MITOCHONDRIAL METABOLISM

MITOCHONDRIAL METABOLISM
An Approach for Disease Management

Edited by

JALAL POURAHMAD

MOHSEN REZAEI

ELSEVIER

ACADEMIC PRESS
An imprint of Elsevier

Academic Press is an imprint of Elsevier
125 London Wall, London EC2Y 5AS, United Kingdom
525 B Street, Suite 1650, San Diego, CA 92101, United States
50 Hampshire Street, 5th Floor, Cambridge, MA 02139, United States
The Boulevard, Langford Lane, Kidlington, Oxford OX5 1GB, United Kingdom

Library of Congress Cataloging-in-Publication Data
A catalog record for this book is available from the Library of Congress

British Library Cataloguing-in-Publication Data
A catalogue record for this book is available from the British Library

ISBN 978-0-12-822416-8

For information on all Academic Press publications
visit our website at https://www.elsevier.com/books-and-journals

Publisher: Andre Gerhard Wolff
Editorial Project Manager: Sara Pianavilla
Production Project Manager: Kiruthika Govindaraju
Cover Designer: Christian J. Bilbow

Typeset by SPi Global, India

Contents

Contributors

Mohammad Abdollahi
Toxicology and Diseases Group (TDG), Pharmaceutical Sciences Research Center (PSRC), The Institute of Pharmaceutical Sciences (TIPS); Department of Toxicology and Pharmacology, School of Pharmacy, Tehran University of Medical Sciences, Tehran, Iran

Siddhesh Aras
Center for Molecular Medicine and Genetics, Wayne State University School of Medicine, Detroit, MI; Perinatology Research Branch, Division of Obstetrics and Maternal-Fetal Medicine, Division of Intramural Research, Eunice Kennedy Shriver National Institute of Child Health and Human Development, National Institutes of Health, U.S. Department of Health and Human Services, Bethesda, MD and Detroit, MI, United States

Luiz Roberto G. Britto
Laboratory of Cellular Neurobiology, Department of Physiology and Biophysics, Biomedical Sciences Institute, University of São Paulo, São Paulo, SP, Brazil

Christian Cortés-Rojo
Instituto de Investigaciones Químico Biológicas, Universidad Michoacana de San Nicolás de Hidalgo, Morelia, Michoacan, Mexico

Majid Dadmehr
Research Institute for Islamic and Complementary Medicine; School of Persian Medicine, Iran University of Medical Sciences, Tehran, Iran

Ana Flávia Fernandes Ferreira
Laboratory of Cellular Neurobiology, Department of Physiology and Biophysics, Biomedical Sciences Institute, University of São Paulo, São Paulo, SP, Brazil

Lawrence I. Grossman
Center for Molecular Medicine and Genetics, Wayne State University School of Medicine, Detroit, MI; Perinatology Research Branch, Division of Obstetrics and Maternal-Fetal Medicine, Division of Intramural Research, Eunice Kennedy Shriver National Institute of Child Health and Human Development, National Institutes of Health, U.S. Department of Health and Human Services, Bethesda, MD and Detroit, MI, United States

Somayeh Handali
Medical Biomaterial Research Center (MBRC), Tehran University of Medical Sciences, Tehran, Iran

Reza Heidari
Pharmaceutical Sciences Research Center, Shiraz University of Medical Sciences, Shiraz, Iran

Maribel Huerta-Cervantes
Instituto de Investigaciones Químico Biológicas, Universidad Michoacana de San Nicolás de Hidalgo, Morelia, Michoacan, Mexico

Zhaleh Jamali
Department of Addiction Studies; Student Research Committee, School of Medicine, Shahroud University of Medical Sciences, Shahroud, Iran

Rocío Montoya-Pérez
Instituto de Investigaciones Químico Biológicas, Universidad Michoacana de San Nicolás de Hidalgo, Morelia, Michoacan, Mexico

Taraneh Mousavi
Toxicology and Diseases Group (TDG), Pharmaceutical Sciences Research Center (PSRC), The Institute of Pharmaceutical Sciences (TIPS); Department of Toxicology and Pharmacology, School of Pharmacy, Tehran University of Medical Sciences, Tehran, Iran

Shilan Mozaffari
Toxicology and Diseases Group (TDG), Pharmaceutical Sciences Research Center (PSRC), The Institute of Pharmaceutical Sciences (TIPS), Tehran University of Medical Sciences, Tehran, Iran

Hossein Niknahad
Pharmaceutical Sciences Research Center, Shiraz University of Medical Sciences, Shiraz; Department of Pharmacology and Toxicology, School of Pharmacy, Shiraz University of Medical Sciences, Shiraz, Fars, Iran

Mohammad Mehdi Ommati
College of Life Sciences, Shanxi Agricultural University, Taigu, Shanxi, China

Donovan J. Peña-Montes
Instituto de Investigaciones Químico Biológicas, Universidad Michoacana de San Nicolás de Hidalgo, Morelia, Michoacan, Mexico

Jalal Pourahmad
Department of Pharmacology & Toxicology, School of Pharmacy, Shahid Beheshti University of Medical Sciences, Tehran, Iran

Neeraja Purandare
Center for Molecular Medicine and Genetics, Wayne State University School of Medicine, Detroit, MI; Perinatology Research Branch, Division of Obstetrics and Maternal-Fetal Medicine, Division of Intramural Research, Eunice Kennedy Shriver National Institute of Child Health and Human Development, National Institutes of Health, U.S. Department of Health and Human Services, Bethesda, MD and Detroit, MI, United States

Mohsen Rezaei
Department of Toxicology, Faculty of Medical Sciences, Tarbiat Modares University, Tehran, Iran

Alfredo Saavedra-Molina
Instituto de Investigaciones Químico Biológicas, Universidad Michoacana de San Nicolás de Hidalgo, Morelia, Michoacan, Mexico

Rafael Salgado-Garciglia
Instituto de Investigaciones Químico Biológicas, Universidad Michoacana de San Nicolás de Hidalgo, Morelia, Michoacan, Mexico

Ahmad Salimi
Department of Pharmacology and Toxicology, School of Pharmacy; Traditional Medicine and Hydrotherapy Research Center, Ardabil University of Medical Sciences, Ardabil, Iran

Monique Patrício Singulani
Laboratory of Cellular Neurobiology, Department of Physiology and Biophysics, Biomedical Sciences Institute, University of São Paulo, São Paulo, SP, Brazil

Bahareh Sadat Yousefsani
Research Institute for Islamic and Complementary Medicine; School of Persian Medicine, Iran University of Medical Sciences, Tehran, Iran

Preface

In recent years, our knowledge about mitochondria has dramatically improved. Studies have revealed the new aspects of capabilities of mitochondria and their involvement in the cellular function, which are as important as their traditional role in ATP production. From direct involvement in cell fate and death to their important role in the remodeling of cell metabolic activities are among these roles. In addition to the significant role that these organelles play in the homeostasis and integrity of cell and organ physiological activities, we now know that in many chronic diseases that affect a large number of the population today, dysfunction of these organelles was identified. Mitochondrial dysfunction has received much attention in cancers, diabetes, neurodegenerative and psychiatric diseases. Numerous studies in this field have identified targets that can be used to treat these diseases. Although in some cases it is not possible to determine with certainty whether mitochondrial dysfunction has caused the disease or whether the complex pathogenesis of that particular disease has ultimately damaged the mitochondria, as in the chicken or the egg causality dilemma, correction and modification of mitochondrial function can be promising in terms of the disease management. This approach becomes more important when we consider that based on the emerging evidence, many environmental pollutants to which we are more or less exposed have a destructive role on mitochondrial activity, and interestingly, on the other hand, contact with these contaminants is associated with those mentioned diseases.

With the increasing evidence that mitochondria play an essential role in the pathogenesis of diseases, targeting of them for the treatment purposes has also been rigorously taken into consideration. To do this, we must begin to know the mitochondrial metabolic irregularities involved in those diseases and then to determine the possibility or feasibility of manipulating them via pharmacological approaches. In the present book, some details regarding the significant role of mitochondria in the chronic diseases and new insights in targeting of mitochondria for therapeutic purposes have been discussed. In view of this and to better achieve our goals, the chapters of the book are arranged in the following order:

Chapter 1: The recent advances in the antioxidant burden of mitochondria and the mechanism behind its sensing and maintaining of its integrity have been discussed in this chapter.

Chapter 2: The role of mitochondria in the metabolism of biomolecules with the focus on molecular pathomechanisms and also targeted therapeutics is a very interesting topic. Finding molecular pathways and their interrelationship, which has the potential to serve as a therapeutic target, has been discussed here.

Chapter 3: This chapter updates information about the pathways behind the mitochondria-related cell deaths and discusses their manipulation to evoke or prevent the cell demise in different disease states.

Chapter 4: Emerging evidence indicates that the environmental pollutants particularly heavy metals including arsenic, cadmium, copper etc. and other nonmetals agents disrupt the mitochondrial normal function and many of them are etiologically linked to some chronic disease states. These toxic mitochondriotropic agents and new insights into their toxic mechanisms are updated here.

Chapter 5: The role of caloric restriction and diet on the mitochondria have been discussed. Calorie restriction improves mitochondrial function and quality along with lowering free radical production and also extends life span in organisms ranging from yeast to mammals. Impaired quality control of mitochondria during the aging process or malnutrition can be modified through the calorie restriction.

Chapter 6: New advances in the identification of altered mitochondria associated signaling pathways in carcinogenesis has been discussed here. Many studies have addressed the alteration of mtDNA, oxidative phosphorylation, and other enzymatic reactions in the context of the carcinogenesis.

Chapter 7: Mitochondrial-targeted cancer therapy strategies have been discussed.

Chapter 8: New advances in the identification and targeting of mitochondria-associated metabolic pathways in diabetes that are highly sought after more effective treatments have been discussed.

Chapter 9: Some of the most common neurologic conditions (Parkinson's disease, Alzheimer's disease, multiple sclerosis, epilepsy, etc.) in which the role of mitochondrial dysfunction in their pathogenesis have been broadly investigated.

Chapter 10: The role of mitochondria in mental disorders has been addressed by many researchers and the molecular link between mitochondrial dynamic (fission and fusion) and psychiatric states (e.g., depression) have attracted much intention. This field is very interesting and some third parties like gut microflora have also been engaged.

Chapter 11: This chapter discusses the different aspects of mitotherapy in terms of current problems, major challenges, and future promises.

We believe that this book will provide new horizons for readers to realize the importance of mitochondria in the present-day research on therapies dealing with mitochondria associated chronic diseases including diabetes, cancer, and neurodegenerative disorders.

Jalal Pourahmad and Mohsen Rezaei
February 20, 2021

CHAPTER 1

Mitochondria as balancers of reduction/oxidation for intracellular environment

Mohsen Rezaei[a], Somayeh Handali[b], and Jalal Pourahmad[c]
[a]Department of Toxicology, Faculty of Medical Sciences, Tarbiat Modares University, Tehran, Iran
[b]Medical Biomaterial Research Center (MBRC), Tehran University of Medical Sciences, Tehran, Iran
[c]Department of Pharmacology & Toxicology, School of Pharmacy, Shahid Beheshti University of Medical Sciences, Tehran, Iran

Contents

1 Introduction

The main endogenous sources of reactive oxygen species (ROS) are cytosolic sources including nicotinamide adenine dinucleotide phosphate (NADPH) oxidases (NOXs), cyclooxygenases, cytochrome P450 enzymes, xanthine oxidase, and mitochondria originated sources including respiratory chain, NOX4, p66shc, and monoamine oxidases (MAOs).[1] Oxidative phosphorylation (OXPHOS) in mitochondria contributes to ATP production via the electron transport chain (ETC) and also the generation of ROS.[2] ROS include superoxide ($O_2^{\bullet-}$), hydrogen peroxide (H_2O_2), hydroxyl radical ($OH^{\bullet}$), hypochlorous acid (HOCl), and peroxynitrite ($ONOO^-$).[3–5] Leakage of electrons at complexes I and III from ETC (the major sites of ROS generation) leads to reduction of oxygen to $O_2^{\bullet-}$ (primary ROS with a very short lifetime, approximately 1×10^{-9} s).[3, 4, 6–8] Then, $O_2^{\bullet-}$ converts to H_2O_2 by superoxide dismutase 1 (SOD1, Cu-ZnSOD) in the mitochondrial intermembrane space and superoxide dismutase 2 (SOD2, MnSOD) in the mitochondrial matrix (Fig. 1).[9, 10] H_2O_2 further reduces to H_2O by catalase (CAT), glutathione peroxidase (GPX), peroxiredoxin (PRDX), and

https://doi.org/10.1016/B978-0-12-822416-8.00002-6

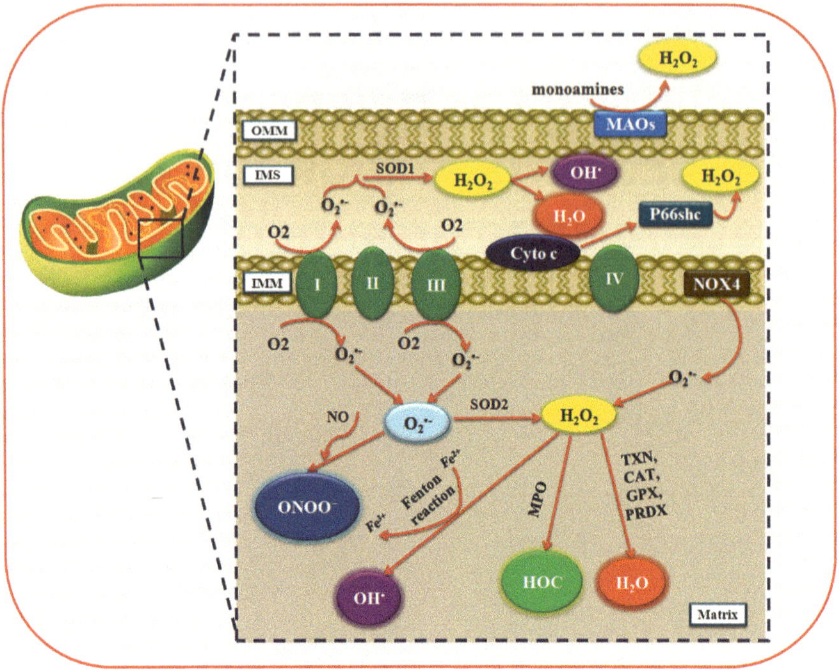

Fig. 1 Mitochondrial ROS generation.

thioredoxin (TXN).[9,11] CAT is mainly found in the peroxisome fraction of cells and very little levels of CAT have been found in mitochondria. CAT is not present in mitochondria of mammalian cells, except for rat heart mitochondria.[12,13] A redox balance on production and neutralization of free radicals is existed in organisms that maintain the ROS level at a steady state and reducing equivalents at optimum capacities.[14] Produced ROS are transferred to the matrix and eliminated by the antioxidant systems; however, high levels of ROS in the matrix reach to the intermembrane space and cytosol which contribute in redox signaling or lead to oxidative damage. ROS can pass the membranes through different ways. For instance, H_2O_2 can transfer by aquaporins and superoxide can travel via anion channels.[15] ROS are involved in reprogramming cells for proliferation and intracellular signaling.[16–18]

Mitochondria are also the main site for scavenging of ROS and some similar antioxidant enzymes exist in the cytosol.[12] Excessive ROS leads to lipid peroxidation (LPO),[19] DNA damage,[20] and protein oxidation.[4] H_2O_2 can react with iron (Fe^{2+}) in the Fenton reaction and produce OH·[11] Peroxidation of lipids is begun by OH overproduction.[4] H_2O_2 can be converted to HOCl by myeloperoxidase (MPO).[5] Moreover, $O_2^{\bullet-}$

contributes to the generation of peroxynitrite ($ONOO^-$) in a reaction with mitochondrial nitric oxide (NO), which may lead to lipid peroxidation (Fig. 1).[4, 11]

ETC can also produce other reactive species such as nitric oxide (NO) and reactive nitrogen species (RNS).[6] It has been suggested that NO may be produced by a mitochondrial NO synthase. This free radical can potentially affect mitochondrial functions including competing with O_2 at complex IV and reacting with O_2 to form peroxynitrite. NO may be produced in nerve terminals, vascular endothelium, or other cytosolic sites.[21]

NOXs are a family of enzymes and another source for intracellular ROS production and are involved in the transfer of electrons from NADPH to O_2 which may result in the generation of $O_2^{\bullet-}$ and H_2O_2. NOX4 is located in the inner mitochondrial membrane of renal cells and cardiomyocytes and generates $O_2^{\bullet-}$ and H_2O_2 in the mitochondrial matrix (Fig. 1).[1] P66shc belongs to the Shc (Src homology 2 domain and collagen-homology region) family of cytosolic adaptor proteins. This protein exists in the mitochondrial intermembrane space and is involved in mitochondrial ROS production through the cytochrome *c* oxidation and triggering of H_2O_2 production (Fig. 1).[1] MAOs are located in the outer mitochondrial membrane and degrade monoamines such as epinephrine and norepinephrine to aldehydes and H_2O_2 (Fig. 1).[1] Intra-mitochondrial ROS production by MAO can induce lipid peroxidation that results in phospholipase C activation and calcium signaling provoked by inositol trisphosphate.[8]

The mitochondrial flavoprotein long-chain acyl-CoA dehydrogenase (LCAD) can also be considered as an intra-mitochondrial source for H_2O_2 production. In vitro experiments revealed that LCAD which contributes to a key step in mitochondrial fatty acid oxidation (FAO) generated H_2O_2 by transferring electrons to O_2. There is evidence that mitochondrial enzyme very long-chain acyl-CoA dehydrogenase (VLCAD) attaches to cardiolipin in the mitochondrial inner membrane and is in close interaction with the ETC in a higher-order FAO complex.[22]

Besides their contribution to oxidative damage, mitochondrial ROS can also trigger several signaling pathways in the cells. Stabilization of hypoxia-inducible transcription factors (HIFs) under hypoxic condition is one of the well-known pathways that require the mitochondrial ROS production. HIF stabilization results in the beginning of transcriptional changes including the alteration of genes involved in angiogenesis. Proliferation of endothelial cells is essential for angiogenesis that is highly depended on the expression of vascular endothelial growth factor (VEGF) which is upregulated by HIF. The proliferative effect of VEGF is directed by the activation of the extracellular-signal-regulated kinase/mitogen-activated protein kinase (ERK/MAPK) pathway.[23]

Mitochondrial ROS are also needed for T-cells activation. Lowered amount of ROS produced by mitochondria complex III resulted in decreased T-cell activation in mice upon the stimulation with CD3 or CD28. In addition, it was observed that translocation of mitochondria to the immunological synapse in a T-cell line and mitochondrial H_2O_2 are both necessary for T-cell receptor (TCR) signal transduction via MAPK signaling. These findings suggest that mitochondrial ROS that is required for T-cell stimulation and proliferation increase the TCR signal transduction following activation of antigen exposure.[23]

2 Mitochondrial ROS in pathological condition

Increasing ROS levels leads to protein denaturation, DNA damage, and lipid peroxidation,[24] which eventually may lead to oxidative stress, mitochondria, and other membranous organelles damaging and neurodegenerative diseases, diabetes, cancer, cardiovascular diseases, and rheumatoid arthritis.[25, 26] ROS oxidizes the disulfide bonds in proteins and lipids which lead to instability of membranes and production of short half-life lipid radicals. This toxic event may propagate the oxidative damage in membranes and develop to chronic diseases.[20] ROS also contribute to the peroxidation of cardiolipin in mitochondria which results in cytochrome *c* release into the cytosol which triggers apoptosis.[27] ROS stimulate DNA damage by oxidation of guanine nucleobases. However, the repair mechanism may not be flawless and these errors can lead to some mutations.[4]

OH can specifically alter the enzymes and structural proteins and oxidize polyunsaturated fatty acids.[28] It was reported that interaction of OH with DNA molecule damages the nitrogenous bases, pyrimidine and purine, and deoxyribose backbone of DNA.[6]

In some general insult like inflammation, ROS production is elevated by the action of certain enzymes, including NADPH oxidase, xanthine oxidase, and cyclooxygenase-2, which activate NF-κB (nuclear factor kappa B), the primary redox-sensitive signaling pathway for inflammation.[29] In the next sections, the role of ROS is discussed in relation to specific chronic diseases.

2.1 Mitochondrial ROS and diabetes

Hyperglycemia is implicated in ROS generation in diabetic patients through mitochondrial respiratory chain enzymes, lipoxygenases, xanthine oxidases, peroxidases, cyclooxygenases, and nitric oxide synthases.[30] The major function of β-cells is to regulate the concentration of glucose in the body by insulin secretion. In β-cells, glucose enters through GLUT2 (glucose transporter-2) and becomes phosphorylated to glucose-6-phosphate and is converted to pyruvate by glycolysis. Pyruvate enters mitochondria and is converted to acetyl coenzyme A through pyruvate dehydrogenase. Acetyl coenzyme A then enters the tricarboxylic acid (TCA) cycle, resulting in the production of NADH. NADH fuels the ETC for ATP production and elevates the ATP/ADP ratio. This in turn results in the closure of ATP-sensitive K^+ channels (K_{ATP}) that depolarizes the plasma membrane and induces the opening of voltage-dependent Ca^{2+} channels and influx of Ca^{2+}; consequently, facilitates insulin secretion through the fusion of insulin granules with the plasma membrane.[31] When β-cells are exposed to high levels of glucose (hyperglycemia), ROS generated by the glucose oxidation pathway activate uncoupling protein-2 (UCP-2). This protein reduces the ATP/ADP ratio through leakage of proton in the β-cells mitochondria, which decreases secretion of insulin.[32, 33] β-Cells possess low antioxidant defenses than other tissues.[31]

Hyperglycemia also increases ROS by stimulating xanthine oxidase, NADPH oxidase, and mitochondrial pathway which leads to oxidative stress.[34] Mitochondrial ROS influence the nuclear DNA which in turn activate poly (ADP-ribose) polymerase (PARP), a DNA repair enzyme. Then, PARP decreases the activity of glyceraldehyde-3 phosphate dehydrogenase (GAPDH) which in turn inhibits the glycolytic pathway and results to the accumulation of glycerol 3-phosphate. Consequently, glycerol 3-phosphate induces other alternative glucose metabolic pathways including hexosamine pathway, polyol pathway, protein kinase C (PKC) pathway, and AGE (advanced glycation end) production.[35–37] Hexosamine pathway converts fructose-6-phosphate to glucosamine-6-phosphate through the transfer of an amide group from glutamine. Excess hexosamines have been shown to induce insulin resistance in vitro and in vivo.[38] The polyol pathway consists of a 2-step metabolic pathway that glucose is reduced to sorbitol and then sorbitol converts to fructose.[39] This pathway has a pathogenic role in diabetic retinopathy.[40] Multiple diabetic problems have been related to the enhancement of PKC activation including alteration in blood flow, vascular permeability, and angiogenesis.[41] In podocytes, activation of calcium channel and inhibition of Ca^{2+}-ATPase contribute to AGE-induced increase in the concentration of intracellular calcium, which results in ER stress and podocyte apoptosis. This demise of podocytes due to apoptosis is characteristic in diabetic nephropathy that results in albuminuria and renal dysfunction.[42] AGEs are produced through the Maillard reaction that is a nonenzymatic amino-carbonyl reaction. This reaction takes places between the carbonyl group of glucose, fructose, galactose, and ribose or glucose metabolism intermediates (glucose-6-phosphate, ribose-5-phosphate, fructose-6-phosphate glyceraldehyde, and deoxyribose-5-phosphate), with an amine group and other molecules. The yield of the later reaction is a reversible Schiff base and consequently, Amadori products, which are early products of the Maillard reaction. Production of the Amadori products is elevated in hyperglycemia. By glycoxidation of biomolecules, these highly reactive compounds interact with amine groups and metal ions to form malondialdehyde (MDA), glyoxal (GO), and methylglyoxal (MGO).[43]

2.2 Mitochondrial ROS and cancer

The mechanism of ROS in tumor development is associated with DNA damaging at 8-hydroxy 2′-deoxyguanosine (8-OHdG) that leads to mutation.[20] In some cancer cells, antioxidant capacity has been upregulated through the elevated expression of enzymes involved in ROS detoxification. Nuclear respiratory factor 2 (NRF2) is a transcriptional activator of genes that alleviate oxidative damage; activity of NRF2 is inhibited by kelch-like ECH-associated protein 1 (KEAP1). In some tumors, NRF2 is activated constitutively, either by gain-of-function mutations or by KEAP1 suppression. NRF2 gain-of-function provides resistance features to tumor cells against the chemotherapies that act by inducing the oxidant stress. Inactivation of KEAP1 can be the result of posttranslational modifications that contribute to the overactivation of NRF2.[15] ROS can also be involved during metastasis that suggesting to have a signaling role within cells for survival, migration, and metastasis promotion.[44] It has been shown that ROS facilitates metastasis in various studies.[4] Anoikis is the process of programmed cell death induces by the detachment of normal cells from the matrix. It was suggested that ROS could induce anoikis resistance and escaping apoptosis in cancer cells during metastasis.[20] The mechanism of anoikis inhibition is still not clear. It has been suggested that attenuation of SOD2 (MnSOD) promotes anoikis resistance and tumor metastasis. NF-κB is transcriptional regulators of SOD2 which activates following cell detachment. In normal cells, matrix detachment enhances ROS levels, results in

$NF_{K}B$ pathway activation, which in turn induces MnSOD expression to detoxify ROS. MnSOD exhibits antianoikis activity. Reduction of MnSOD or NF-κB enhances the toxic effect of superoxide free radical in suspended cells; therefore, increases the sensitivity of cells to anoikis. It has been suggested that two parallel pathways including the decrease in glucose oxidation (reduced mitochondrial ROS generation) and increase of antioxidant capacity to scavenge ROS take place in response to matrix detachment which permit cells to cope with oxidative stress.[45] In normal and cancerous cell lines, the role of ROS in cell death and viability has been controversial and presumably depended on the relative amount of the ROS. Moreover, increasing expression of NOX4 confers anoikis inhibition in lung cancer cells through ROS-mediated activation of epidermal growth factor receptor (EGFR), which enhances cell survival. It is likely that upregulation of NOX4 increases EGFR phosphorylation through H_2O_2 since NOX4 mainly produces H_2O_2 rather than superoxide anions. Overexpression of EGFR inhibits anoikis by downregulation of Bim (pro-apoptotic protein).[46]

HIF-1 and its targets including VEGF and metalloproteinases (MMPs) that are contributed to angiogenesis and matrix remodeling may be promoted by ROS. ROS stabilize HIF-1a that is necessary for metastasis and evident in stromal remodeling and connective tissue degradation through the upregulation of MMPs. This action provides an optimum environment for tumor cells to enter and spread into adjacent tissue. Furthermore, HIF-1a upregulates VEGF that in turn increases tumor angiogenesis and facilitates the migration of tumor cells into the bloodstream.[47]

Impaired mitochondrial respiration is linked to the high level of ROS generation and excessive mutations that may progress to tumor formation. Mutations in complex I promote tumorigenesis via ROS production; as a result, DNA damage and mutation that may activate oncogenes (gain of function) will be occurred. Besides this, enhancing glycolytic metabolism and lactate production promotes metastasis.[44] Complex II has a close relationship to TCA cycle and mitochondrial ETC and all of its subunits are expressed by nucleus genes. Electrons from complex I and also complex II (oxidation of succinate) enter into the respiratory chain.[44] Mutation in SDH leads to the accumulation of succinate. Enhanced level of succinate results in improved histone methylation through inhibition of histone demethylase Jumonji D3, which increases epigenetic changes and oncogenic transformation. Moreover, mutations in subunits A-D of succinate dehydrogenase (SDH) are related to pheochromocytoma and paragangliomas. Accumulation of succinate enhances the ROS generation through α-ketoglutarate-dependent enzyme blockade and reversing of electrons flow from complex II to complex I and leads to pheochromocytoma and paragangliomas by the stabilization action of HIF.[48, 49]

2.3 Mitochondrial ROS and brain

Oxidative stress has been etiologically related to many brain diseases including Alzheimer's disease (AD) and Parkinson's disease (PD).[27] Relatively, elevated oxygen consumption, excess amount of membrane polyunsaturated fatty acids, and lower antioxidant capacity including reduced glutathione, CAT, SOD, and GPX activities make the central nervous system (CNS) very vulnerable and sensitive to ROS devastating effects.[27]

AD is related to deposits of amyloid β (Aβ) and tau which results in inflammation, increasing iron and enhancing ROS production.[50] Intracellular tau oligomers interact with the mitochondrial respiratory chain, triggering the release of cytochrome *c* and inducing ROS generation.[51] ROS production is an important intracellular event that leads to activation of p38-MAPK, tau hyperphosphorylation, and destabilization of microtubules which may induce neurodegeneration and cell death.[52] Increasing Ca^{2+} transport from endoplasmic reticulum (ER) to mitochondria also results in reduced ATP production and enhanced ROS generation which triggers apoptosis.[53] In AD, anomalous mitochondrial electron transfer is evident which mostly occurred in complexes I and IV, resulting in declined mitochondrial membrane potential, reduced ATP production (complex V), and increased ROS generation. Dysfunction of complex I is mainly contributed to tau effects; whereas, dysregulation of complex IV is Aβ-dependent and these damages happened at both protein structures and activities.[54] Other damages related to oxidative stress in AD are glucose autoxidation that leads to AGE_S production and Aβ-toxicity induction.[55]

In the PD, the activity of ETC complex I decreases that leads to enhanced ROS production, mitochondrial DNA damage, ETC and bioenergetics disruption, release of cytochrome *c* into the cytosol, and finally induction of mitochondria-dependent apoptotic pathway.[27] Many gene products are related to PD pathogenesis including α-synuclein, LRRK2 (leucine-rich repeat kinase 2), DJ-1, parkin, and PINK1. Parkin (an E3 ubiquitin ligase) is associated with the outer mitochondrial membrane and prevents the release of cytochrome *c*.[56] Parkin can attach covalently to ubiquitin on different misfolded protein substrates to help their degradation.[57] Parkin is also involved in mitophagy and removal of impaired mitochondria.[58] PINK1 (PTEN-induced putative kinase-1) is a serine–threonine kinase in the mitochondria and its overexpression enhances mitochondrial fission. Protection against oxidative damage and reduction of mitochondrial cytochrome *c* release as well as apoptosis are among its functions.[27, 57] Overexpression of α-synuclein (α-syn) is associated with mitochondrial membrane instability and cytochrome *c* release. α-Syn accumulation leads to complex I impairment, mitochondrial inter-membrane potential disruption, and increase in ROS production.[59, 60] It has been suggested that α-syn may enter in the mitochondrial matrix or the intermembrane space through channel proteins including TOM34 and TIM13A and inhibit the complex I.[61] Furthermore, α-syn enhances Ca^{2+} transport to the mitochondria which leads to oxidative stress and triggers apoptosis.[62] SIRT3 is present in the cytosol in its inactive form and transports to the mitochondria during oxidative stress. It has been suggested that TOM20 (translocase of outer membrane) carries SIRT3 to the mitochondria. α-Syn reduces the mitochondrial SIRT3 levels which results in the ROS production dysregulation. SIRT3 deacetylates and activates the ROS neutralizing enzymes, including SOD2.[63] In oxidative stress insult, DJ-1 as an oxidative stress sensor chaperone relocalized into the mitochondria and takes its protective role.[27, 64] Generally, the DJ1 protein is found in the cytosol, nucleus, and mitochondria.[58] Insufficient DJ-1 increases α-syn accumulation by accelerating the LAMP2A (lysosome-associated membrane protein type-2A) degradation in lysosomes.[65] It has been shown that α-syn and DJ-1 interact with mitochondria-associated membranes (MAM) through the chaperone Grp75. These interactions stimulate MAM assembly and activity by controlling ER-mitochondria Ca^{2+} and lipid homeostasis. These findings indicated that MAM disruption contributes in the pathogenesis of PD.[64] In PD patients, there is impairment for reuptake of dopamine (DA) into the synaptic vesicles; so that a high level of cytosolic DA is present. DA can be auto-oxidized to form cytotoxic radicals that may impair mitochondria and activate intrinsic apoptotic pathway or inhibit the respiratory chain. The produced DA quinines or semiquinones can modify some vital cellular proteins such as α-syn, DJ-1, SOD2, parkin, and UCH-L1. The DA quinones are also undergoing oxidation to form aminochrome with redox-cycling activity that may in turn contribute to the generation of the superoxide radicals, antioxidants depletion, and the degenerative outcome (Fig. 2).[66]

3 Mitochondrial antioxidant defense

Mitochondria possess numerous defense mechanisms to counteract the elevated ROS production. Glutathione (GSH) is a tripeptide antioxidant that is produced by the amino acids glutamate, cysteine, and glycine in two

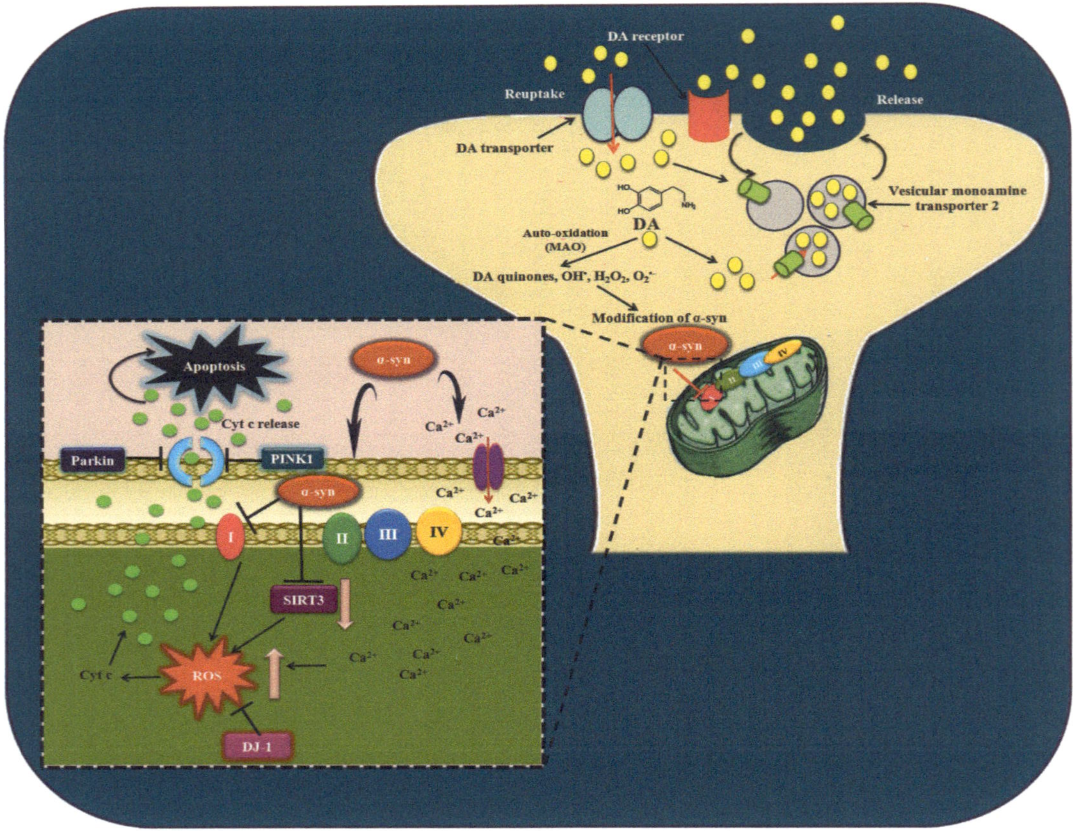

Fig. 2 Role of mitochondrial dysfunction in PD.

ATP-dependent steps in the cytoplasm (is present in millimolar concentrations) (Fig. 3).[15, 67–69] The first reaction is γ-glutamyl-cysteine formation from glutamate and cysteine catalyzed by the glutamine-cysteine ligase (GCL).[70] Then, the addition of glycine to γ-glutamyl-cysteine forms GSH by glutathione synthetase (GS).[67] Cysteine along with glutamate and glycine are the rate-limiting metabolites for GSH synthesis.[4] GSH is oxidized by glutathione peroxidase (GSH-PX), a selenium-based enzyme for the neutralization of ROS that leads to the formation of glutathione-disulfide (GSSG). GSH can also be regenerated by the reduction of GSSG via the glutathione reductase (GSR) enzyme. This reaction needs NADPH for producing two GSH molecules from one GSSG molecule.[70] GSR is a homodimer enzyme and found both in the cytosol and in mitochondria.[13] GSH synthesis in mitochondria is not possible owing to the lack of relevant enzymes.[71] Cytosolic GSH can transport into mitochondria via TTC (tricarboxylate carrier), DIC (dicarboxylate carrier), KGC (a-KG carrier), and 2-oxoglutarate (2-OGC) GSH carriers.[72, 73]

Different isoenzymes of GPX are located at distinct subcellular organelles with diverse tissue-specific expression patterns. GPX is the main scavenging enzyme for H_2O_2 when ascorbate peroxidases become inhibited at persistent oxidative stress. In addition to H_2O_2, other toxic organic hydroperoxides are also detoxified by GPX to neutral substances like water or alcohols.[74] Similar to GSR, GPX is located both in the cytosol and in the mitochondria.[13]

At physiological pH, GSH behaves like an anion and it was suggested that inner mitochondria membrane organic anion carriers may act as GSH transports.[69] Overloading of mitochondrial cholesterol content reduces mitochondrial GSH by impairment of these cytosolic GSH transports and subsequently the entrance of GSH into mitochondria.[75] GSH is predominantly found in its reduced form in the mitochondria in such a way that GSH:GSSG ratio is normally greater than 100:1. This ratio keeps the mitochondrial protein thiols in their reduced state.[76] Under oxidative conditions and apoptosis, the ratio of GSH:GSSG decreased to less than 4:1.[72] Lower ratio of GSH:GSSG in cuprizone-induced demyelinated mice confirmed a ROS over production in the muscle tissues.[77]

Other H_2O_2 scavenging enzyme is CAT that is mostly present in peroxisomes and converts H_2O_2 to oxygen and water.[68] CAT considerably decreases the oxidative stress and prevents the damage to the mitochondrial structure by increasing the mitochondrial membrane potential ($\Delta\psi m$).[74] TXN and SOD exhibit high catalytic activity for H_2O_2.[4, 25] In comparison

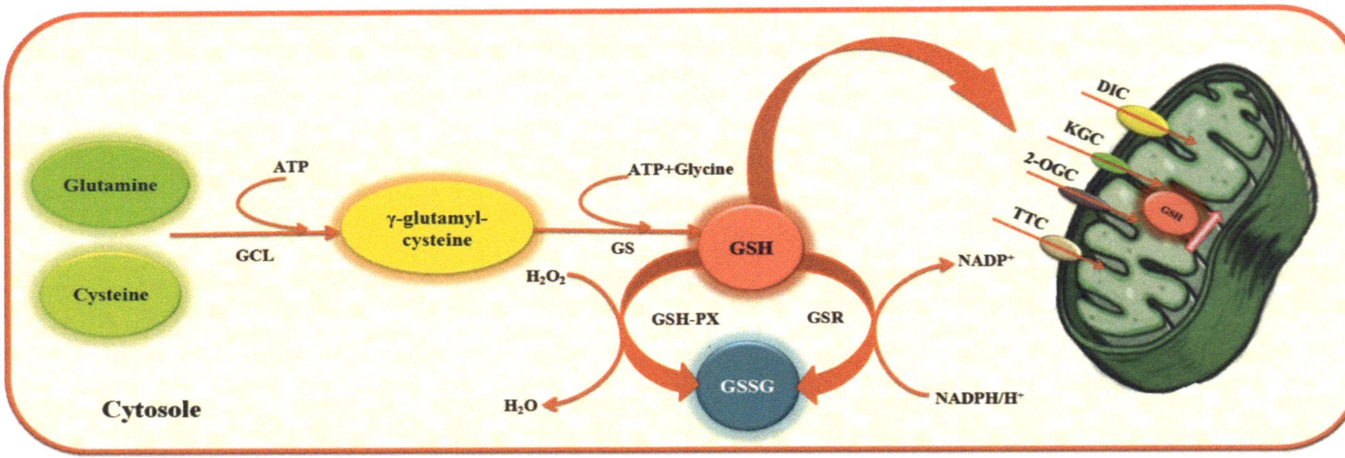

Fig. 3 Synthesis and metabolism of GSH.

with GSH, TXNs, as small antioxidant proteins, are less abundantly found in the mitochondria.[4]

Coenzyme Q10 (CoQ10) is an essential endogenous compound for electron transporting in the respiratory chain and the production of ATP. It is an important antioxidant that protects the mitochondrial biomolecules from oxidation by decreasing ROS/RNS production and suppressing of proinflammatory mediators production through the diminution of nuclear factor κB (NF-κB) gene expression.[13]

The ability of MnSOD to decrease ROS is increased by the mitochondrial NAD^+-dependent deacetylase sirtuin (SIRT3). SIRT3 decreases ROS levels by deacetylating MnSOD and increases its ROS neutralization capacity by upregulating the expression of MnSOD and other genes involved in oxidative stress responsiveness. Posttranslational modification of SIRT3 by carbonylation can result in the loss of its activity. Carbonylation of SIRT3 leads to spinal MnSOD inhibition in carrageenan-induced thermal hyperalgesia in rats. Furthermore, SIRT3 carbonylation inhibited during the application of natural and synthetic antioxidants by restoring of MnSOD enzymatic activity and finally blocking the development of thermal hyperalgesia. These findings provide a therapeutic approach for inhibition of posttranslational modifications of SIRT3 aimed to alleviate the pain states when the central sensitization is evidently contributed to ROS overproduction.[29,78]

Some other antioxidants including lipoic acid are also important for proper mitochondrial function. Lipoic acid is an organosulfur compound with capacity to exchange the thiol-disulfide and ROS scavenging. Lipoic acid is a mitochondrial nutrient and can directly or indirectly protect mitochondria from oxidative damage. It can restore glutathione to a normal redox statement and its application promoted mitochondrial function and inhibited capillary cell apoptosis resulting in the prevention of diabetic retinopathy in rats.[79] Among the tocopherol (vitamin E) isomers, α-tocopherol is the most abundant isomer found in mammalian membranes. During oxidative stress, vitamin E is markedly oxidized and mitochondrial membrane lipid peroxidation may occur in such a way that a close correlation has been suggested between the amount of mitochondrial lipid peroxidation and α-tocopherol. In vitro studies indicated that vitamin E is an excellent protective against lipid peroxidation in both mitochondrial and microsomal membranes.[80]

Along with natural antioxidants, exogenous molecules can also be beneficial to mitochondria. SS-Peptides, the Szeto-Schiller (SS) compounds, have a tendency to accumulate more than 1000-fold in the inner mitochondrial membrane compared to the cytosol. They are tetrapeptides with an alternating aromatic-cationic amino acids motif. Accumulation of these positively charged peptides is mainly related to the mitochondrial potential not solely to their positive charges since even the depolarized mitochondria are capable to uptake and concentrate these molecules. SS peptides can scavenge H_2O_2 and $ONOO^-$ and protect against lipid peroxidation. This effect is associated with improved mitochondrial membrane integrity, inhibition of cytochrome *c* release, and prevention of ROS-induced cell death. Administration of these peptides recovered cardiac function in an animal ischemia/reperfusion model. SS peptides reduced the mitochondrial H_2O_2 levels during high-fat diet and maintained insulin sensitivity in skeletal muscle. Further studies suggested the usefulness of these molecules in ischemia/reperfusion injury and neurodegenerative diseases.[13]

References

1. Peoples JN, Saraf A, Ghazal N, Pham TT, Kwong JQ. Mitochondrial dysfunction and oxidative stress in heart disease. *Exp Mol Med.* 2019;51:1–13.
2. Lian G, Gnanaprakasam JR, Wang T, et al. Glutathione de novo synthesis but not recycling process coordinates with glutamine catabolism to control redox homeostasis and directs murine T cell differentiation. *elife.* 2018;7, e36158.
3. Bigarella CL, Liang R, Ghaffari S. Stem cells and the impact of ROS signaling. *Development (Cambridge, England).* 2014;141:4206–4218.
4. Harris IS, DeNicola GM. The complex interplay between antioxidants and ROS in Cancer. *Trends Cell Biol.* 2020;30:440–451.
5. Ambrosone CB, Ahn J, Singh KK, et al. Polymorphisms in genes related to oxidative stress (MPO, MnSOD, CAT) and survival after treatment for breast cancer. *Cancer Res.* 2005;65:1105–1111.
6. Bhatti JS, Bhatti GK, Reddy PH. Mitochondrial dysfunction and oxidative stress in metabolic disorders—a step towards mitochondria based therapeutic strategies. *Biochimica et Biophysica Acta (BBA)-Molecular Basis of Disease.* 2017;1863:1066–1077.
7. Blajszczak C, Bonini MG. Mitochondria targeting by environmental stressors: implications for redox cellular signaling. *Toxicology.* 2017;391:84–89.
8. Angelova PR, Abramov AY. Role of mitochondrial ROS in the brain: from physiology to neurodegeneration. *FEBS Lett.* 2018;592:692–702.
9. Kim YS, Gupta Vallur P, Phaëton R, Mythreye K, Hempel N. Insights into the dichotomous regulation of SOD2 in cancer. *Antioxidants.* 2017;6:86.
10. Zorov DB, Juhaszova M, Sollott SJ. Mitochondrial reactive oxygen species (ROS) and ROS-induced ROS release. *Physiol Rev.* 2014;94:909–950.
11. Choi TG, Kim SS. Physiological functions of mitochondrial reactive oxygen species. In: *Free radical medicine and biology.* IntechOpen; 2019.
12. Holzerová E, Prokisch H. Mitochondria: much ado about nothing? How dangerous is reactive oxygen species production? *Int J Biochem Cell Biol.* 2015;63:16–20.
13. Morón ÚM, Castilla-Cortázar I. Protection against oxidative stress and "IGF-I deficiency conditions". In: *Antioxidant enzyme.* vol. 89. IntechOpen; 2012.

14. Georgieva E, Ivanova D, Zhelev Z, Bakalova R, Gulubova M, Aoki I. Mitochondrial dysfunction and redox imbalance as a diagnostic marker of "free radical diseases". *Anticancer Res.* 2017;37:5373–5381.
15. Sabharwal SS, Schumacker PT. Mitochondrial ROS in cancer: initiators, amplifiers or an Achilles' heel? *Nat Rev Cancer.* 2014;14:709–721.
16. Jang S, Javadov S. Association between ROS production, swelling and the respirasome integrity in cardiac mitochondria. *Arch Biochem Biophys.* 2017;630:1–8.
17. Ballard JWO, Towarnicki SG. Mitochondria, the gut microbiome and ROS. *Cell Signal.* 2020;109737.
18. Chowdhury AR, Zielonka J, Kalyanaraman B, Hartley RC, Murphy MP, Avadhani NG. Mitochondria-targeted paraquat and metformin mediate ROS production to induce multiple pathways of retrograde signaling: a dose-dependent phenomenon. *Redox Biol.* 2020;101606.
19. Al-Menhali AS, Banu S, Angelova PR, et al. Lipid peroxidation is involved in calcium dependent upregulation of mitochondrial metabolism in skeletal muscle. *Biochim Biophys Acta Gen Subj.* 2020;1864, 129487.
20. Yang H, Villani RM, Wang H, et al. The role of cellular reactive oxygen species in cancer chemotherapy. *J Exp Clin Cancer Res.* 2018;37:266.
21. Stowe DF, Camara AK. Mitochondrial reactive oxygen species production in excitable cells: modulators of mitochondrial and cell function. *Antioxid Redox Signal.* 2009;11:1373–1414.
22. Zhang Y, Bharathi SS, Beck ME, Goetzman ES. The fatty acid oxidation enzyme long-chain acyl-CoA dehydrogenase can be a source of mitochondrial hydrogen peroxide. *Redox Biol.* 2019;26:101253.
23. Weinberg F, Ramnath N, Nagrath D. Reactive oxygen species in the tumor microenvironment: an overview. *Cancer.* 2019;11:1191.
24. Hahn HJ, Kim KB, An IS, Ahn KJ, Han HJ. Protective effects of rosmarinic acid against hydrogen peroxide-induced cellular senescence and the inflammatory response in normal human dermal fibroblasts. *Mol Med Rep.* 2017;16:9763–9769.
25. Stefanatos R, Sanz A. The role of mitochondrial ROS in the aging brain. *FEBS Lett.* 2018;592:743–758.
26. Di Meo S, Napolitano G, Venditti P. *Physiological and pathological role of ROS: benefits and limitations of antioxidant treatment.* Multidisciplinary Digital Publishing Institute; 2019.
27. Bhat AH, Dar KB, Anees S, et al. Oxidative stress, mitochondrial dysfunction and neurodegenerative diseases; a mechanistic insight. *Biomed Pharmacother.* 2015;74:101–110.
28. Henkler F, Brinkmann J, Luch A. The role of oxidative stress in carcinogenesis induced by metals and xenobiotics. *Cancer.* 2010;2:376–396.
29. Ji LL, Yeo D, Kang C, Zhang T. The role of mitochondria in redox signaling of muscle homeostasis. *J Sport Health Sci.* 2020;9:386–393.
30. Volpe CMO, Villar-Delfino PH, dos Anjos PMF, Nogueira-Machado JA. Cellular death, reactive oxygen species (ROS) and diabetic complications. *Cell Death Dis.* 2018;9:1–9.
31. Burgos-Morón E, Abad-Jiménez Z, Martínez de Marañón A, et al. Relationship between oxidative stress, ER stress, and inflammation in type 2 diabetes: the battle continues. *J Clin Med.* 2019;8:1385.
32. Remedi M, Nichols C, Koster J. The mitochondria and insulin release: Nnt just a passing relationship. *Cell Metab.* 2006;3:5–7.
33. Fernández-Mejía C. Oxidative stress in Diabetes mellitus and the role of vitamins with antioxidant actions. In: *Oxidative stress and chronic degenerative diseases-a role for antioxidants.* 209. IntechOpen; 2013.
34. Zhu W, Chen M, Shou Q, Li Y, Hu F. Biological activities of Chinese propolis and Brazilian propolis on streptozotocin-induced type 1 diabetes mellitus in rats. *Evid Based Complement Alternat Med.* 2011;2011.

35. Brownlee M. The pathobiology of diabetic complications: a unifying mechanism. *Diabetes.* 2005;54:1615–1625.
36. Ighodaro OM. Molecular pathways associated with oxidative stress in diabetes mellitus. *Biomed Pharmacother.* 2018;108:656–662.
37. Yan L-J. Pathogenesis of chronic hyperglycemia: from reductive stress to oxidative stress. *J Diabetes Res.* 2014;2014.
38. Daniels MC, McClain DA, Crook ED. Transcriptional regulation of transforming growth factor β1 by glucose: investigation into the role of the hexosamine biosynthesis pathway. *Am J Med Sci.* 2000;319:138–142.
39. Jang S-M, Kim M-J, Choi M-S, Kwon E-Y, Lee M-K. Inhibitory effects of ursolic acid on hepatic polyol pathway and glucose production in streptozotocin-induced diabetic mice. *Metabolism.* 2010;59:512–519.
40. Lorenzi M. The polyol pathway as a mechanism for diabetic retinopathy: attractive, elusive, and resilient. *Exp Diabesity Res.* 2007;2007.
41. Evcimen ND, King GL. The role of protein kinase C activation and the vascular complications of diabetes. *Pharmacol Res.* 2007;55:498–510.
42. Pathomthongtaweechai N, Chutipongtanate S. AGE/RAGE signaling-mediated endoplasmic reticulum stress and future prospects in non-coding RNA therapeutics for diabetic nephropathy. *Biomed Pharmacother.* 2020;131:110655.
43. Amorim RG, Guedes GDS, Vasconcelos SML, Santos JCDF. Kidney disease in diabetes mellitus: cross-linking between hyperglycemia, redox imbalance and inflammation. *Arq Bras Cardiol.* 2019;112:577–587.
44. Tan AS, Baty JW, Berridge MV. The role of mitochondrial electron transport in tumorigenesis and metastasis. *Biochim Biophys Acta Gen Subj.* 2014;1840:1454–1463.
45. Kamarajugadda S, Cai Q, Chen H, et al. Manganese superoxide dismutase promotes anoikis resistance and tumor metastasis. *Cell Death Dis.* 2013;4, e504.
46. Kim H, Sung JY, Park E-K, et al. Regulation of anoikis resistance by NADPH oxidase 4 and epidermal growth factor receptor. *Br J Cancer.* 2017;116:370–381.
47. Giampazolias E, Tait SW. Mitochondria and the hallmarks of cancer. *FEBS J.* 2016;283:803–814.
48. Moreno C, Santos RM, Burns R, Zhang WC. Succinate dehydrogenase and ribonucleic acid networks in cancer and other diseases. *Cancer.* 2020;12:3237.
49. Bardella C, Pollard PJ, Tomlinson I. SDH mutations in cancer. *Biochim Biophys Acta Biomembr.* 2011;1807:1432–1443.
50. Ashok A, Singh N, Chaudhary S, et al. Retinal degeneration and Alzheimer's disease: an evolving link. *Int J Mol Sci.* 2020;21:7290.
51. Shafiei SS, Guerrero-Muñoz MJ, Castillo-Carranza DL. Tau oligomers: cytotoxicity, propagation, and mitochondrial damage. *Front Aging Neurosci.* 2017;9:83.
52. Liu Z, Li P, Wu J, Yi W, Ping L, Xinxin H. The cascade of oxidative stress and Tau protein autophagic dysfunction in Alzheimer's disease. In: *Alzheimer's disease—challenges future.* vol. 2. IntechOpen; 2015.
53. Wang X, Zheng W. Ca2+ homeostasis dysregulation in Alzheimer's disease: a focus on plasma membrane and cell organelles. *FASEB J.* 2019;33:6697–6712.
54. Eckert A, Schmitt K, Götz J. Mitochondrial dysfunction-the beginning of the end in Alzheimer's disease? Separate and synergistic modes of tau and amyloid-β toxicity. *Alzheimers Res Ther.* 2011;3:1–11.
55. Ahmad W, Ijaz B, Shabbiri K, Ahmed F, Rehman S. Oxidative toxicity in diabetes and Alzheimer's disease: mechanisms behind ROS/RNS generation. *J Biomed Sci.* 2017;24:76.
56. Caldwell GA, Caldwell KA. Traversing a wormhole to combat Parkinson's disease. *Dis Model Mech.* 2008;1:32–36.
57. Maiti P, Manna J, Dunbar GL. Current understanding of the molecular mechanisms in Parkinson's disease: targets for potential treatments. *Translational Neurodegener.* 2017;6:28.

58. Barnhill LM, Murata H, Bronstein JM. Studying the pathophysiology of Parkinson's disease using zebrafish. *Biomedicine*. 2020;8:197.
59. Di Rita A, Strappazzon F. Mitophagy could fight Parkinson's disease through antioxidant action. *Rev Neurosci*. 2019;30:729–742.
60. Guerra de Souza AC, Prediger RD, Cimarosti H. SUMO-regulated mitochondrial function in Parkinson's disease. *J Neurochem*. 2016;137:673–686.
61. Witt SN, Flower TR. α-Synuclein, oxidative stress and apoptosis from the perspective of a yeast model of Parkinson's disease. *FEMS Yeast Res*. 2006;6:1107–1116.
62. Larsen S, Hanss Z, Krüger R. The genetic architecture of mitochondrial dysfunction in Parkinson's disease. *Cell Tissue Res*. 2018;373:21–37.
63. Park J-H, Burgess JD, Faroqi AH, et al. Alpha-synuclein-induced mitochondrial dysfunction is mediated via a sirtuin 3-dependent pathway. *Molecular Neurodegener.* 2020;15:1–19.
64. Zaichick SV, McGrath KM, Caraveo G. The role of Ca2+ signaling in Parkinson's disease. *Dis Model Mech*. 2017;10:519–535.
65. Xu C-Y, Kang W-Y, Chen Y-M, et al. DJ-1 inhibits α-synuclein aggregation by regulating chaperone-mediated autophagy. *Front Aging Neurosci*. 2017;9:308.
66. Ortiz GG, Pacheco-Moisés FP, Mireles-Ramírez MA, et al. Oxidative stress and Parkinson's disease: effects on environmental toxicology. In: *Free radicals and diseases*. Intechopen; 2016:183–209.
67. Jozefczak M, Remans T, Vangronsveld J, Cuypers A. Glutathione is a key player in metal-induced oxidative stress defenses. *Int J Mol Sci*. 2012;13:3145–3175.
68. Ježek J, Cooper KF, Strich R. Reactive oxygen species and mitochondrial dynamics: the yin and yang of mitochondrial dysfunction and cancer progression. *Antioxidants*. 2018;7:13.
69. Lash LH. Mitochondrial glutathione transport: physiological, pathological and toxicological implications. *Chem Biol Interact*. 2006;163:54–67.
70. Guilford FT, Hope J. Deficient glutathione in the pathophysiology of mycotoxin-related illness. *Toxins*. 2014;6:608–623.
71. Alboghobeish S, Pashmforosh M, Zeidooni L, Samimi A, Rezaei M. High fat diet deteriorates the memory impairment induced by arsenic in mice: a sub chronic in vivo study. *Metab Brain Dis*. 2019;34:1595–1606.
72. Conde de la Rosa L, García-Ruiz C, Fernández-Checa JC. *Glutathione in mammalian biology*. Springer; 2014.
73. Xiao W, Loscalzo J. Metabolic responses to reductive stress. *Antioxid Redox Signal*. 2020;32:1330–1347.
74. Su L-J, Zhang J-H, Gomez H, et al. Reactive oxygen species-induced lipid peroxidation in apoptosis, autophagy, and ferroptosis. *Oxid Med Cell Longev*. 2019;2019.
75. Torres S, Matías N, Baulies A, et al. Mitochondrial GSH replenishment as a potential therapeutic approach for Niemann pick type C disease. *Redox Biol*. 2017;11:60–72.
76. Garcia J, Han D, Sancheti H, Yap L-P, Kaplowitz N, Cadenas E. Regulation of mitochondrial glutathione redox status and protein glutathionylation by respiratory substrates. *J Biol Chem*. 2010;285:39646–39654.
77. Khodaei F, Rashedinia M, Heidari R, Rezaei M, Khoshnoud MJ. Ellagic acid improves muscle dysfunction in cuprizone-induced demyelinated mice via mitochondrial Sirt3 regulation. *Life Sci*. 2019;237:116954.
78. Ilari S, Giancotti LA, Lauro F, et al. Antioxidant modulation of sirtuin 3 during acute inflammatory pain: the ROS control. *Pharmacol Res*. 2020;104851.
79. Kowluru RA, Mishra M. Oxidative stress, mitochondrial damage and diabetic retinopathy. *Biochim Biophys Acta (BBA) - Mol Basis Dis*. 2015;1852:2474–2483.
80. Maguire JJ, Wilson DS, Packer L. Mitochondrial electron transport-linked tocopheroxyl radical reduction. *J Biol Chem*. 1989;264:21462–21465.

CHAPTER 2

Mitochondria as biosynthetic centers and targeted therapeutics

Reza Heidari[a], Mohammad Mehdi Ommati[b], and Hossein Niknahad[a,c]
[a]Pharmaceutical Sciences Research Center, Shiraz University of Medical Sciences, Shiraz, Iran
[b]College of Life Sciences, Shanxi Agricultural University, Taigu, Shanxi, China
[c]Department of Pharmacology and Toxicology, School of Pharmacy, Shiraz University of Medical Sciences, Shiraz, Fars, Iran

Contents

1 Introduction

Mitochondria have long been known as the organelles responsible for energy (ATP) supply to the cell. However, extensive studies on this organelle have shown that mitochondria play a key role not only in the production of ATP but also in the synthesis of substances that are vital for cell survival and growth. Besides, mitochondria have been shown to play a critical role in cell death. On the other hand, various disorders related to the biosynthesis of different molecules in mitochondria have been identified to lead to disease. Therefore, focusing on these biosynthetic pathways could be considered potential therapeutic targets in various complications (e.g., cancer). This requires accurate and extensive knowledge about the mechanisms of the synthesis of these compounds in mitochondria.

In this chapter, energy (ATP) metabolism as the fundamental function of cellular mitochondria is first discussed, and the diseases in which ATP production is impaired are described. Afterward, the synthesis of other vital biomolecules such as nucleotides, heme, citrate, and amino acids

Mitochondrial Metabolism
https://doi.org/10.1016/B978-0-12-822416-8.00008-7

by mitochondria are highlighted. In each section, the related pathological conditions are also described. At the end of the chapter, the pharmacological and ancillary treatments used in mitochondrial diseases are mentioned. Finally, newly emerged methods of treating mitochondrial disorders such as gene therapy and mitochondrial replacement therapy are noted.

The data collected in this chapter could provide a better insight into the complex function of mitochondria in the cell. Eventually, novel strategies for the treatment of human diseases could be suggested.

2 Role of mitochondria in the metabolism of biomolecules

2.1 Energy metabolism: The fundamental role of mitochondria

The production of "adenosine triphosphate" (ATP) in cells plays a vital role in maintaining cellular homeostasis, and enough ATP levels guarantee pivotal processes such as the function of different transporters. ATP production is the final stage of a multistage process; hence, factors that disrupt energy metabolism reactions at any stage will eventually disrupt cell homeostasis[1,2] (Fig. 1). Several interrelated mechanisms should be balanced for maintaining normal mitochondrial function and ATP production (Fig. 1).

Pyruvate (produced from glucose metabolism) and fatty acids are the substrates for ATP production in mitochondria. These substrates will be oxidized in the mitochondrial matrix. However, pyruvate and fatty acids should be transported to the mitochondrial matrix via specific carriers. The outer mitochondrial membrane is almost permeable to any nonpolar compound (smaller than 50 kDa). Thus, foreign materials easily pass through the outer membrane (Fig. 1). Unlike the outer membrane, the mitochondria's inner membrane (IMM) does not allow compounds to pass easily.[1,14] The IMM contains particular carriers responsible for transporting various substances to the matrix. For example, long-chain fatty acids first bind to coenzyme A (CoA) (Fig. 1), then, the fatty acids-CoA complex binds to carnitine to form acylcarnitine. Acylcarnitine passes through the inner membrane of the mitochondria. Finally, fatty acids are released in the mitochondrial matrix and broken down into smaller compounds through β-oxidation (Fig. 1). Pyruvate is also transported to the mitochondrial matrix via specific carriers. In the mitochondrial matrix, both pyruvate and fatty acids are converted to acetyl-CoA. Acetyl-CoA is the primary substrate for the Krebs cycle (citrate cycle). An essential function of the citrate cycle is the production of reducing equivalents, such as nicotinamide adenine dinucleotide (NADH)[14] (Fig. 1).

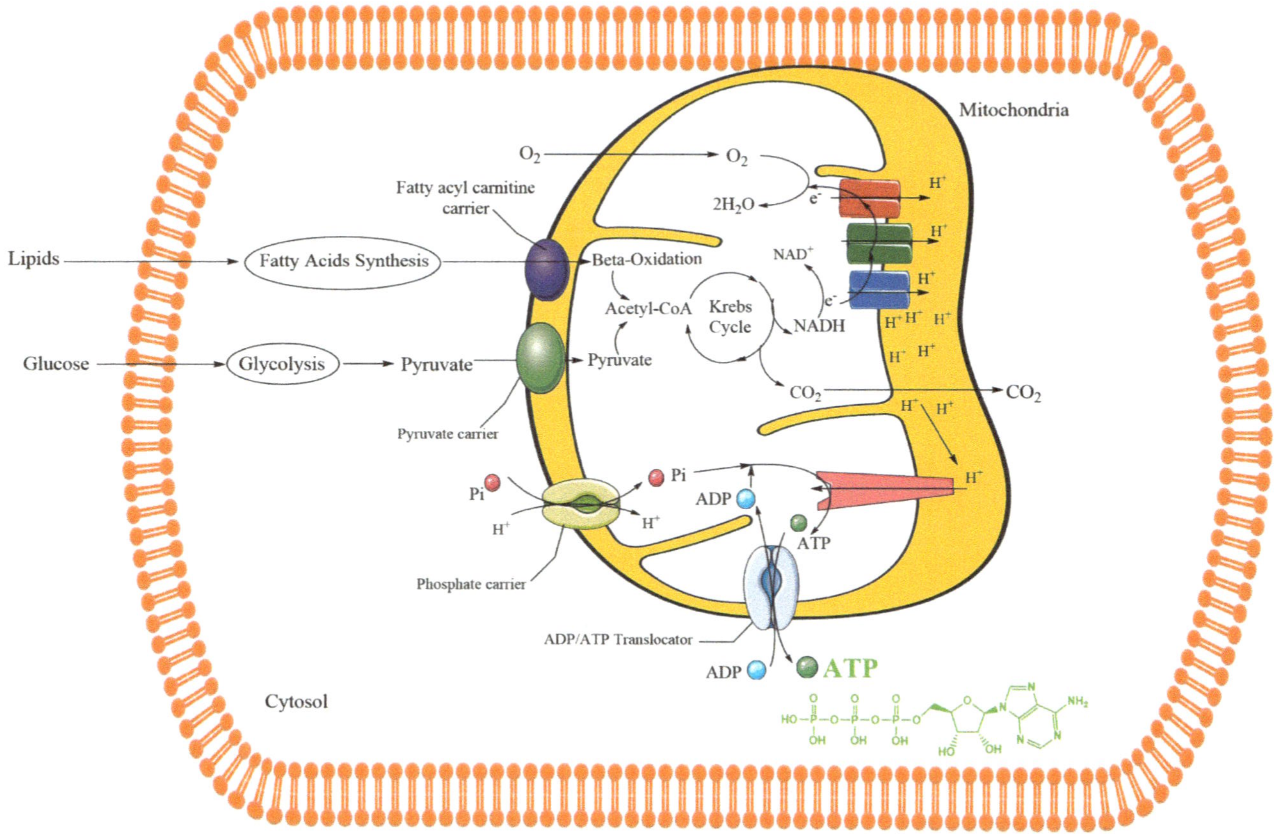

Fig. 1 Energy metabolism is the main feature of cellular mitochondria. Adenosine triphosphate (ATP) production might be impaired by a wide range of diseases or xenobiotics. Any significant change in cellular ATP levels could lead to deleterious consequences. Many pharmacological interventions have been developed regarding targeting mitochondria and enhancing cellular ATP levels (Table 2). These ancillary treatments could be applied to a wide range of human diseases (Table 2).

NADH is oxidized by the mitochondrial electron transport chain (ETC), consisting of four protein complexes (Fig. 1). The energy from the oxidation of NADH is used to pump protons (H^+) from the mitochondrial matrix to the intramembrane space. The electrons produced by NADH oxidation are finally transferred to molecular oxygen. Due to the positive charge of H^+, and its immense accumulation in the intermembrane area, a potential difference ($\Delta\Psi$) is generated between the two sides of the inner mitochondrial membrane (Fig. 1). Finally, the energy created from the H^+ concentration gradient activates the ATP synthase complex. ATP synthase facilitates the interaction of adenosine diphosphate (ADP) and phosphate (Pi) to produce ATP (Fig. 1). Because ATP production is a complex process and involves various enzymes and transporters, its disruption at any stage can lead to impaired ATP synthesis.[1,14] Multiple diseases and many xenobiotics have been identified that can decrease the function of mitochondria as the cellular power plants by interacting with stages of ATP production. The following is a discussion of disorders caused by reduced ATP production in human diseases.

2.2 Diseases related to impaired mitochondria energy (ATP) metabolism

There is a plethora of clinical evidence on mitochondria-associated disorders.[15] Pathogenic mutation of mtDNA or nuclear DNA could lead to a wide range of heterogeneous mitochondrial diseases. High energy demand organs such as the brain, heart, skeletal muscle, kidney, and liver are the most influenced tissues when mitochondrial function is impaired. Hence, the clinical manifestations of mitochondrial diseases related to impaired ATP synthesis include cognitive impairment, locomotor dysfunction, seizure, heart diseases, kidney injury, and hepatic impairment.[15] It is worth to mention that cells contain thousands of mitochondria. Hence, the severity of mitochondrial cytopathies is associated with the percentage of mitochondria in each cell with a particular genetic mutation. Mutation in nuclear or mtDNA genes and, consequently, the impairment in the synthesis of proteins such as ETC components is the leading cause of mitochondria metabolic disorders[16] (Fig. 2).

Presently, the clinical focus is on treating the symptoms of mitochondrial diseases that induce energy crisis in different tissues (Table 1). For example, cardiac complications or seizures are managed in these patients by appropriate drugs. However, these strategies could not prevent the disease's progression or alleviate severe symptoms such as neurodegeneration. Therefore,

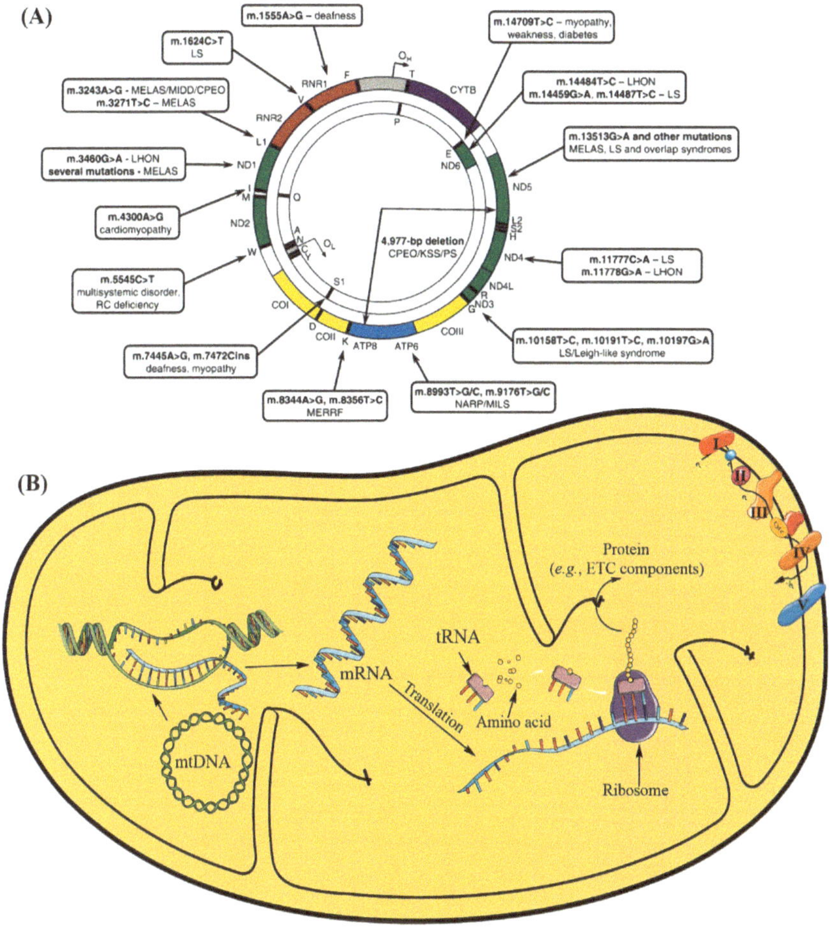

Fig. 2 Mitochondrial DNA (mtDNA) plays a crucial role in the synthesis of mitochondrial respiratory chain components. Several mitochondrial cytopathies are associated with a mutation in mtDNA genes. Leber hereditary optic neuropathy (LHON), myoclonic epilepsy with ragged red fibers (MERRF), and mitochondrial encephalopathy, lactic acidosis, and stroke-like episodes (MELAS) (A) are well-known mitochondrial cytopathies associated with nDNA or mtDNA (or both) mutations (Table 1). mtDNA mutations could lead to impaired protein synthesis (e.g., ETC components). Eventually, mitochondrial depolarization, energy (ATP) crisis, and cell death will occur (B). Gene therapy (Section 3.1) might provide a promising option for mitochondrial cytopathies. On the other hand, ancillary treatments (antioxidants and mitochondria protecting agents; Table 2). *(Panel (A) was adapted from https://doi.org/10.1016/j.bbabio.2009.09.005.)*

Table 1 Several mitochondrial cytopathies have been identified in association with mutation in nuclear or mitochondrial DNA.

Disorder	Main cause	Phenotype	Currently available therapeutic options
Leber hereditary optic neuropathy (LHON)	Mutation in mtDNA gene responsible for encoding the NADH dehydrogenase (Complex I) protein	Degeneration of retinal ganglion cells and their axons, acute or subacute loss of vision	Antioxidants treatment for the alleviation of disease symptoms. Idebenone is under clinical trial for LHON
Mitochondrial encephalomyopathy, lactic acidosis, and stroke-like episodes (MELAS)	Mutations in the genes in mtDNA responsible for NADH dehydrogenase (Complex I) and mitochondrial tRNA synthesis	Muscle weakness and pain, recurrent headaches, loss of appetite, vomiting, and seizures	Managing symptoms, pharmacological intervention to correct tRNA structures
Myoclonic epilepsy with ragged red fibers (MERRF)	Mutations in the genes in mtDNA responsible for mitochondrial tRNA synthesis	Myoclonus seizure, generalized epilepsy, ataxia, weakness, dementia	Managing symptoms by antioxidants and mitochondrial protecting agents
Kearn-Sayre syndrome and chronic progressive external ophthalmoplegia (CPEO)	Nuclear or mtDNA defects, impaired synthesis of ETC components, depletion of mtDNA	Ptosis and ophthalmoplegia, pigmentary retinal degeneration, cardiac, endocrine, and neurological abnormalities	No specific treatment, managing symptoms
Neuropathy, ataxia, and retinitis pimentosa (NARP)	Mutation in mtDNA gene involved in encoding ATP synthase protein	Numbness, tingling, sensory neuropathy, muscle weakness, ataxia, learning disabilities, developmental delays, vision loss	No specific treatment, managing symptoms
Leigh syndrome (subacute necrotizing encephalomyopathy; SNEM)	Mutation in mtDNA gene involved in encoding cytochrome *c* oxidase and/or pyruvate dehydrogenase complex (PDHC)	The progressive loss of mental and movement abilities, dystonia, nystagmus	No specific treatment, managing symptoms

CoQ deficiency	Mutation in the gene (Nuclear) involved in coenzyme Q10 synthesis	Decreased activity of ETC, skeletal muscle disorders	Managing symptoms by the administration of COQ10
Mitochondrial encephalopathy, lactic acidosis, and stroke-like episodes (MELAS)	Mutation in mtDNA genes involved in ETC subunits synthesis	Mitochondrial respiratory complex deficiency, cellular energy crisis	No specific treatment, managing symptoms
Glutaric acidemia type 2	Mutation in the nucleic DNA gene involved in flavoprotein dehydrogenase	Accumulation of by-products of fatty acids and amino acids metabolism	No specific treatment
mtDNA depletion syndrome	Mutation in genes (nuclear) involved in mtDNA polymerase γ enzyme synthesis	mtDNA depletion, symptoms of MNGIE	No specific treatment
Mitochondrial myopathy and lactic acidosis	Mutation in mtDNA genes involved in tRNA synthesis	Respiratory chain deficiency, energy crisis	No specific treatment, managing symptoms by the administration of compounds for boosting ATP production
Mitochondrial neurogastrointestinal encephalomyopathy (MNGIE)	Mutation in the nuclear DNA gene involved in thymidine phosphorylase enzyme synthesis	Accumulation of thymidine and deoxyuridine in mitochondria, impaired mtDNA transcription, and replication	No specific treatment

Pathogenic mutation of mtDNA could lead to impaired ATP synthesis in high energy demand tissues. To date, the therapeutic options against these disorders are almost symptomatic.

despite some pharmacological interventions (Table 2), mitochondrial diseases lead to severe disability and death. However, emerging strategies such as gene therapy and mitochondrial donation (see Section 3.1) much-resolved worries induced by mitochondrial diseases.

2.3 Biosynthesis of biomolecules other than ATP by mitochondria

As mentioned, cellular mitochondria are well-known for their crucial role in ATP production. Meanwhile, these organelles are responsible for the biosynthesis of a variety of other molecules. Mitochondria-dependent mechanisms synthesize heme, several amino acids, cholesterol, fatty acids, and nucleotides. Dysfunctional synthesis of biomolecules in the cellular mitochondria is associated with several disorders. Therefore, these biosynthesis pathways could be therapeutic targets for the management of human diseases. The mitochondria-mediated biosynthesis of these compounds and their related pathologies and potential therapeutic points are discussed in the next sections.

2.3.1 Heme biosynthesis in mitochondria

Heme is a metallocofactor presented in the structure of many hemoproteins[17] and also acts as a signaling molecule that regulates diverse physiological processes.[17] Different types of heme (*b*, *c*, *o*, *a*) have been identified in eukaryotic cells. Heme *b* is synthesized in the mitochondria and used as a precursor for other heme types.[17] As mentioned, heme incorporates numerous vital proteins, and any disturbances in heme metabolism could severely affect cellular function. Therefore, identifying the pathological conditions associated with mitochondrial heme biosynthesis is an exciting field of investigation and could help the development of novel therapeutic options.

Heme biosynthesis reactions involve mitochondrial and cytoplasmic steps (Fig. 3). Glycine, succinyl CoA, molecular oxygen, and ferrous iron (Fe^{2+}) are required for heme synthesis. Iron is transported to the mitochondrial matrix through several routes.[17] The best-known iron transport mechanism to the mitochondria is through the mitoferrin transporter[17] (Fig. 3). Dysregulation of mitoferrin function has been reported in some pathological conditions associated with mitochondrial iron accumulation,[17] which will be discussed later in this chapter. It has been found that the mitochondria-located enzyme aminolevulinate synthase (ALAS) is an essential enzyme in the heme synthesis process (Fig. 3). Aminolevulinate (ALA) is exported to the cytoplasm via specific transporters.[17] ALA is converted to porphobilinogen in the cytoplasm (Fig. 3). Porphobilinogen is converted via several intermediates into coproporphyrinogen III (CPG III) in the cytosol (Fig. 3). CPG III is then transported to the mitochondrial intermembrane space, where it is converted to coproporphyrinogen IX by coproporphyrinogen oxidase (Fig. 3). Coproporphyrinogen IX is transported to the mitochondrial matrix and converted to protoporphyrin IX by protoporphyrin oxidase (Fig. 3). The final step of heme synthesis occurs in the mitochondrial matrix, where the enzyme ferrochelatase catalyzes the insertion of Fe^{2+} into the heme structure (Fig. 3). Once it was synthesized, heme is trafficked to the cytoplasm. On the other hand, heme also incorporates different components of ETC (Complexes I–IV) (Fig. 3). Hence, any changes in heme synthesis could disturb proper mitochondrial oxidative phosphorylation and energy metabolism.[17]

Several pathological conditions associated with disturbed heme synthesis have been identified, including some types of anemia. Anemia is a very heterogeneous group of acquired (e.g., iron deficiency) or inherited diseases. Iron deficiency could impair heme synthesis. This complication could significantly improve by iron supplementation. On the other hand, impaired heme synthesis in the bone marrow erythroblasts is associated with different forms of hereditary anemia.[18] Erythroblasts are involved in the development of mature red blood cells in the erythropoiesis process. Sideroblastic anemia and erythropoietic porphyria are inheritable diseases associated with a defect in heme biosynthesis in erythroblastic mitochondria.[18] Mutation in genes encoding heme synthesis enzymes such as ferrochelatase and ALA dehydrogenase (Fig. 3) are involved in the pathogenesis of these kinds of anemia.[18] Chronic erythrocyte transfusions and bone marrow or hematopoietic stem cell transplantation are therapeutic options for these diseases. Gene therapy (see Section 3.1.1) of the enzymes involved in the pathogenesis of these diseases (e.g., ferrochelatase) could also be used to manage these kinds of anemia.

Heme also incorporates essential proteins of vascular endothelial cells.[19] Some investigations reported that heme synthesis is directly connected with pathological angiogenesis.[19] Although the precise mechanisms of how heme derives angiogenesis are not precisely known, the effects of heme synthesis on mitochondrial function and energy metabolism seem to play a role in this process.[19] Therefore, the enzymes involved in heme synthesis could be potential targets for managing pathological conditions such as proliferative diabetic retinopathy, wet age-related macular degeneration, and retinopathy of prematurity.[19] It has been found that the inhibition of

Table 2 Some pharmacological interventions and ancillary treatments are used for the enhancement of mitochondrial function and energy production.

Option	Primary mechanism(s) of action	Reference(s)
Betaine	Incorporation in the synthesis of vital molecules in mitochondria, boosting mitochondrial antioxidant system, osmolyte	3
Carnosine	Ameliorating mitochondria-mediated oxidative stress, enhancing cellular ATP levels, mitochondrial matrix buffering agent	4
Coenzyme Q_{10}	Vitamins and cofactors	5–7
Curcumin	Alleviating mitochondria-originated oxidative stress, boosting cellular and mitochondrial antioxidant defense	8
Elamipretide	Significant decreasing of ROS, stabilize cardiolipin	9
Folic/folinic acid	Nucleotides synthesis, preventing mtDNA depletion	10
Glycine	Incorporation in the synthesis of vital molecules in mitochondria, boosting mitochondrial antioxidant system	11
Histidine	Incorporation in the synthesis of vital molecules in mitochondria, boosting mitochondrial antioxidant system	11
Idebenone	Significant decreasing of mitochondria-originated ROS	12
Taurine	Regulation of proper ETC components synthesis, preventing mitochondria-mediated ROS formation, enhancing ATP synthesis	13
Vitamin E	Vitamins and cofactors	5–7

enzymes, such as FECH in the mitochondria (Fig. 3), significantly suppressed the angiogenesis process.[19] The inhibition of heme synthesis could increase mitochondria-mediated ROS formation, depolarize mitochondria, and deplete ATP stores. These events could finally suppress the angiogenesis process. Griseofulvin, an inhibitor of FECH,[20, 21] significantly inhibited heme synthesis and suppressed angiogenesis in the in vitro experiments or animal models.[20, 21] Drugs such as salicylic acid are also reported as inhibitors of heme synthesis in mitochondria.[22] These drugs might be useful against pathological neovascularization. However, the clinical significance of the effects of these drugs on angiogenesis remains to be investigated.

Lack of heme synthesis also affects several cytoplasmic signaling pathways.[19] It has been found that eNOS activity is significantly decreased when heme synthesis is inhibited.[19] eNOS inhibition could reduce nitric oxide (NO) synthesis and suppress vascular relaxation.[19] It has been found that NO is a pro-angiogenic agent.[19] Prevention of angiogenesis could be beneficial in the abovementioned vascular diseases (e.g., proliferative diabetic retinopathy) or preventing tumor growth in various cancer types.

The role of mitochondrial heme synthesis is also widely investigated in neurodegenerative diseases.[23, 24] It has been found that mitochondrial dysfunction and iron accumulation are evident in several neurological disorders such as Alzheimer's disease, Friedrich ataxia, Parkinsonism, and Huntington's disease.[24] Some of these disorders (e.g., Friedrich ataxia) are directly connected with heme biosynthesis and storage impairment. Iron-induced mitochondrial impairment and oxidative stress seem to play a role in the pathogenesis of other neurodegenerative diseases such as Parkinsonism, Alzheimer's disease, and Huntington's disease.[24, 25] Physiologically, heme regulates many biological processes such as gene expression, ion channel function, and energy production in the nervous system.[24] Thus, heme is essential for nerve survival and proper operation. In the next part, some neurodegenerative disorders associated with heme synthesis dysregulation and mitochondrial impairment are discussed.

Friedreich's ataxia (FRDA) is a well-known neurodegenerative disease with impaired heme biosynthesis and storage.[25] Severe motor deficiency and cognitive dysfunction are the most prominent clinical manifestations of FRDA.[25] Patients of FRDA die at young ages due to neurological disorders and cardiac complications.[26] This condition is called cardiac neuropathy because of damage to the nerves that innervate the heart.[25, 26] It has been found that severe mitochondrial iron loading is occurred in FRDA.[25] Defected mitoferrin function or decreased protein level named frataxin (Fig. 3) is possibly involved in the heme biosynthesis dis-homeostasis in FRDA.[25] Despite the etiology of iron accumulation in mitochondria and disturbances of heme synthesis in FRDA, these events could have serious consequences such as severe mitochondria-mediated ROS formation, mitochondrial depolarization, depletion of cellular ATP stores, and the release of cell death mediators from mitochondria.[25, 27] Moreover, it has been found that the activity of mitochondrial respiratory chain complexes is severely suppressed in FRDA patients.[27] These events could be connected with iron-induced mitotoxicity.[25] Finally, neuronal cell death and brain injury will occur in FRDA.

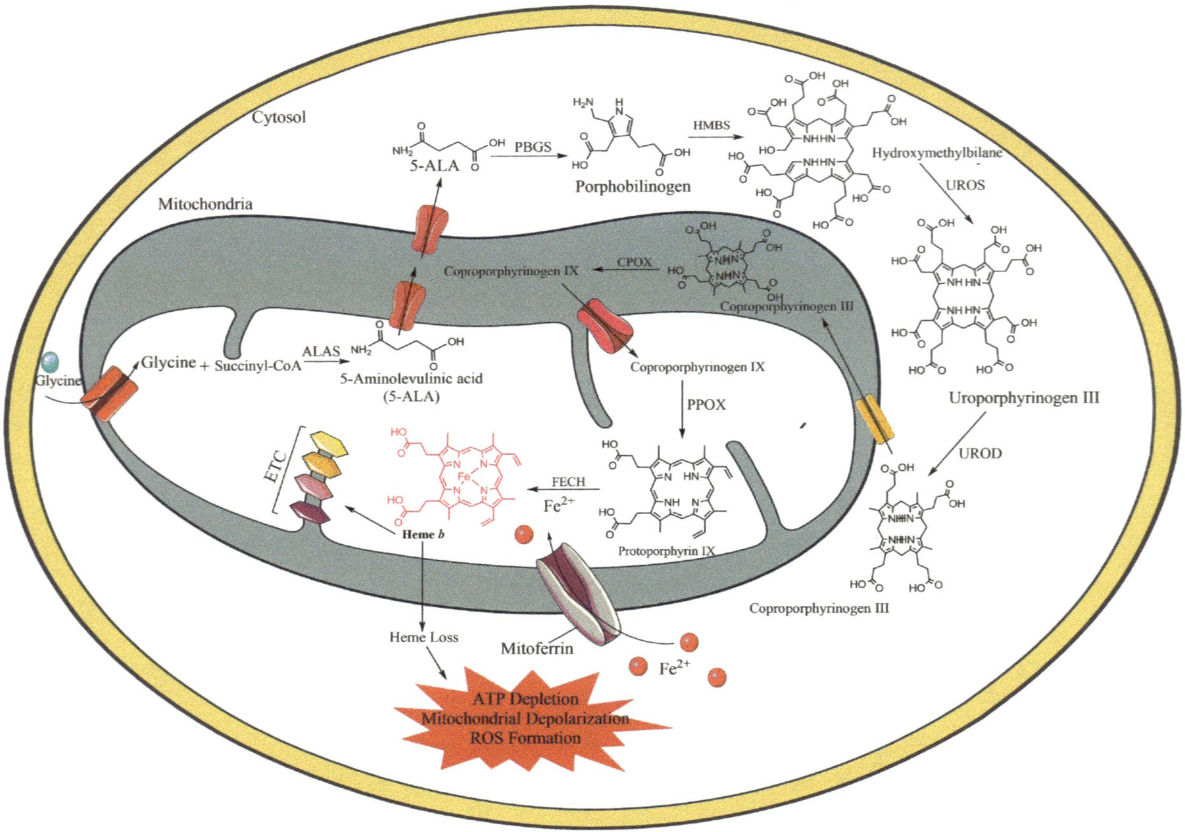

Fig. 3 Mitochondria play a critical role in heme biosynthesis. *ETC*, electron transport chain. Several human diseases, such as various forms of anemia, neurodegenerative disorders, or cardiovascular diseases, are associated with impaired heme metabolism and storage in cellular mitochondria. *ALAS*, aminolevulinic acid synthase; *CPOX*: coproporphyrinogen oxidase; *ETC*, electron transport chain; *FECH*, ferrochelatase; *HMBS*, hydroxymethylbilane synthase; *PBGS*, porphobilinogen synthase; *PPOX*, protoporphyrinogen oxidase; *UROD*, uroporphyrinogen decarboxylase; *UROS*, uroporphyrinogen III synthase.

Increased markers of oxidative stress are evident in different biological specimens taken from FRDA cases.[28] Moreover, primary antioxidant signaling is defected in the CNS of FRDA patients and the experimental models of this disease.[25, 27–29] Impaired antioxidant defense accelerates the neurodegeneration process in FRDA. As cellular mitochondria are the major site of ROS formation, the proximity of iron molecule and oxygen radicals could lead to the formation of hydroxyl radical (OH) as a noxious oxygen species. The level of other oxygen species is also significantly increased in FRDA[25]; therefore, severe mitochondrial impairment and mitochondria-mediated ROS formation are the primary mechanisms proposed for neuronal and other cells (e.g., astrocytes) injury in FRDA.[25] Based on these data, several studies evaluated mitochondria-targeted and antioxidant therapies for FRDA.[5–7] In this regard, compounds such as coenzyme Q_{10} and vitamin E have been administered.[5–7] Hart et al. reported that these treatments significantly improved mitochondrial energy synthesis and sustained the progression of neurodegeneration in FRDA patients.[6] In another study by Lodi and others, it has been found that antioxidants (vitamin E and Coenzyme Q_{10}) also improved skeletal muscle mitochondrial function as well as cardiac functionality in FRDA.[5] They suggested targeting mitochondria and antioxidant therapy as a disease-modifying strategy in FRDA.[5]

Other than FRDA, several types of neuropathic porphyria as hereditary diseases have also been identified. Aminolevulinic acid dehydratase (ALAD) deficiency, acute intermittent porphyria, hereditary coproporphyria, variegate porphyria, posterior column ataxia, and retinitis pigmentosa, nonsyndromic retinitis pigmentosa, hereditary sensory and autonomic neuropathy, and Fowler syndrome are other neurodegenerative disorders associated with impaired heme biosynthesis.[23] Psychiatric disturbances, peripheral neuropathies, progressive cardiac complications, retinitis pigmentosa, lack of pain perception, and severe hydrocephaly are clinical features of these diseases.[23] Mutations in the genes encoding mitochondrial heme transporters or enzymes of heme biosynthesis path are involved in the pathogenesis of these neurodegenerative disorders.[23] As same as FRDA syndrome, oxidative stress and severe mitochondrial impairments have been identified in the clinical cases and experimental models of these disorders.[23, 30] These events could contribute to the mechanism of neurodegeneration. Targeting heme metabolism or the administration of antioxidants and mitochondria protecting agents have been tested to blunt the complications of these hereditary diseases.[30, 31] However, there is no specific cure for these complications to date. Future investigations on strategies to restore mitochondrial function or correct the mutation in the heme biosynthesis process (e.g., gene therapy; see Section 3.1.2) could provide promising tools for managing neurodegenerative diseases associated with dis-homeostasis of heme.

As mentioned, several other neurodegenerative disorders such as Alzheimer's disease, Huntington's disease, and Parkinsonism are connected with iron dis-homeostasis. These diseases may not directly be associated with heme synthesis, but the role of iron-induced mitochondrial injury and energy crisis has been widely reviewed in these disorders. On the other hand, many studies targeted cellular mitochondria and antioxidant defense systems to alleviate these neurodegenerative disorders.[23] However, there is scarce clinical evidence for the therapeutic potential of such treatments.

Collectively, it should be mentioned that there is no specific therapeutic against diseases associated with the dis-homeostasis of iron and mitochondrial heme biosynthesis. Iron chelating drugs or the induction of heme degradation paths (e.g., heme oxygenase; HO-1) are therapeutic strategies tested in the neurodegenerative disorders associated with impaired heme metabolism.[24] On the other hand, as previously mentioned, several antioxidant agents such as coenzyme Q_{10} and vitamin E (Table 2) have been administered in complications associated with dysregulation of heme biosynthesis. Future studies on the precise mechanisms of heme dis-homeostasis in mitochondria could enhance our understanding of related diseases' pathophysiology and provide great clinical value.

2.3.2 *Citrate: A regulator of cellular anabolic reactions*

Citrate is a carbon source for the biosynthesis of different molecules such as fatty acids (FAs), ubiquinone, dolichol, and steroids (Fig. 4). Citrate is synthesized in the mitochondrial matrix by citrate synthase (CS) (a TCA cycle enzyme) (Fig. 4) and transported to the cytoplasm. In the cytoplasm, citrate is converted to acetyl-CoA (Fig. 4). Acetyl-CoA is a significant precursor for synthesizing vital molecules such as steroids, fatty acids, and ubiquinone. Acetyl-CoA also participates in essential signalings, such as histone acetylation (Fig. 4). Therefore, changes in citrate synthesis or transport could disturb cellular homeostasis. On the other hand, it seems that the citrate path intervention could target some diseases such as dyslipidemias and cancer.

Pathogenic mutations in the citrate carrier (SLC25A1; Fig. 4) have been identified.[32, 33] These mutations could impair mitochondrial citrate export and profoundly affect the cytoplasm's biosynthesis processes.[32] It has been found that SLC25A1 dysfunction could lead to severe pathophysiology. Neurometabolic disorders, severe muscular weakness, respiratory distress,

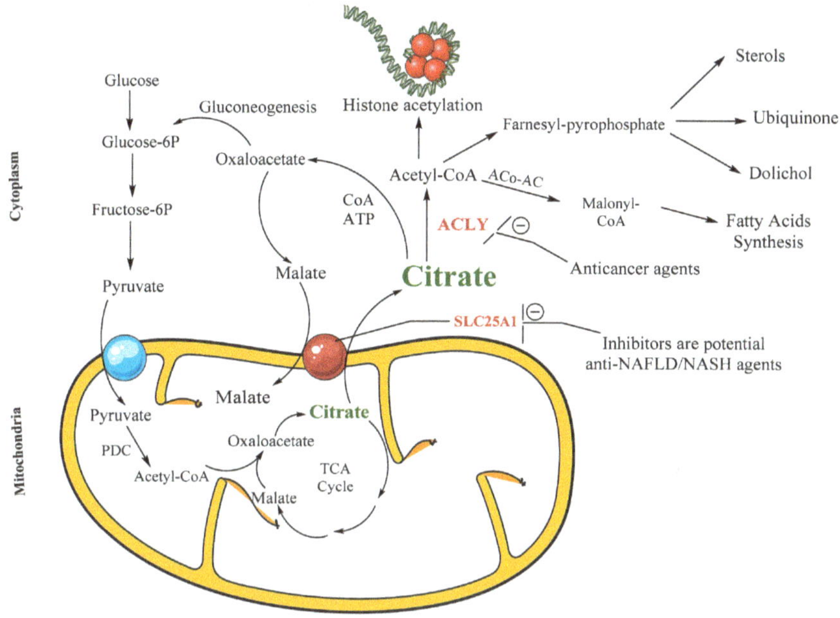

Fig. 4 Citrate acts as an essential mediator for synthesizing biomolecules such as fatty acids, ubiquinone, steroids, and dolichol. *ACo-AC*, acetyl-CoA carboxylase; *ACLY*, ATP citrate lyase; *PDC*, pyruvate dehydrogenase complex; *SLC25A1*, mitochondrial citrate carrier; *TCA*, tricarboxylic acid. Enzymes and transporters in the citrate biosynthesis and metabolism pathway are potential candidates for targeting diseases such as cancer or dyslipidemia.

seizure, and the lack of psychomotor development have been reported in patients with mutated SLC25A1.[32] On the other hand, the downstream synthesis of vital molecules such as bile acids, hormones, and cholesterol is immensely influenced. Molecules such as sphingolipids and dolichols are vital for brain development and function. This might explain the pathophysiology of severe brain injury in patients with mutated mitochondrial citrate transporter.[32] On the other hand, low cytoplasmic citrate could suppress ubiquinone synthesis and impairs mitochondrial function[32] (Fig. 4).

There is no specific treatment for patients with SLC25A1 mutation. On the other hand, reductive carboxylation of 2-oxoglutarate or acetate metabolism might partially compensate for disturbing mitochondrial citrate export.[32] Hence, acetic acid, with care, could be considered as a dietary supplement in patients with SLC25A1 mutation.[32]

Pharmacological inhibition of SLC25A1 has been proposed as a strategy to revert some pathological conditions such as steatosis, glucose intolerance, and inflammation in nonalcoholic fatty liver (NAFLD) and nonalcoholic steatohepatitis (NASH).[34] Manipulating the cytoplasmic citrate pool by SLC25A1 inhibitors could significantly affect the lipogenic and gluconeogenic pathways.[34] These events could finally affect liver tissue lipid profile and alleviate NAFLD and NASH.[34] Tan et al. reported that animals treated with a SLC25A1 inhibitor could afford a high-fat diet without liver injury, weight gain, or glucose metabolism disruption.[34] These data suggest the potential application of SLC25A1 inhibitors as promising agents in pathological conditions involving the disruption of lipid metabolism.

An important cytoplasmic enzyme named ATP citrate lyase (ACLY) facilitates citrate as a carbon source for synthesizing fatty acids and steroids (Fig. 4). Therefore, the inhibition of enzymes responsible for citrate metabolism has been investigated as therapeutic targets for several human diseases. It has been reported that ACLY is overexpressed in different types of cancer.[35, 36] Interestingly, many cancer cells rely on the ACLY for proliferation. Severe ACLY overexpression in tumor cells is associated with poor prognosis.[37] Fatty acids are widely used in rapidly developing cells (e.g., cancer cells) for membrane composition, signal transduction, and protein modification.[36] Therefore, the inhibition of ACLY could be a target for anticancer drugs. It has been found that the inhibition of ACLY significantly inhibited the growth of cancer cells both in vitro and in vivo.[37] The development of agents with more potency and selectivity for ACLY could lead to effective cancer chemotherapy strategies. Many researchers are believed that the investigation of the ACLY enzyme could lead to the development of new drugs for different types of cancer.[37]

Acetyl-CoA is used as a building block for lipogenesis and cholestrologenesis (Fig. 4); thus, the ACLY enzyme inhibition has been investigated to treat complications such as dyslipidemia and hepatic steatosis.[34] The inhibition of ACLY could also significantly prevent lipogenesis.[35, 38] ACLY inhibitors revealed promising antilipidemic effects in high-fat diet models of hyperlipidemia.[35, 38] Thus, these compounds could be potential future antihyperlipidemic compounds with promising therapeutic efficacy.

Cytoplasmic acetyl-CoA is also required for essential reactions such as histone acetylation[39] (Fig. 4). Histones are proteins that pack DNA molecules in structures called nucleosomes. Histone acetylation occurs when DNA transcriptional activity is increased. Therefore, DNA becomes more accessible for transcription. In tumor cells that need rapid DNA transcription, histone acetylation widely occurs, and acetyl-CoA is an essential compound used for histone acetylation in the nucleus of cancer cells.[39] As a metabolic source of acetyl-CoA, mitochondria-originated citrate plays a crucial role in histone acetylation and the proliferation of cancer cells (Fig. 4). Hence, eliminating this carbon pool for histone acetylation in cancer cells might have clinical value. For this purpose, mitochondrial citrate transporter inhibitors or ACLY enzyme inhibition could be considered potential targets (Fig. 4).

The data presented in this section unravel the importance of intervention in mitochondria-mediated cellular citrate synthesis as a therapeutic target. Future investigations will provide insight into the role of mitochondrial citrate metabolism and carrier in the pathogenesis of human diseases.

2.3.3 Folate cycle and nucleotides synthesis

Purine and pyrimidine nucleotides are building blocks required for vital cellular functions such as cell proliferation and growth. The folate cycle (FC) is a crucial mechanism involved in the biosynthesis of nucleotides. FC provides nucleotides for DNA synthesis (mitochondrial and nuclear) (Fig. 5). FC occurs in both cytoplasm and mitochondria. Cytoplasmic and mitochondrial FC are interconnected by tetrahydrofolate (THF) and formate (Fig. 5). THF is transported to the mitochondrial matrix through specific carriers. THF is converted to glycine and 5,10-methylene-THF by serine hydroxymethyltransferase enzyme (SHMT2) in the mitochondrial matrix (Fig. 5). 5,10-Methylene-THF is used by the enzyme thymidylate synthase to produce nucleotides (e.g., deoxythymidine monophosphate; dTMP) (Fig. 5). Some other mitochondria-located enzymes are also involved in nucleotides synthesis. Dihydroorotate dehydrogenase (DHODH) is an important enzyme involved in the de novo synthesis of pyrimidine nucleotides. DHODH is located on the surface of the inner mitochondrial membrane. This enzyme is responsible for uridine monophosphate (UMP) synthesis. UMP is the precursor of pyrimidine nucleotides (Fig. 5).

It has been well-established that folate deficiency leads to the instability of mtDNA.[40] It has been well-established that folate deprivation significantly depletes mtDNA in several tissues.[40] Hence, folate deficiency could lead to decreased expression of mitochondrial respiratory chain components. Consequently, the mitochondrial function will be impaired (Fig. 5). Finally, mitochondria-mediated oxidative stress and cell death will occur. Interestingly, it has been revealed that FC metabolites, such as 5,10-CH_2-THF (Fig. 4), play a crucial role in mitochondrial protein translation[41] (Fig. 5). 5,10-CH_2-THF is needed for proper mitochondrial tRNA modification[41] (Fig. 5). Dysfunctional tRNA cannot participate in the appropriate protein synthesis process (e.g., ETC components synthesis) in mitochondria.

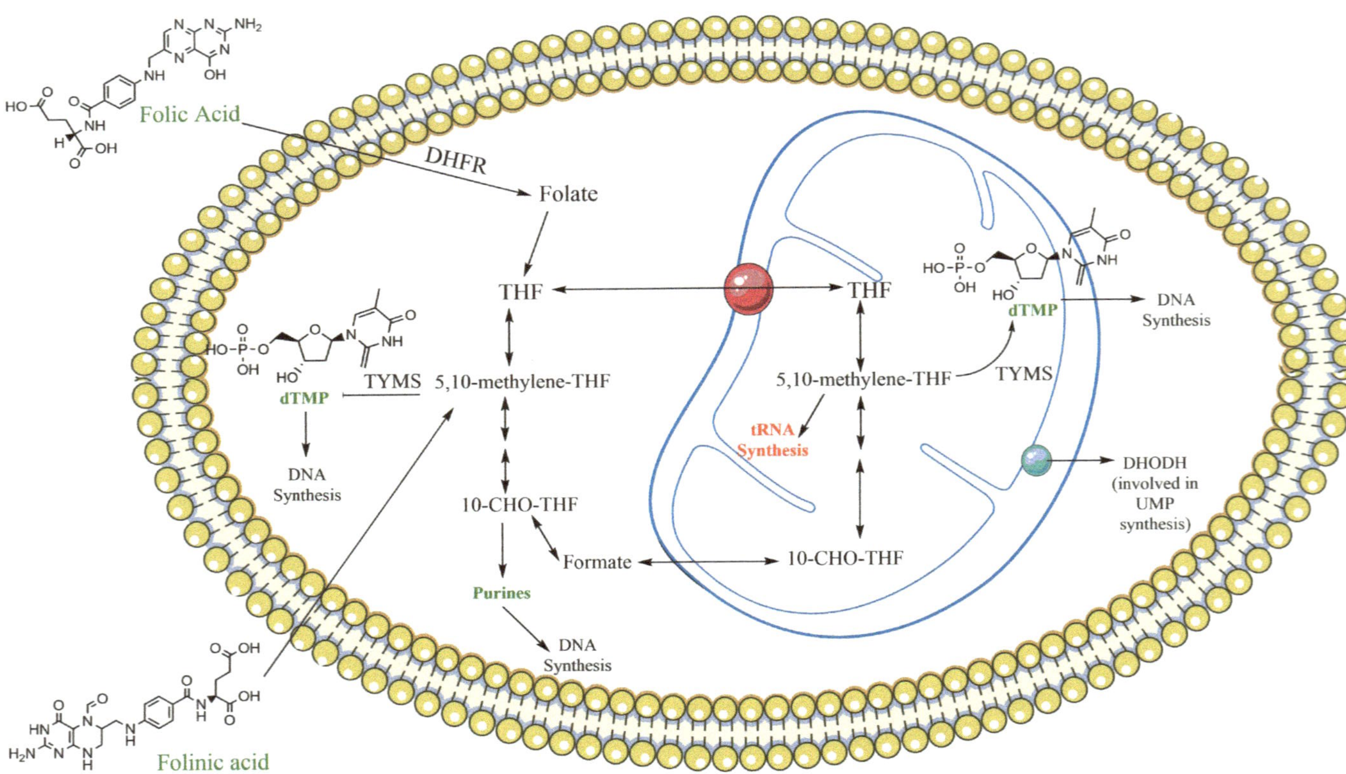

Fig. 5 Cytoplasmic and mitochondrial compartmentalization of the folate cycle. Nucleotide synthesis relies on carbon metabolism in the folate cycle. Anticancer and immune suppressor drugs target folate biosynthesis. Impaired mitochondrial metabolism of folate could lead to mtDNA depletion and cytotoxicity. On the other hand, as 5,10-CH_2-THF plays a pivotal role in mitochondrial tRNA synthesis, the production of several mtDNA-encoded proteins could also be impaired upon folate biosynthesis dysfunction. *DHFR*, dihydrofolate reductase; *dTMP*, deoxythymidine monophosphate; *THF*, tetrahydrofolate; *TYMS*, thymidylate synthase; *UMP*, uridine monophosphate.

Impaired mitochondrial tRNA has been detected in mitochondrial diseases such as MERRF and MELAS[42] (Table 1).

Several mitochondrial diseases associated with folate deficiency have been identified.[43, 44] It has been found that some neurological disorders such as Aicardi-Goutieres, Rett, and Kearns-Sayre syndromes are associated with folate deficiency and consequent mitochondrial impairment.[44, 45] Some other mitochondrial cytopathies such as NARP and Leigh syndrome also are associated with a low level of folate metabolites (e.g., 5-MTHF) in biological samples (e.g., in cerebrospinal fluid).[44] Actually, cerebral folate deficiency (CFD) is a frequent event observed in several central nervous system-associated mitochondrial disorders.[10, 44, 46] Based on these studies, folic/folinic acid supplementation could have a role in the management of CFD-associated mitochondrial diseases (Table 2). There is clinical evidence on the efficiency of folic/folinic acid in managing patients with mitochondrial cytopathies associated with disturbed folate metabolism.[47]

Immune cells, such as T cells, depend on the FC-based synthesis of nucleotides for their differentiation and activation.[48] Hence, several therapeutic agents that suppress the immune response and modulate autoimmune diseases target the FC.[49, 50] Methotrexate, a suppressor of FC, is a drug used to manage immune response (e.g., in rheumatoid arthritis) or suppressing cancer cell growth.[51] This drug inhibits the dihydrofolate reductase (DHFR) enzyme in FC and prevents the production of tetrahydrofolate (THF) (Fig. 5). 5-Fluorouracil (5-FU) is another drug used to suppress FC-mediated nucleotides synthesis. 5-FU is an inhibitor of the enzyme thymidylate synthase (TYMS) (Fig. 5). It has also been found that T cells need DHODH and pyrimidine synthesis during their differentiation into mature and effective cells.[52] Hence, DHODH inhibitors also could suppress the immune response (Fig. 5).

Targeting one-carbon metabolism and de novo synthesis of nucleotides through FC is a hot topic in cancer therapy.[53] It has been well-documented that FC's proper function guarantees the high cell survival rate and proliferation required for tumor growth, and many metabolites from FC are vital for cancer cell growth and expansion.[53] In addition to molecules such as nucleotides, which are building blocks for DNA synthesis in rapidly growing cancer cells, molecules such as NADPH are important metabolites produced during FC. As previous studies have shown, oxidative stress occurs much more frequently in cancer cells than in normal tissues[54]; therefore, cancer cells need more robust defense mechanisms to counteract the harmful effects of oxidative stress. NADPH is an essential cofactor for encountering oxidative stress and reconstruction of damaged proteins.

Interestingly, it has been reported that the inhibition of mitochondrial enzymes involved in the FC could enhance the response of a tumor to chemotherapy.[55] Lee et al. reported that the inhibition of mitochondrial methylenetetrahydrofolate dehydrogenase 1 (MTHFD1L) significantly sensitized the hepatocarcinoma to sorafenib.[55] It has been found that MTHFD1L is significantly upregulated in cancer (e.g., hepatocellular carcinoma).[55] These data indicate the importance of this enzyme in cancer. The inhibition of MTHFD1L could lead to severe oxidative stress in tumors that eventually lead to cell death.[55] Based on these data, the FC's pharmacological inhibition (e.g., methotrexate) could be an excellent option to decrease cancer cell survival synergistically.[31]

Leflunomide is a well-known inhibitor of DHODH and interferes with the synthesis of pyrimidine nucleotides[52] (Fig. 5). Thus, this drug inhibits the multiplication of fast-dividing cells.[52] The inhibition of DHODH also inhibits the differentiation of some types of cancer cells (e.g., AML).[52] Leflunomide is clinically administered as an immunosuppressant in diseases such as rheumatoid arthritis, psoriatic arthritis, Crohn's disease, prostate cancer, sarcoidosis, or prevent organ transplant rejection.[52]

It could be concluded that the FC and the enzymes involved in this process are potential therapeutic targets in the treatment of various diseases such as cerebral complications, cancer, and immune system-related disorders. Further studies in this field and the extension of pharmacological interventions could lead to the development of more effective drugs against mitochondrial cytopathies.

2.3.4 Amino acids synthesis and metabolism in the mitochondria

The association of mitochondria with the synthesis of various amino acids (AAs) has long been known. Several AAs are totally synthesized in the mitochondrial matrix. Glutamine, alanine, glutamate, proline, and aspartate are synthesized explicitly in the mitochondria.[56] Some AAs are not synthesized in the mitochondria, but they are significantly compartmentalized in this organelle.[56] On the other hand, it has been found that the metabolism of more than 17 AAs utilizes mitochondrial enzymes.[56] However, the significance of the localization and synthesis of AAs in mitochondria is mainly unknown.

AAs also contribute to the synthesis of vital molecules in mitochondria. For example, glutamate is involved in α-ketoglutarate synthesis; and isoleucine, phenylalanine, and tyrosine could incorporate into acetyl-CoA metabolism. Isoleucine, valine, and methionine could also form succinyl-CoA.[56] Over 40 human diseases have been identified in the association of disturbed AAs synthesis and degradation in the mitochondria.[56] Phenylketonuria (disruption in phenylalanine metabolism), maple syrup urine disease (disturbed valine, leucine, and isoleucine degradation), and ornithine transcarbamylase deficiency (impaired arginine metabolism, impaired urea cycle) are some well-known examples of these diseases. Readers could find a complete list of these disorders in a study by Guda et al.[56]

Unfortunately, little is known about targeting mitochondrial AAs metabolism for the treatment of human diseases to date. This issue might be associated with the complexity of cellular AAs metabolism. Actually, the metabolic path of AAs span across several subcellular compartments other than mitochondria.[56] This issue complicates the investigation about targeting AAs metabolic pathways.

The synthesis of glutamine is one of the most investigated subjects related to AAs metabolism in mitochondria. Glutamine synthetase uses glutamate and ammonia to produce glutamine. This pathway is a critical mechanism of ammonia detoxification in tissues that lack the urea cycle (e.g., the brain).[57] In clinical complications such as hepatic encephalopathy, the brain level of ammonia is dramatically increased.[57] On the other hand, glutamine is used for nucleotides synthesis in some types of cancer cells (e.g., glioblastoma).[58] Hence, targeting mitochondria, amino acid synthesis could be a target for managing several human diseases. Unlike glutamine, the mechanisms and importance of compartmentalization of AAs, such as glutamate in mitochondria, are vague.

Interestingly, several studies revealed the positive effects of AAs on mitochondrial function.[11, 59] It has been found that the administration of some AAs and their derivatives could significantly prevent mitochondria-mediated cell death.[3, 11, 60–62] The role of AAs in blunting mitochondria-mediated ROS formation is an exciting feature of these compounds. The effects of AAs on oxidative stress could be mediated through their direct interaction with oxygen radicals or by interfering with mitochondrial redox reactions.[3, 11, 60–62]

The data presented in the current section highlight the importance of more investigation in AAs metabolism. As there is a lack of data in this field, future works could enhance our understanding of diseases caused by abnormal amino acid metabolism in mitochondria. Moreover, more studies are warranted to clarify the mechanisms of mitochondria protective properties provided by several AAs.

3 Pharmacological options to target mitochondrial function

With tremendous advances in molecular biology and genetics, a substantial and in-depth understanding of mitochondrial diseases has been provided. However, little is known about the therapeutic interventions against these complications (Table 1). In this section, pharmacological solutions to alleviate the symptoms of mitochondrial diseases are discussed. It should be mentioned that although these strategies may well control the symptoms of the disease, they are not definitive and permanent treatments.

Despite the etiology of mitochondrial impairment, enhancing cellular energy metabolism could improve cell and organ function. Several pharmacological approaches have been suggested for alleviating the impairment of mitochondrial oxidative phosphorylation. Decreasing mitochondria-mediated ROS formation and oxidative stress is a widely tested strategy to blunt mitochondrial cytopathies symptoms. In this regard, several synthetic and natural antioxidants have been studied. Also, several agents also have been used to boost mitochondrial ATP production. An overview of the effects of these compounds on mitochondrial function is given below.

Idebenone is a synthetic analog of ubiquinone as a vital antioxidant and electron transport chain component. This drug is now approved for the management of LHON in Europe[12] (Table 2). Mitochondria-targeted antioxidants also have been widely investigated. Elamipretide is a mitochondrially targeted peptide that profoundly decreased mitochondrial-mediated ROS formation and protected mitochondrial cardiolipin structure[9] (Table 2). Clinical studies revealed promising results for elamipretide in the management of mitochondrial diseases.[63] Curcumin is another powerful antioxidant agent used for the alleviation of symptoms of mitochondrial disorders.[64] Clinical trials are also in progress to evaluate curcumin's effects on mitochondrial function in diseases such as LHON[8] (Table 2). Our previous studies on some antioxidant amino acids and peptides revealed that these compounds could effectively preserve mitochondrial indices of functionality in different experimental models (Table 2).[3, 11, 60–62, 65] As mentioned in previous sections, COQ_{10} and vitamin E are also clinically administered against mitochondrial diseases[5–7] (Table 2). These compounds could enhance mitochondrial ATP production in patients with mitochondrial cytopathies.[5–7] Creatine is another compound investigated for its positive effects on mitochondrial diseases (Table 2). Creatine-bond phosphate (phosphocreatine) is the primary source of energy phosphate.[5–7] Brain and skeletal muscle tissues have a high level of phosphocreatine. It has been found that phosphocreatine content is significantly decreased in the skeletal muscle and brain of patients with mitochondrial disease.[5–7] Creatinine therapy in these patients enhanced considerably motor ability and aerobic activity.[5–7] Nevertheless, a very high dose is needed to administer daily (e.g., 2–10 g/day).

Several other compounds could also be named for their positive effects on mitochondria. However, it is important to mention that all the aforementioned pharmacological options symptomatically alleviate mitochondrial diseases. Therefore, suboptimal control of the progress of the mitochondrial disease could frequently occur. The next section highlights the importance of permanent and newly emerged therapeutic strategies for mitochondrial disease.

3.1 Novel strategies in targeting mitochondrial diseases

Pharmacological interventions are only supportive and ancillary treatments for patients with mitochondrial diseases. Although the administration of several antioxidants and mitochondrial cofactors positively improved the quality of life in mitochondrial disorders (Table 2), this is not a permanent strategy and often leads to heterogeneous results. On the other hand, there are scarce clinical trials for the effectiveness of pharmacotherapy in managing mitochondrial disorders. Pharmacological interventions could also profoundly force considerable economic pressure on families as some of these drugs are expensive and should be consumed for the whole life. Some studies also mentioned that patients' perception of their quality of life is poor during drug therapy of mitochondrial diseases.[12] Therefore, in addition to all the advantages of drug therapy methods to improve the lives of patients with mitochondrial diseases, the existence of such disadvantages led researchers to find permanent treatments for these complications.

Recently, new strategies for the treatment of mitochondrial diseases have emerged. Gene therapy and mitochondrial replacement therapy are new options for the management of mitochondrial diseases. The main difference between these types of therapies compared with pharmacological interventions is their tremendous effectiveness, which could even wholly treat the condition. On the other hand, despite many advances in these fields, some problems are still associated with these modern therapies, which are mentioned in the following sections.

3.1.1 Gene therapy: A promising therapeutic intervention in mitochondrial disease?

Gene therapy (GT) is an evolving field of investigation that locally replaced functional genes with mutated ones to treat genetic disorders. It has been found that GT dramatically improved some mitochondrial disorders in both experimental models and clinical cases.[12] Several strategies have been developed for editing mutated genes in both mitochondria and the nucleus. In this regard, using adeno-associated viruses (AAVs) is a good option as AAVs have long persistence in cells.[66] GT provided a promising tool for treating several inherited genetic diseases[67]; however, using this technology to manage mitochondrial disorders interfaced with several obstacles.

It is imperative to mention that GT is currently applied for mitochondrial disorders associated with a mutation in nuclear DNA genes encoding mitochondrial components.[12] Specific biological features of mitochondria complicate the genetic manipulation of this organelle. Correcting mutations in mtDNA is a big challenge compared to the GT of nuclear DNA.[12] However, induction of the expression of mtDNA encoded proteins by nuclear DNA is a fascinating strategy that could overcome this obstacle. These nuclear DNA-encoded proteins could have a mitochondria-targeting sequence added to them in the cellular nucleus.[12] Some experimental models and clinical trials test this strategy in mitochondrial diseases, such as LHON[68] (Table 1).

Another exciting field of study associated with GT of mitochondrial disease is shifting mtDNA mutation heteroplasmy.[12] Heteroplasmy implies multiple individual copies of mtDNA in a given cell or tissue. Due to high mtDNA copies in a cell, low mtDNA changes could not significantly improve the clinical effect. However, if the number of mutated mitochondrial DNAs exceeds a certain level, mitochondrial disease symptoms will develop. Interestingly, shifting mitochondrial mutant heteroplasmy could serve as a therapeutic option in mitochondrial diseases. Novel strategies have been developed to reduce the mitochondrial level of mutated DNA.[12] These techniques include mitochondria-targeted restriction endonucleases, zinc finger endonucleases, transcription activator-like nucleases, and clustered regulatory interspaced palindrome repeat (CRISPER)/Cas9.[12] These strategies could reduce the mtDNA mutation load and increase the capacity to produce ATP.[12]

Despite the tremendous advances in GT, this technique has not yet been widely used to treat mitochondrial diseases. This may be due to the complexity of mitochondrial diseases or the inconvenience of this technique to date. However, it seems that GT will play a vital role in the future treatment of mitochondrial diseases caused by DNA mutations.

3.1.2 Mitochondrial replacement therapy

The mitochondrial genome is maternally inherited to the next generation. Therefore, all inherited mitochondrial disorders come from oocytes. The goal of mitochondria replacement therapy (MRT) or mitochondrial donation is to prevent transferring mutant mtDNA from affected oocytes to the next generation.[69] Thus, mothers with inherited mitochondrial diseases find a chance to have a healthy child. It is worth noting that MRT is not a treatment for existed mitochondrial diseases, but it abolishes the inheritance of mitochondrial impairments from mothers to their progenies.

MRT refers to a range of techniques in which the donated mitochondria from a healthy female donor is transplanted through IVF.[69, 70] Pronuclear transfer, maternal spindle transfer, and polar body genome transfer are techniques used in MRT.[69] For more comprehensive information and how to do these techniques, readers are referred to comprehensive reviews in this field (e.g., Refs. 69, 70).

It should be noted that MRT is mistakenly considered a way to germline modification in many countries.[70] Therefore, this technique might be strictly prohibited. Consequently, it is the scientific community's responsibility to provide accurate and enlightening explanations in this regard.

According to the data presented in this section, MRT could provide a suitable solution to prevent mutated mtDNA transmission to the next generation. However, there is still a long way to go before this method becomes convenient.

4 Prospects

Mitochondria are not only the hub of ATP synthesis but also contain numerous enzymes and transporters that play a critical role in maintaining cellular homeostasis. Impairment in proper mitochondrial function is associated with a wide range of diseases. Today, remarkable progress has been achieved in mitochondrial medicine due to tremendous progress in identifying and understanding mitochondrial disorders' pathogenesis. On the other hand, targeting mitochondria could provide promising therapeutic strategies for many human diseases.

Advanced techniques such as GT and MRT are also a breakthrough in targeting mitochondrial diseases. However, due to the economic, social, legal, and ethical barriers to this technique, its widespread use is currently limited. mtDNA mutation affects a wide range of organs. Hence, gene therapy of mitochondria for a specific tissue often fails. Many vectors are needed to target various tissues, leading to complications such as exaggerated immune response. This issue also considerably increases the cost of treatment. Other problems, such as crossing physiological barriers (e.g., BBB), are another challenge for gene therapy of mitochondrial diseases. Indeed, further research in this field could significantly resolve many of these issues.

Finally, it should be mentioned that by studying the synthesis pathways of biomolecules, good outcomes have been achieved in the treatment of mitochondrial diseases. However, more investigations are needed to develop effective clinical outcomes.

References

1. Osellame LD, Blacker TS, Duchen MR. Cellular and molecular mechanisms of mitochondrial function. *Best Pract Res Clin Endocrinol Metab.* 2012;26(6):711–723.
2. Duchen MR, Szabadkai G. Roles of mitochondria in human disease. *Essays Biochem.* 2010;47:115–137.
3. Heidari R, Niknahad H, Sadeghi A, et al. Betaine treatment protects liver through regulating mitochondrial function and counteracting oxidative stress in acute and chronic animal models of hepatic injury. *Biomed Pharmacother.* 2018;103:75–86.
4. Heidari R, Ghanbarinejad V, Ommati MM, Jamshidzadeh A, Niknahad H. Regulation of mitochondrial function and energy metabolism: a primary mechanism of cytoprotection provided by carnosine. *Trends Pharmacol Sci.* 2018;4(1):41–50.
5. Lodi R, Hart PE, Rajagopalan B, et al. Antioxidant treatment improves in vivo cardiac and skeletal muscle bioenergetics in patients with Friedreich's ataxia. *Ann Neurol.* 2001;49(5):590–596.
6. Hart PE, Lodi R, Rajagopalan B, et al. Antioxidant treatment of patients with Friedreich ataxia: four-year follow-up. *Arch Neurol.* 2005;62(4):621–626.
7. Cooper JM, Schapira AHV. Friedreich's ataxia: coenzyme Q10 and vitamin E therapy. *Mitochondrion.* 2007;7(Suppl):S127–S135.
8. Mahidol U, Randomized A. *Double-Blind, Placebo-Controlled Trial of Curcumin in Leber's Hereditary Optic Neuropathy (LHON)*; 2012. clinicaltrials.gov.
9. Mitchell W, Ng EA, Tamucci JD, et al. Molecular mechanism of action of mitochondrial therapeutic SS-31 (Elamipretide): membrane interactions and effects on surface electrostatics. *bioRxiv.* 2019;1–36.
10. Ormazabal A, Casado M, Molero-Luis M, et al. Can folic acid have a role in mitochondrial disorders? *Drug Discov Today.* 2015;20(11):1349–1354.
11. Heidari R, Ghanbarinejad V, Ommati MM, Jamshidzadeh A, Niknahad H. Mitochondria protecting amino acids: application against a wide range of mitochondria-linked complications. *PharmaNutrition.* 2018;6(4):180–190.
12. Hirano M, Emmanuele V, Quinzii CM. Emerging therapies for mitochondrial diseases. *Essays Biochem.* 2018;62(3):467–481.
13. Mousavi K, Niknahad H, Ghalamfarsa A, et al. Taurine mitigates cirrhosis-associated heart injury through mitochondrial-dependent and antioxidative mechanisms. *Clin Exp Hepatol.* 2020;6(3):207–219.

14. Svensson OL. *Mitochondria: Structure, Functions and Dysfunctions.* 1st ed. Hauppauge, NY: Nova Science Publishers, Inc; 2011.
15. Russell OM, Gorman GS, Lightowlers RN, Turnbull DM. Mitochondrial diseases: hope for the future. *Cell.* 2020;181(1):168–188.
16. Mustafa MF, Fakurazi S, Abdullah MA, Maniam S. Pathogenic mitochondria DNA mutations: current detection tools and interventions. *Gene.* 2020;11(2):192.
17. Swenson SA, Moore CM, Marcero JR, Medlock AE, Reddi AR, Khalimonchuk O. From synthesis to utilization: the ins and outs of mitochondrial heme. *Cells.* 2020;9(3).
18. Iolascon A, Falco LD, Beaumont C. Molecular basis of inherited microcytic anemia due to defects in iron acquisition or heme synthesis. *Haematologica.* 2009;94(3):395–408.
19. Shetty T, Corson TW. Mitochondrial heme synthesis enzymes as therapeutic targets in vascular diseases. *Front Pharmacol.* 2020;11.
20. Basavarajappa HD, Sulaiman RS, Qi X, et al. Ferrochelatase is a therapeutic target for ocular neovascularization. *EMBO Mol Med.* 2017;9(6):786–801.
21. Corson TW, Sulaiman RS, Alkhairy S, Gupta K, Basavarajappa HD. Griseofulvin inhibits choroidal neovascularization. *Invest Ophthalmol Vis Sci.* 2016;57(12):2122.
22. Gupta V, Liu S, Ando H, et al. Salicylic acid induces mitochondrial injury by inhibiting ferrochelatase heme biosynthesis activity. *Mol Pharmacol.* 2013;84(6):824–833.
23. Fiorito V, Chiabrando D, Tolosano E. Mitochondrial targeting in neurodegeneration: a heme perspective. *Pharmaceuticals (Basel).* 2018;11(3):1–21.
24. Chiabrando D, Fiorito V, Petrillo S, Tolosano E. Unraveling the role of heme in neurodegeneration. *Front Neurosci.* 2018;12:1–14.
25. Llorens JV, Soriano S, Calap-Quintana P, Gonzalez-Cabo P, Moltó MD. The role of iron in Friedreich's ataxia: insights from studies in human tissues and cellular and animal models. *Front Neurosci.* 2019;13:1–16.
26. Hanson E, Sheldon M, Pacheco B, Alkubeysi M, Raizada V. Heart disease in Friedreich's ataxia. *World J Cardiol.* 2019;11(1):1–12.
27. Calabrese V, Lodi R, Tonon C, et al. Oxidative stress, mitochondrial dysfunction and cellular stress response in Friedreich's ataxia. *J Neurol Sci.* 2005;233(1–2):145–162.
28. Lupoli F, Vannocci T, Longo G, Niccolai N, Pastore A. The role of oxidative stress in Friedreich's ataxia. *FEBS Lett.* 2018;592(5):718–727.
29. La Rosa P, Bertini ES, Piemonte F. The NRF2 signaling network defines clinical biomarkers and therapeutic opportunity in Friedreich's ataxia. *Int J Mol Sci.* 2020;21(3):1–16.
30. Homedan C, Laafi J, Schmitt C, et al. Acute intermittent porphyria causes hepatic mitochondrial energetic failure in a mouse model. *Int J Biochem Cell Biol.* 2014;51:93–101.
31. Ferrer MD, Tauler P, Sureda A, Palacín C, Tur JA, Pons A. Antioxidants restore protoporphyrinogen oxidase in variegate porphyria patients. *Eur J Clin Invest.* 2013;43(7):668–678.
32. Majd H, King MS, Smith AC, Kunji ERS. Pathogenic mutations of the human mitochondrial citrate carrier SLC25A1 lead to impaired citrate export required for lipid, dolichol, ubiquinone and sterol synthesis. *Biochim Biophys Acta Bioenerg.* 2018;1859(1):1–7.
33. Edvardson S, Porcelli V, Jalas C, et al. Agenesis of corpus callosum and optic nerve hypoplasia due to mutations in SLC25A1 encoding the mitochondrial citrate transporter. *J Med Genet.* 2013;50(4):240–245.
34. Tan M, Mosaoa R, Graham GT, et al. Inhibition of the mitochondrial citrate carrier, Slc25a1, reverts steatosis, glucose intolerance, and inflammation in preclinical models of NAFLD/NASH. *Cell Death Differ.* 2020;1–15.
35. Enache LS. ATP citrate lyase–biology and implication in human pathology. *Rev Rom Med Lab.* 2008;12(3):17–30.

36. Devi Khwairakpam A, Singh Shyamananda M, Lalduhsaki Sailo B, et al. ATP citrate lyase (ACLY): a promising target for cancer prevention and treatment. *Curr Drug Targets.* 2015;16(2):156–163.
37. Migita T, Narita T, Nomura K, et al. ATP citrate lyase: activation and therapeutic implications in non–small cell lung cancer. *Cancer Res.* 2008;68(20):8547–8554.
38. Feng X, Zhang L, Xu S, Shen A-Z. ATP-citrate lyase (ACLY) in lipid metabolism and atherosclerosis: an updated review. *Prog Lipid Res.* 2020;77:101006.
39. Feron O. The many metabolic sources of acetyl-CoA to support histone acetylation and influence cancer progression. *Ann Transl Med.* 2019;7(Suppl 8):1–6.
40. Chang C-M, Yu C-C, Lu H-T, Chou Y-F, Huang R-FS. Folate deprivation promotes mitochondrial oxidative decay: DNA large deletions, cytochrome c oxidase dysfunction, membrane depolarization and superoxide overproduction in rat liver. *Br J Nutr.* 2007;97(5):855–863.
41. Zheng Y, Cantley LC. Toward a better understanding of folate metabolism in health and disease. *J Exp Med.* 2019;216(2):253–266.
42. El-Hattab AW, Adesina AM, Jones J, Scaglia F. MELAS syndrome: clinical manifestations, pathogenesis, and treatment options. *Mol Genet Metab.* 2015;116(1–2):4–12.
43. Froese DS, Fowler B, Baumgartner MR. Vitamin B12, folate, and the methionine remethylation cycle—biochemistry, pathways, and regulation. *J Inherit Metab Dis.* 2019;42(4):673–685.
44. Garcia-Cazorla A, Quadros EV, Nascimento A, et al. Mitochondrial diseases associated with cerebral folate deficiency. *Neurology.* 2008;70(16):1360–1362.
45. Blau N, Bonafe L, Krägeloh-Mann I, et al. Cerebrospinal fluid pterins and folates in Aicardi-Goutieres syndrome: a new phenotype. *Neurology.* 2003;61(5):642–647.
46. Agodi A, Barchitta M, Quattrocchi A, Marchese AE, Boffetta P. Folate deficiency is not associated with increased mitochondrial genomic instability: results from dietary intake and lymphocytic mtDNA 4977-bp deletion in healthy young women in Italy. *Mutagenesis.* 2014;29(2):101–106.
47. Quijada-Fraile P, O'Callaghan M, Martín-Hernández E, et al. Follow-up of folinic acid supplementation for patients with cerebral folate deficiency and Kearns-Sayre syndrome. *Orphanet J Rare Dis.* 2014;9:1–9.
48. Ron-Harel N, Santos D, Ghergurovich JM, et al. Mitochondrial biogenesis and proteome remodeling promotes one carbon metabolism for T cell activation. *Cell Metab.* 2016;24(1):104–117.
49. Navarro-Peran E, Cabezas-Herrera J, Sanchez-Del-Campo L, García-Cánovas F, Rodríguez-López JN. The anti-inflammatory and anti-cancer properties of epigallocatechin-3-gallate are mediated by folate cycle disruption, adenosine release and NF-κB suppression. *Inflamm Res.* 2008;57(10):472–478.
50. Schaer DA, Geeganage S, Amaladas N, et al. The folate pathway inhibitor pemetrexed pleiotropically enhances effects of cancer immunotherapy. *Clin Cancer Res.* 2019;25(23):7175–7188.
51. Rider BJ. Methotrexate. In: Enna SJ, Bylund DB, eds. *xPharm: The Comprehensive Pharmacology Reference.* New York: Elsevier; 2007:1–5.
52. Greene S, Watanabe K, Braatz-Trulson J, Lou L. Inhibition of dihydroorotate dehydrogenase by the immunosuppressive agent leflunomide. *Biochem Pharmacol.* 1995;50(6):861–867.
53. Shuvalov O, Petukhov A, Daks A, Fedorova O, Vasileva E, Barlev NA. One-carbon metabolism and nucleotide biosynthesis as attractive targets for anticancer therapy. *Oncotarget.* 2017;8(14):23955–23977.
54. Hayes JD, Dinkova-Kostova AT, Tew KD. Oxidative stress in cancer. *Cancer Cell.* 2020;38(2):167–197.
55. Lee D, Wong CC-L. The folate cycle is a new metabolic weakness of cancer. *Mol Cell Oncol.* 2017;4(6):1–3.

56. Guda P, Guda C, Subramaniam S. Reconstruction of pathways associated with amino acid metabolism in human mitochondria. *Genomics Proteomics Bioinformatics.* 2007;5(3–4):166–176.
57. Heidari R. Brain mitochondria as potential therapeutic targets for managing hepatic encephalopathy. *Life Sci.* 2019;218:65–80.
58. Obara-Michlewska M, Szeliga M. Targeting glutamine addiction in gliomas. *Cancers (Basel).* 2020;12(2).
59. Ommati MM, Farshad O, Jamshidzadeh A, Heidari R. Taurine enhances skeletal muscle mitochondrial function in a rat model of resistance training. *PharmaNutrition.* 2019;9:100161.
60. Mohammadi H, Ommati MM, Farshad O, et al. Taurine and isolated mitochondria: a concentration-response study. *Trends Pharmacol Sci.* 2019;5(4):197–206.
61. Ommati MM, Heidari R, Ghanbarinejad V, Abdoli N, Niknahad H. Taurine treatment provides neuroprotection in a mouse model of manganism. *Biol Trace Elem Res.* 2019;190(2):384–395.
62. Ommati MM, Farshad O, Azarpira N, Shafaghat M, Niknahad H, Heidari R. Betaine alleviates cholestasis-associated renal injury by mitigating oxidative stress and enhancing mitochondrial function. *Biologia.* 2021;76:351–365.
63. Karaa A, Haas R, Goldstein A, Vockley J, Weaver WD, Cohen BH. Randomized dose-escalation trial of elamipretide in adults with primary mitochondrial myopathy. *Neurology.* 2018;90(14):e1212–e1221.
64. Zhu Y-g, Chen X-c, Chen Z-z, et al. Curcumin protects mitochondria from oxidative damage and attenuates apoptosis in cortical neurons. *Acta Pharmacol Sin.* 2004;25(12):1606–1612.
65. Ommati MM, Jamshidzadeh A, Heidari R, et al. Carnosine and histidine supplementation blunt lead-induced reproductive toxicity through antioxidative and mitochondria-dependent mechanisms. *Biol Trace Elem Res.* 2019;187(1):151–162.
66. Wang D, Tai PWL, Gao G. Adeno-associated virus vector as a platform for gene therapy delivery. *Nat Rev Drug Discov.* 2019;18(5):358–378.
67. Gonçalves GAR, Paiva RDMA. Gene therapy: advances, challenges and perspectives. *Einstein (Sao Paulo).* 2017;15(3):369–375.
68. Zhang Y, Tian Z, Yuan J, et al. The progress of gene therapy for Leber's optic hereditary neuropathy. *Curr Gene Ther.* 2018;17(4):320–326.
69. Amato P, Tachibana M, Sparman M, Mitalipov S. Three-parent IVF: gene replacement for the prevention of inherited mitochondrial diseases. *Fertil Steril.* 2014;101(1):31–35.
70. Sharma H, Singh D, Mahant A, Sohal SK, Kesavan AK, Samiksha. Development of mitochondrial replacement therapy: a review. *Heliyon.* 2020;6(9), e04643.

CHAPTER 3

Mitochondrial contribution to cell death

Shilan Mozaffari and Mohammad Abdollahi
Toxicology and Diseases Group (TDG), Pharmaceutical Sciences Research Center (PSRC), The Institute of Pharmaceutical Sciences (TIPS), Tehran University of Medical Sciences, Tehran, Iran

Contents

1 Role of mitochondria in cell death

Mitochondria, double-membrane dynamic organelles in cells, have long been thought to be the cells' powerhouse. The pivotal role of mitochondria in cells is to generate adenosine triphosphate (ATP) through respiration and metabolism regulation. Mitochondrial regular performance is crucial for survival and cell function. It is considered as quality control for cell survival. Various ATP levels at different cycles of cells indicate mitochondria's implication in each level's regulation and initiation. It has recently been acknowledged that mitochondria are involved in some mechanisms of cell death.[1, 2] Based on robust biological studies, mitochondria are considered as deciding crossroads for cell death or survival.[3–5] It is demonstrated to contribute to death pathways. Mitochondria could play a vital role in either programmed (apoptosis) or accidental death process.[2, 6] Autophagy is another path in which mitochondria have an essential part in its conduction. Autophagy

Mitochondrial Metabolism
https://doi.org/10.1016/B978-0-12-822416-8.00004-X

sometimes could be categorized as a death pathway. Mitochondria's autophagy role could vary by either facilitating the autophagosome biogenesis as a membrane or being a target organelle (mitophagy) of the autophagy process.[2]

Despite the complexity of different signal networking in death pathways, a broad range of essential factors and signals and the aftereffect of alteration of each element are identified already.[1] This chapter will explain several cell death categories, and crucial associated determinants in each pathway will be described further. We will review the prominent components and how they could induce or inhibit the pathways in different situations. However, the main emphasis will be on mitochondrial contribution to cell death pathways.

There is compelling evidence on mitochondrial dysfunction's association with various pathological states and diseases.[7–9] Of these, cancers and malignancies, metabolic syndromes, cardiomyopathies, and neurodegenerative diseases such as Parkinson's disease and multiple sclerosis are highlighted. Therefore, knowing the exact critical players of each pathway and their regulatory forces, intracellular factors might promise new therapeutic opportunities.[10]

2 Cell death pathways

Cell death is a crucial physiological process due to its role in early development and removing dysfunctional damaged cells with inefficient energy. Each mammalian cell has a cycle, including distinct stages of life and functioning. Regarding the regulation programs, cells undergo the death pathway when required. Overall, two dominant cell death types are described in mammalian cells, programmed or regulated cell death versus unwanted cell loss.[11] Programmed cell death (apoptosis) includes extrinsic death receptor pathway and intrinsic mitochondrial accompanied pathway.[11, 12] Necrosis and autophagy are identified as nonprogrammed death pathways.[11, 13, 14] Deregulated and interrupted apoptosis is demonstrated to be linked with immunodeficiency, autoimmune disorders, and cancers. It has appeared that the death receptor pathway and mitochondrial-mediated death are mediating both programmed and necrosis death.

In conclusion, four types of cell death are taken into account as follows: mitochondrial-associated apoptosis, mitochondrial-associated necrosis, death receptor-mediated apoptosis, and death receptor-mediated necrosis. In this context, we will review detailed biology of these four mitochondrial-contributed death extensively understood so far among 12 identified cell death pathways in mammalian cells.[11]

2.1 Apoptosis

Apoptosis is demonstrated as an active energy-dependent process initiated by cells. As we mentioned before, apoptosis, a programmed death pathway, consists of two types of extrinsic and intrinsic signaling.[11] In both of these death pathways, death is promoted by proteases called caspases. The following step consists of protein degradation, which provokes and upregulates cell death extensively. Other significant influencing factors are B-cell lymphoma-2 (Bcl-2) protein family members that will be further introduced in detail.

2.1.1 Extrinsic apoptosis

In the death receptor-mediated pathway, extrinsic death factors bind to their specific receptors on the cell surface, bringing about protein complex formation (complexes I and II). In complex II, caspase-8 and/or -10 are key responsible actors for activating other caspases-3 and -7. The activated caspases target multiple proteins that culminate in apoptosis.[1] In addition to the direct caspase-3 activation mechanism, caspase-8 can cleave BH3-only proteins and Bid. This is followed by Bax and Bak proteins' translocation into the outer mitochondrial membrane (OMM) and altering its permeability. Cytochrome c releases in the cell, forming the apoptosome complex with apoptosis-activating factor (Apaf-1) and procaspase-9. Activated caspase-9 eventually stimulates caspase-3 leading to apoptosis.[1, 15]

2.1.2 Intrinsic (mitochondrial-mediated) apoptosis

The initiator death signals affect mitochondria during the intrinsic death pathway. There is emerging evidence that mitochondria have a remarkable role in the activation of caspase-induced apoptosis. However, they are involved in the caspase-independent pathway (mediated by apoptosis-inducing factor (AIF) and endonuclease G protein).[1, 16, 17] Heat, radiation, infections, lack of nutrients, oxidative stress, and hypoxia are some identified stressors that hypothesized to initiate cell death process. In response to stress situations, intracellular signals promote the intrinsic death pathway. These signals are the consequences of the release of proteins from the intermembrane zone of mitochondria. These proteins can either increase the numbers of outer membrane pores or their permeability. Subsequently, apoptotic factors release is raised. Cytochrome c is an apoptotic factor released after Bcl-2-associated x protein, Bax, and Bcl-2 antagonist/killer 1 and Bak proteins activation. Within this pathway, caspases are activated following cytochrome c stimulation.[15, 17] Cytochrome c activates caspase-9 via apoptosome formation.[2] Altogether, this pathway is believed as a balancing pathway between mitochondria and factors within the cytosol.

2.2 Necrosis

According to biomedical research, a traumatic cell injury has been reported as a driving force of necrosis. A metabolic dysfunction, energetic collapse, ion-gradient breakdown, and cell swelling are evolved in this death pathway. The key mechanism in necrosis is mitochondrial permeability transition pore (mPTP) opening. Conversely to apoptosis, there may remain some bodies of cells that are harmful to other cells in a tissue and can activate inflammation as the next step. Necrosis may be involved in several disorders, including cerebral or heart ischemia–reperfusion injuries, trauma, and some neurodegenerative diseases.[18, 19]

2.2.1 Necroptosis

Although necrosis is categorized as nonprogrammed cell death, particular signals are regulating this pathway. According to the literature, both apoptosis and programmed necrosis (necroptosis) are initiated by the same stimulants and activation of death receptors. However, the strength of the exposure will lead the cell toward necrosis or apoptosis. Necroptosis may occur either during the development process or pathological cell damages. Conversely to apoptosis, necroptosis occurs due to energy depletion in cells.[1, 18, 20] Caspase independence mechanism is another distinction of necroptosis.[1, 18, 20] Necroptosis is correlated with activation of serine/tyrosine kinases receptor and interaction with protein kinase 1, receptor-interacting protein 1 (RIPK1), and RIPK3, as fundamental factors in this pathway.[21] RIPK3 activates the mixed lineage kinase-like domain (MLKL) by phosphorylation. MLKL is identified as an essential mediator in the necrosis pathway. MLKL is delocalized in the plasma membrane and increases Ca^{2+} influx. The ultimate stage is to disrupt the plasma membrane's permeability, through which necroptosis factors are released.[18, 21]

2.2.2 Mitochondrial-contributed necrosis

Growing evidence in the literature suggests that during mitochondrial-mediated necrosis, Ca^{2+} ions are the leading players altering the permeability of the inner mitochondrial membrane (IMM).[2] Proton gradient increase in the inner membrane leads to ATP synthesis, followed by lack of energy as the final consequence. The plasma membrane dysfunction is not clearly illustrated so far. Permeability transition in the IMM is reported to be the result of an overload of Ca^{2+}, exposure to reactive oxygen species (ROS), and nitrogen species that promote IMM pores opening and influx of solutes into the mitochondria. Loss of mitochondrial membrane potential and equilibrium ionic gradients inhibit ATP synthesis, leading to matrix expansion, swelling, and membrane rupture. Cytochrome c release into the cytoplasm, and finally, cell death occurs.[2, 18, 22, 23] Moreover, some evidence has shown a mitochondrial phosphatase, phosphoglycerate mutase family member 5 (PGAM5), as a significant factor in necrosis.[2]

3 Caspases

Caspases are enzymes from subclasses of the cysteine-dependent protease family. Their proposed mechanism is to hydrolyze peptide bonds. Caspases are regarded as proenzymes, and their performance depends on activation. Therefore, activation of caspases via mitochondria is of significant importance.[1, 16] It has been implied that mitochondria affect the regulation of caspases indirectly. Second mitochondria-derived activator of caspases (SMACs) are proteins that are released from mitochondria. SMACs bind to inhibitor apoptosis proteins (IAP) and block their inhibitory effect. It is noted that IAP proteins are the main suppressors of caspases activation. Caspases have two primary types, including long (2, 8–10) and short (3, 6, 7) prodomain groups. Short prodomain types are known as effectors, and long prodomain enzymes are initiator caspases.[1, 24]

4 Apoptotic proteins

4.1 Cytochrome c

Cytochrome c is one of the critical apoptotic factors. It is involved in the electron transport chain, and its release from mitochondria depends on Bax and Bak proteins activation. As a result, apoptosome is developed. Cytochrome c is released into the cytosol following the increased permeability of the OMM. It binds with apoptotic protease-activating factor-1 (Apaf-1) and ATP.

Consequently, this complex binds to procaspase-9. The final complex is called apoptosome. Procaspase-9 is activated, and caspase-9 is formed. Then, as a result, the effector caspase-3 is developed.[19, 22]

4.2 Bcl-2 proteins family

The family of various proteins—mostly inhibit cell death pathway—are known as Bcl-2 family. It has been indicated that Bcl-2 proteins are contributed to several death pathways, including mitochondrial-contributed death.[1, 25] This family has three proteins based on their primary function, and the Bcl-2 homology domain (BH) exists in protein.[16] The prosurvival or antiapoptotic group involves Bcl-2, Bcl-xl, Bcl-w, and MCL-1. This group of proteins conserves BH domains 1, 2, 3, and 4 in the structure. The proapoptotic group consists of Bax and Bak. These proteins provoke apoptosis and contain BH domains 1, 2, and 3. The third type consists of Bok, Bik, Bim, Bad, Noxa, and Puma. These proteins are known as BH3-only proteins due to their BH domain 3.[16, 25] The indispensable role of proapoptotic proteins has been emphasized when their cytochrome c release contribution was corroborated.[16] Bak and Bax proteins are involved in the permeabilization of OMM and promoting cell death. Oligomerized Bax and Bak proteins are responsible for transitional pore generation on OMM.

Apoptotic proteins, including cytochrome c, release into cytosol and apoptosome complex forms. The result is activating caspase-dependent death pathway.[26, 27] It has been stipulated that activation of these proteins depends on several steps, including oligomerization, and they might have a synergistic effect together. Nevertheless, the detailed mechanism is not determined yet.[16] On the other hand, Bcl-2, Bcl-xL, MCL-1, and Bcl-w inhibit Bak and Bax via binding to these proteins and hindering their oligomerization. Thus, the balance between Bcl-2 family members conducts cells toward death or survival.[16, 25] BH3-only proteins are inactive and rely on different activators.[16] They are demonstrated to be responsible for the activation of Bak and Bax proteins. The exact steps are still controversial, but binding and blocking antiapoptotic proteins is an indirect mechanism. After death pathway induction, some selective BH3-only proteins are direct causative factors for Bak and Bax activation.[28, 29] By translocation to OMM, BH3-only proteins interact with Bax and Bak proteins culminating in cell death. Simultaneously, with BH3-only proteins relocation, oligomerization of Bak and Bax at the surface of mitochondria occurs.[30, 31] Furthermore, other than BH3-only members, a few non-Bcl-2 family proteins are indicated to promote Bak and Bax activation.[32, 33] Altogether, relying on these hypotheses, intriguing potential targets will be identified to modulate cell death within pathological situations.

4.3 Apoptosis-inducing factor

AIF has been introduced as one of the proapoptotic proteins in mitochondria. It appears to be responsible for the death of particular cell types such as neurons and some tumor cells.[1, 15, 34] Besides, it is demonstrated that specific death stimulants are determinants for the involvement of AIF during the death pathway. On the occasion of Ca^{2+} homeostasis perturbation, AIF is the most common preferred moderator.[1, 34] The exact mechanism of AIF release from mitochondria to cytosol and then the nucleus is not fully described so far.[1] Ca^{2+} increase within the intermembrane zone of mitochondria activates calpain. This leads to AIF cleavage and its release into the cytosol.[1, 15] AIF causes condensation of chromatin in the nucleus and DNA fragmentation. Moreover, it promotes the release of cytochrome c from mitochondria. It has been revealed that Bcl-2 inhibits the AIF discharge from mitochondria.[34]

5 Mitochondrial permeability transition (MPT)

The outer and inner membranes cover mitochondria. According to the detailed description of mitochondrial-mediated apoptosis and necrosis, the outer and inner membrane permeability in mitochondria is considered prominent in death pathways.[35] There is substantial evidence on the link of Ca^{2+} and pore opening of mitochondrial membrane.[36] Also, diminished ATP, oxidative stress, low pH, and inorganic phosphates have been introduced to have a role.[37] Furthermore, prosurvival Bcl-2 proteins are involved in the OMM protection during apoptosis pathway.[25, 37] The proapoptotic members of Bcl-2 family lead to OMM perturbation and cytochrome c release. According to the mechanisms mentioned above, cytochrome c facilitates the activation of caspases.[25, 35] SMACs are the other stimulators of caspases via suppressing IAPs.[16] Permeabilization leads to membrane collapse and ATP depletion, leading to cell death. New pore generation is also observed due to transformation in the membrane proteins such as adenine nucleotide translocase (ANT). During this process, mitochondrial proteins such as cytochrome c release into the cell cytosol.[22, 35] Mitochondrial permeability is a critical trigger to regulate mitochondrial cell death to find therapeutic options in several diseases such as neurodegenerative disorders, autoimmune diseases, and several types of cancers.[5, 37] Cyclophilin D is a modulator of mitochondrial permeability transition (MPT) via blocking cytochrome c release. Regarding the fact that several diseases have shown the involvement of pore opening pathology, cyclosporine A has been suggested as a management trigger. Cyclophilin D has been reported to bind with cyclosporine A inhibiting the pore opening.[35,37]

6 Mitophagy

Autophagy is a kind of cell response to stress. It has been revealed that cells entering the autophagy process might recover after the removal of stressors. Hence, the exact return point is not identified so far. Therefore, autophagy might be considered as a survival mechanism in some cells.[1] Some literature indicates MPT as the no return point in death pathway.[2] Aging, stress, and other detrimental factors contribute to damage and dysfunction of mitochondria. It is observed that some acute injuries initiate and promote apoptosis or necrosis by altering the permeability of mitochondrial membrane.[8] Arranged autophagy of mitochondria and mitophagy consists of degradation and recycling pathways. It is occurring when the function of mitochondria is impaired.[1,7] Through dynamic fission and fusion, components of damaged mitochondria are segregated and equilibrated. Mitochondrial fission and fusion are programmed processes with related regulators and promoters. Dynamic-related protein 1 (DrP1) is a critical regulator in mitochondrial fission. DrP1 is separated from cytosol and transport to the OMM to regulate fission.

GTPases mitofusins 1 and 2 (Mfn1/2) are involved in OMM fusion, and optic atrophy 1 (OPA1) is demonstrated to dedicate IMM fusion.[38–40] When mitophagy takes place, the fission process enables dysfunctional mitochondria to detach from the network. Remaining mitochondrial fragments are cleaned via lysosome-driven autophagy.[41] Alteration in mitophagy and accumulation of impaired mitochondria bring about the loss of proper responsiveness to the cellular signals after lack of energy. Cell death is usually the final step in this situation.[3] Increased mitochondrial fragments have been observed in acute ischemia–reperfusion injuries.[7,42] Several degenerative diseases, including Alzheimer's disease, Parkinson's disease, multiple sclerosis, and Huntington's disease, have been linked to diminished gene expression of mitophagy regulator (inhibition of mitochondrial division).[8]

7 Calcium^{2+} role in the cell death pathway

Calcium $^{2+}$ (Ca^{2+}) is essentially required in many cell functions. It is recognized as a critical signaling mediator within cells. Mitochondrial function regulation and Ca^{2+} dynamic are tightly associated with each other.[43–45] Mitochondria involve in regulating Ca^{2+} homeostasis due to their ion-retaining capacity. Moreover, Ca^{2+} has a multifaceted role within ion-dependent channels by facilitating mitochondrial functions and signaling.[44–47] The main Ca^{2+} signaling is maintaining the counterbalance of producing energy by mitochondria relevant to cell demand.[19, 45] Growing evidence suggests the implication of malfunction of mitochondria and/or Ca^{2+} dynamic in a range of cardiac pathological states. Cancers and metabolic diseases are also suggested to originate from mitochondrial dysfunction associated with interrupted homeostasis of Ca^{2+}. Several studies have shown intertwined function of Ca^{2+} and mitochondria both in cell death and survival.[43, 44] The contribution of Ca^{2+} in different death pathways is well recognized.[1] Impaired regulation of Ca^{2+} and mitochondria brings about cell dysfunction or death.[43] Similar to the plasma membrane, IMM contains Ca^{2+}-activated K^+ channels (Kca). The most prominent roles for these channels are regulating mitochondrial ROS production and maintaining Ca^{2+} homeostasis in mitochondria. Altogether, Kca channels are revealed to be responsible players in cell death.[48] Altering Ca^{2+} homeostasis in mitochondria is associated with several metabolic diseases and cancers.[43, 44] During necrosis, Ca^{2+} overload and accumulation in mitochondria lead to permeability transition pore in IMM (known as mPTP) and collapse following morphological changes.[1] Excessive mitochondrial Ca^{2+} combined with ROS has been increasingly linked to several pathological states in the brain, heart, and muscles.[19, 23, 44]

8 Conclusion

Mitochondria affect cells physiology via influencing their functions and communications. Hence, mitochondrial contribution to cell death and finding critical players in death pathway is of great important. Biomedical research is currently emphasized to investigate potential beneficial keys based on emerging evidence on cell death's mechanism. For instance, in cancerous cells, the death pathway is inhibited extensively. Due to these organs' high energy demand, brain, heart, and muscles have been noted as the most affected organs when mitochondrial dysfunction occurs. Several pathological conditions are linked to mitochondrial impaired performance, including cancers, metabolic disorders, neurodegenerative diseases, cardiomyopathies, and ischemia–reperfusion injuries in the heart and brain.[49] Thus, studying the effects of mitochondrial dysfunctions on health status is required. However, current biological studies have not fully met the question of whether this approach can be helpful in managements of diseases.[50]

References

1. Orrenius S, Gogvadze V, Zhivotovsky B. Calcium and mitochondria in the regulation of cell death. *Biochem Biophys Res Commun*. 2015;460(1):72–81.
2. Lee MS. Role of mitochondrial function in cell death and body metabolism. *Front Biosci (Landmark Ed)*. 2016;21:1233–1244.
3. Sedlackova L, Korolchuk VI. Mitochondrial quality control as a key determinant of cell survival. *Biochim Biophys Acta Mol Cell Res*. 2019;1866(4):575–587.
4. Schapira AHV. Mitochondrial diseases. *Lancet*. 2012;379(9828):1825–1834.
5. Bock FJ, Tait SWG. Mitochondria as multifaceted regulators of cell death. *Nat Rev Mol Cell Biol*. 2020;21(2):85–100.
6. Dawson TM, Dawson VL. Mitochondrial mechanisms of neuronal cell death: potential therapeutics. *Annu Rev Pharmacol Toxicol*. 2017;57:437–454.
7. Anzell AR, Maizy R, Przyklenk K, Sanderson TH. Mitochondrial quality control and disease: insights into ischemia-reperfusion injury. *Mol Neurobiol*. 2018;55(3):2547–2564.
8. Green DR, Galluzzi L, Kroemer G. Mitochondria and the autophagy-inflammation-cell death axis in organismal aging. *Science*. 2011;333(6046):1109–1112.
9. Karami-Mohajeri S, Abdollahi M. Mitochondrial dysfunction and organophosphorus compounds. *Toxicol Appl Pharmacol*. 2013;270(1):39–44.
10. Manayi A, Saeidnia S, Gohari AR, Abdollahi M. Methods for the discovery of new anti-aging products- -targeted approaches. *Expert Opin Drug Discov*. 2014;9(4):383–405.
11. Galluzzi L, Vitale I, Aaronson SA, et al. Molecular mechanisms of cell death: recommendations of the nomenclature committee on cell death 2018. *Cell Death Differ*. 2018;25(3):486–541.
12. Brenner D, Mak TW. Mitochondrial cell death effectors. *Curr Opin Cell Biol*. 2009;21(6):871–877.
13. Guseva NV, Taghiyev AF, Rokhlin OW, Cohen MB. Contribution of death receptor and mitochondrial pathways to Fas-mediated apoptosis in the prostatic carcinoma cell line PC3. *Prostate*. 2002;51(4):231–240.
14. Goldenthal MJ. Mitochondrial involvement in myocyte death and heart failure. *Heart Fail Rev*. 2016;21(2):137–155.
15. Tsujimoto Y, Shimizu S. Bcl-2 family: life-or-death switch. *FEBS Lett*. 2000;466(1):6–10.
16. Pradelli LA, Bénéteau M, Ricci J-E. Mitochondrial control of caspase-dependent and -independent cell death. *Cell Mol Life Sci*. 2010;67(10):1589–1597.
17. Tait SW, Green DR. Caspase-independent cell death: leaving the set without the final cut. *Oncogene*. 2008;27(50):6452–6461.
18. Golstein P, Kroemer G. Cell death by necrosis: towards a molecular definition. *Trends Biochem Sci*. 2007;32(1):37–43.
19. Osellame LD, Blacker TS, Duchen MR. Cellular and molecular mechanisms of mitochondrial function. *Best Pract Res Clin Endocrinol Metab*. 2012;26(6):711–723.
20. Leist M, Single B, Castoldi AF, Kühnle S, Nicotera P. Intracellular adenosine triphosphate (ATP) concentration: a switch in the decision between apoptosis and necrosis. *J Exp Med*. 1997;185(8):1481–1486.
21. Sun L, Wang H, Wang Z, et al. Mixed lineage kinase domain-like protein mediates necrosis signaling downstream of RIP3 kinase. *Cell*. 2012;148(1–2):213–227.
22. Waterhouse NJ, Goldstein JC, von Ahsen O, Schuler M, Newmeyer DD, Green DR. Cytochrome c maintains mitochondrial transmembrane potential and ATP generation after outer mitochondrial membrane permeabilization during the apoptotic process. *J Cell Biol*. 2001;153(2):319–328.
23. Joseph SK, Booth DM, Young MP, Hajnóczky G. Redox regulation of ER and mitochondrial ca(2+) signaling in cell survival and death. *Cell Calcium*. 2019;79:89–97.

24. Du C, Fang M, Li Y, Li L, Wang X. Smac, a mitochondrial protein that promotes cytochrome c-dependent caspase activation by eliminating IAP inhibition. *Cell.* 2000;102(1):33–42.
25. Youle RJ, Strasser A. The BCL-2 protein family: opposing activities that mediate cell death. *Nat Rev Mol Cell Biol.* 2008;9(1):47–59.
26. Kuwana T, Mackey MR, Perkins G, et al. Bid, Bax, and lipids cooperate to form supramolecular openings in the outer mitochondrial membrane. *Cell.* 2002;111(3):331–342.
27. Nechushtan A, Smith CL, Lamensdorf I, Yoon SH, Youle RJ. Bax and Bak coalesce into novel mitochondria-associated clusters during apoptosis. *J Cell Biol.* 2001;153(6):1265–1276.
28. Lovell JF, Billen LP, Bindner S, et al. Membrane binding by tBid initiates an ordered series of events culminating in membrane permeabilization by Bax. *Cell.* 2008;135(6):1074–1084.
29. Gavathiotis E, Suzuki M, Davis ML, et al. BAX activation is initiated at a novel interaction site. *Nature.* 2008;455(7216):1076–1081.
30. Huang DC, Strasser A. BH3-only proteins-essential initiators of apoptotic cell death. *Cell.* 2000;103(6):839–842.
31. Korsmeyer SJ, Wei MC, Saito M, Weiler S, Oh KJ, Schlesinger PH. Pro-apoptotic cascade activates BID, which oligomerizes BAK or BAX into pores that result in the release of cytochrome c. *Cell Death Differ.* 2000;7(12):1166–1173.
32. Leu JI, Dumont P, Hafey M, Murphy ME, George DL. Mitochondrial p53 activates Bak and causes disruption of a Bak-Mcl1 complex. *Nat Cell Biol.* 2004;6(5):443–450.
33. Zheng JH, Viacava Follis A, Kriwacki RW, Moldoveanu T. Discoveries and controversies in BCL-2 protein-mediated apoptosis. *FEBS J.* 2016;283(14):2690–2700.
34. Susin SA, Lorenzo HK, Zamzami N, et al. Molecular characterization of mitochondrial apoptosis-inducing factor. *Nature.* 1999;397(6718):441–446.
35. Kroemer G, Galluzzi L, Brenner C. Mitochondrial membrane permeabilization in cell death. *Physiol Rev.* 2007;87(1):99–163.
36. Roy SS, Madesh M, Davies E, Antonsson B, Danial N, Hajnóczky G. Bad targets the permeability transition pore independent of Bax or Bak to switch between Ca2+–dependent cell survival and death. *Mol Cell.* 2009;33(3):377–388.
37. Zhang D, Armstrong JS. Bax and the mitochondrial permeability transition cooperate in the release of cytochrome c during endoplasmic reticulum-stress-induced apoptosis. *Cell Death Differ.* 2007;14(4):703–715.
38. Chan DC. Fusion and fission: interlinked processes critical for mitochondrial health. *Annu Rev Genet.* 2012;46:265–287.
39. Pernas L, Scorrano L. Mito-Morphosis: mitochondrial fusion, fission, and cristae remodeling as key mediators of cellular function. *Annu Rev Physiol.* 2016;78:505–531.
40. Xie L-L, Shi F, Tan Z, Li Y, Bode AM, Cao Y. Mitochondrial network structure homeostasis and cell death. *Cancer Sci.* 2018;109(12):3686–3694.
41. Youle RJ, Narendra DP. Mechanisms of mitophagy. *Nat Rev Mol Cell Biol.* 2011;12(1):9–14.
42. Eisner V, Picard M, Hajnóczky G. Mitochondrial dynamics in adaptive and maladaptive cellular stress responses. *Nat Cell Biol.* 2018;20(7):755–765.
43. Bravo-Sagua R, Parra V, López-Crisosto C, Díaz P, Quest AF, Lavandero S. Calcium transport and signaling in mitochondria. *Compr Physiol.* 2017;7(2):623–634.
44. Duchen MR. Mitochondria and calcium: from cell signaling to cell death. *J Physiol.* 2000;529(Pt 1):57–68.
45. Robb-Gaspers LD, Burnett P, Rutter GA, Denton RM, Rizzuto R, Thomas AP. Integrating cytosolic calcium signals into mitochondrial metabolic responses. *EMBO J.* 1998;17(17):4987–5000.
46. Rizzuto R, Duchen MR, Pozzan T. Flirting in little space: the ER/mitochondria Ca2+ liaison. *Sci STKE.* 2004;2004(215), re1.

47. Nicholls DG, Chalmers S. The integration of mitochondrial calcium transport and storage. *J Bioenerg Biomembr.* 2004;36(4):277–281.
48. Krabbendam IE, Honrath B, Culmsee C, Dolga AM. Mitochondrial Ca^{2+}-activated K^{+} channels and their role in cell life and death pathways. *Cell Calcium.* 2018;69:101–111.
49. Magner M, Kolářová H, Honzik T, Švandová I, Zeman J. Clinical manifestation of mitochondrial diseases. *Dev Period Med.* 2015;19(4):441–449.
50. Nunnari J, Suomalainen A. Mitochondria: in sickness and in health. *Cell.* 2012;148(6):1145–1159.

CHAPTER 4

Mitochondrial response to environmental toxicants

Ahmad Salimi[a,b] **and Zhaleh Jamali**[c,d]
[a]Department of Pharmacology and Toxicology, School of Pharmacy, Ardabil University of Medical Sciences, Ardabil, Iran
[b]Traditional Medicine and Hydrotherapy Research Center, Ardabil University of Medical Sciences, Ardabil, Iran
[c]Department of Addiction Studies, School of Medicine, Shahroud University of Medical Sciences, Shahroud, Iran
[d]Student Research Committee, School of Medicine, Shahroud University of Medical Sciences, Shahroud, Iran

Contents

Mitochondrial Metabolism
https://doi.org/10.1016/B978-0-12-822416-8.00003-8

Abbreviations

ALS	amyotrophic lateral sclerosis
ANT	adenine nucleotide translocator
BaP	benzo[*a*]pyrene
CypD	cyclophilin D
DBF	dibenzofurans
ETC	electron transport chain
IARC	International Agency for Research on Cancer
IMM	inner mitochondrial membrane
IMS	intermembrane space
ISM	interspace membrane
MMP	mitochondrial membrane potential
MPP^+	l-methyl-4-phenylpyridinium
MPTP	mitochondrial permeability transition pore
mtDNA	mitochondrial DNA
NOS	nitric oxide species
OMM	outer membrane
OXPHOS	oxidative phosphorylation
PAHs	polycyclic aromatic hydrocarbons
PBR	peripheral benzodiazepine receptor
PCBs	polychlorinated biphenyls
POPs	persistent organic pollutants
ROS	reactive oxygen species
TCDD	2,3,7,8-tetrachlorodibenzodioxin
TPP	triphenylphosphine
VDAC	voltage-dependent anion channel
$\Delta\Psi m$	mitochondrial membrane potential

1 Introduction

As mitochondria are involved in several biological activities, mitochondrial health is vital to organismal fitness and survival.[1] Mitochondria have some number of biological activities such as energy generation, initiation and amplification of apoptosis, synthesis of macromolecular structures ion exchange, copper and iron homeostasis, calcium signaling and calcium homeostasis, nonshivering thermogenesis, metabolite transport, and protein-activating iron–sulfur cluster assembly.[1] Mitochondria are associated with the cytoplasm and nucleus through signaling nitric oxide species (NOS) and reactive oxygen species (ROS) and cell cycle arrest pathway.[2] Considering the important roles and complex of mitochondria in the cell, maintaining a healthy population of these organelles are vital for organismal fitness and survival.[3] Particularly, mitochondrial plasticity, the capacity to sustain optimum mitochondrial integrity and function, is critical when exposing these organelles to environmental perturbations.[4] A number of studies show that environmental toxicants have significant deleterious effects on mitochondrial function and structure in exposed organisms.[5] Previous studies have shown that exposure to pharmaceuticals, heavy metals, complex hydrocarbon mixtures, nanomaterials, pesticides, UV radiation, and various other inorganic and organic toxicants can disrupt mitochondrial structure and function.[5] For example, we can refer to studies conducted on aquatic and terrestrial invertebrates, fish from the tropics to the polar regions and amphibians.[6] Also, there are examples of environmental toxicant effects on plant mitochondria.[7] However, environmental toxicants have been less well investigated than pharmaceutical agents from the mitochondrial toxicity aspect; there are nevertheless some well-understood instances.[8] For example, rotenone as a complex I inhibitor can result in increased ROS formation, mitochondrial DNA (mtDNA) damage, and cell death through apoptosis or necrosis.[9] Paraquat (*N*, *N*′-dimethyl-4,4′-bipyridinium dichloride) also appears to act largely by redox cycling in a fashion analogous to doxorubicin.[10] There is a well-known example of an environment–environment interaction, the exposure to either cyanide or methanol, and the unfortunate cooccurrence of folate deficiency which result in mitochondrial toxicity-mediated optic neuropathy in exposed humans.[5] Cyanide and carbon monoxide act as complex IV inhibitors and result in severe toxicity.[11] Also, there are many other environmental toxicants that are less well described as mitochondrial toxic agents but are reported to cause at least some of their effects via mitochondrial toxicity.[5] These include lipopolysaccharide, particulate matter, copper and cadmium, dioxin, acrolein, PAH quinones, acrylamide, usnic acid, perfluorinated compounds, methoxychlor, cigarette smoke, 3-nitropropionate, pentachlorophenol, lead, mercury, manganese, and arsenic.[5] Of course, it should be noted that many if not all of the environmental chemicals are likely to affect other targets than mitochondria, and in some cases, mitochondrial toxicity may be secondary to effects elsewhere.[12]

Mitochondria are composed of outer membrane (OMM), interspace membrane (ISM), inner mitochondrial membrane (IMM), and matrix.[13] The matrix encompasses mtDNA, which encodes important proteins for oxidative phosphorylation (OXPHOS).[14] Mitochondrial DNA lacks self-repair ability and introns, making it more susceptible to damage and mutation than nuclear DNA (nDNA).[15] Mutation in mitochondrial DNA can increase ROS generation and affect respiratory chain, whereas overgeneration of ROS in turn damage mtDNA as well as cellular nDNA, proteins, and lipids.[15] Long-time exposure to environmental toxicants leads to ROS formation in mitochondria and can induce mutations in mtDNA and nDNA. It is possible that the mutated cells survive and trigger tumorigenesis.[16] Many studies have shown that different mutations of mtDNA lead to a variety of malignancies, including prostate cancer, colon cancer, and breast cancer.[17] Interaction of environmental toxicants with components of mitochondrial permeability transition pore (MPTP) such as the voltage-dependent anion channel (VDAC), the adenine nucleotide translocator (ANT), peripheral benzodiazepine receptor (PBR), matrix proteins (cyclophilin D [CypD]), cytosolic proteins (glycerol kinase and hexokinase II), and proteins of the intermembrane space (creatine kinase) can result in mitochondrial dysfunction.[18] It has been reported that prolonged MPTP opening induced by environmental agents such as heavy metals can result in organelle swelling, rupture, mitochondrial energetic dysfunction, and cell death.[19] Also, one of the main targets in mitochondria for environmental toxicants is mitochondrial electron transport chain (ETC).[20] Also, many evidences have illustrated that mitochondrial malformation was closely associated with the development of several disorders, including Parkinson's disease, Alzheimer's disease, renal diseases, retinal diseases, amyotrophic lateral sclerosis (ALS), aging, liver diseases, glaucoma, Huntington's disease, autism, cardiovascular and cardiotoxicity disorders, nonalcoholic fatty liver disease (NAFLD), diabetes, epilepsy, kidney injury, bipolar, and schizophrenia disorder.[20]

The effects of environmental contaminants on mitochondrial health have received less attention.[12] In particular, long-term exposure of humans, animals, plants, and microorganisms with sublethal concentrations of environmental contaminants could affect mitochondrial structure and function.[5] In this chapter, we discuss mitochondrial disturbance induced by environmental contaminants with a particular focus on those toxicants that target mitochondrial DNA, electron transport chain, mitochondrial permeability transition pore (MPTP) components, the principles of mitochondrial vulnerability to environmental toxicants, accumulation of environmental toxicants in mitochondria, and mitochondrial responses to environmental toxicants in various organs.

2 Accumulation of environmental toxicants in mitochondria

The effect of environmental contaminants on mitochondrial function requires molecules with suitable physicochemical properties that allow entry into the human circulation and delivery to the intended site for effective binding to the target molecule in the mitochondria.[12] When the small molecule arrives to mitochondria, it can accumulate and lead to deleterious consequences on mitochondrial function and structure.[21] Environmental small molecules that are distributed in the environment have been observed as having effects on mitochondrial function through inhibiting to crucial functions such as electron transport or increasing formation of ROS in the mitochondria.[22] These effects are especially significant when considering organs with a heavy demand for activity of mitochondria, such as the brain and heart, or those having roles in general metabolism, such as the liver.[23] In these tissues, mitochondrial functions are impacted either directly or indirectly.[1] Small molecules directly bind to specific mitochondrial targets and impact directly on a particular mitochondrial function, such as uncoupling of the electron transport chain, modification of mitochondrial resident enzymes, and inhibition of mitochondrial DNA replication.[1] Indirect mechanisms involve small molecules modulating processes within the cell that finally impact mitochondrial structure and function such as inhibition of HMG-CoA reductase, an important enzyme in the biosynthesis of the mitochondrial electron carrier coenzyme Q, or inhibition of fatty acid metabolism by carnitine sequestration in the cytoplasm by environmental toxicants.[24] Therefore, a certain subset of environmental toxicants with mitochondrial toxicity potential act by entrance and accumulation in mitochondria, while other toxicants only need to arrive to the cytoplasm to exert their toxic effects.[25] Delivery of environmental toxicants to mitochondria depend on the physicochemical properties of toxicant and the molecular architecture and composition of mitochondria.[26] The optimal properties of a chemical mostly promote the deleterious effects associated with mitochondrial toxicity. Small molecules are optimally lipophilic in nature, which allows for passage through the circulation by binding to plasma proteins and transit through cell membranes to provide access to intracellular targets such as mitochondria and lysosomes.[27] Also, the molecular architecture and composition of mitochondria are responsible for absorbing some small molecules that leads to accumulation and impact on mitochondrial function and structure.[28] The outer and inner mitochondrial membranes are unique because of highly folded membrane structures particularly in the inner mitochondrial membrane where specific geometry exists with given structures such as the tips of the inner membrane cristae protrusions, as well as the presence of specific phospholipids, such as cardiolipin.[13] These particular properties will allow for certain molecules to accumulate due to fitting into a specific geometry associated with inherent mitochondrial membrane arrangements and/or affinity for cardiolipin.[29] The function of mitochondrial is also dependent on an electrochemical gradient across the inner mitochondrial membrane.[29] This electrochemical gradient is responsible for driving ion and metabolite transport across the inner mitochondrial membrane to power processes such as ATP production.[30] Heme synthesis precursors as anabolic substrates and other metabolites are transported across the inner mitochondrial membrane through special transporters that are indirectly or directly driven by the mitochondrial membrane potential (MMP).[31] Therefore, some lipophilic molecules are capable to transit the inner mitochondrial membrane by passive diffusion due to their chemical properties, and this is driven by the pH and charge differences across the inner mitochondrial membrane.[32] It should be noted that the matrix of mitochondria is slightly alkaline compared with the intermembrane space and cytosol as well as it is an overall negatively charged environment, with a $-180\,mV$ potential across the inner mitochondrial membrane.[33] These features attract positively charged small molecules to the matrix such as triphenylphosphine (TPP), a common organophosphorus compound, with lipophilic and cationic properties.[34] In addition, several mitochondrial toxicants are weak acids that are lipid soluble when protonated. These chemicals act as protonophores and result in dissipation of the mitochondrial membrane potential and mitochondrial dysfunction.[22] These protonophore agents increase the rate of respiration, ROS formation, and heat dissipation in mitochondria, which lead to breakdown of the mitochondrial antioxidant defense.[22] ROS are capable to damage macromolecules such as mitochondrial DNA, proteins, and phospholipids in the mitochondria.[22] These protonophore agents exist in their ionized anion state in the cytoplasm; in the intermembrane space (IMS) of mitochondria, the concentration gradient of protons through the IMM constructs an acidic environment with a lower pH than that in the mitochondrial matrix.[35] Due to the high concentration of protons in the IMS, the ionized form becomes protonated, rendering the small molecule unionized and hence much more lipophilic than the ionized form. In this status, the unionized small molecule can transit via the IMM through diffusion.[36] The small molecule deprotonates in the mitochondrial matrix as protons are used by the electron transport chain to restore the mitochondrial membrane potential.[37] The fast transposition of the protons by the small molecule disintegrates the MMP and prevents proton transposition via ATP synthase, thereby diminishing ATP production potential that is not maintained with an overall outcome of cellular dysfunction and tissue failure.[37] In other words, positively charged molecules are capable of entering the matrix once they enter the mitochondrial intermembrane space.[38] Due to positive charge of trapped small molecules and difference in charge across the membrane, these molecules can accumulate in the mitochondrial matrix.[38] It has been reported that short- to medium-chain fatty acids up to 12 carbons in length are able to pass the inner mitochondrial membrane freely to enter into the mitochondrial matrix without the aid of protein transporters.[39] Other small molecules that resemble these types of fatty acids are able to cross via the IMM because of the shared properties.[39] Therefore, the high lipid content of mitochondrial membranes facilitates accumulation of lipophilic compounds such as polycyclic aromatic hydrocarbons (PAHs), alkylating agents, and cationic metals such as cadmium, lead, manganese, and mercury.[20] Also, the mitochondrial matrix has slightly alkaline pH and a negative charge, which facilitates accumulation of amphiphilic xenobiotics, l-methyl-4-phenylpyridinium (MPP^+), paraquat, and ethidium bromide.[40] It has been reported that the mitochondrial cytochrome P450 system may activate relatively nonreactive chemicals, such as polycyclic aromatic hydrocarbons (PAHs) and mycotoxins.[40]

3 Vulnerability mitochondria to environmental toxicants

Mitochondria are susceptible to certain environmental toxicants due to the negative charge of the mitochondrial matrix that facilitates accumulation of cationic metals and the high lipid content in membrane that facilitates accumulation of lipophilic compounds.[5] Different classes of environmental toxicants have appeared as potent mitochondrial toxicants.[12] These contaminants contain resistant organic pollutants (POP) such as 2,3,7,8-tetrachlorodibenzo-p-dioxin (TCDD), bisphenol A (BPA), and polycyclic aromatic hydrocarbons (PAH); different metals such as copper, cadmium, lead, mercury, arsenic selenium, manganese, zinc, and methyl mercury; as well as toxicants designed such as rotenone to target mitochondrial complexes.[41] In addition, a comprehensive analysis of Comparative Toxicogenomics Database that includes manually curated information about gene-disease and chemical-disease relationships and chemical-gene/protein interactions showed that over 6000 molecules from various classes of chemicals affect mitochondrial oxidative phosphorylation.[42]

Due to enhanced bioavailability in the environment, particularly in aquatic ecosystems, certain mitochondrial toxicants could show a higher toxicity for wildlife compared with human health.[43] For example, the toxicity of selenium has gained more attention as waste product of coal combustion released into the environment and is displayed to affect fish mitochondrial membrane potential leading to toxicity and cell death.[44] Phthalates and BPA, commonly well known for their endocrine-disrupting attributes, are ubiquitous environmental toxicants, particularly in the aquatic environment.[45] Several studies suggest that these environmental toxicants affect mtDNA copy number, mtDNA mutations, mitochondrial membrane potential, ATP content, and mitochondrial mass.[12] Besides, interactive effects of environmental factors and chemical mixtures are a considerable concern in an ecological context.[46] For example, exposure to UV radiation causes mitochondrial dysfunction and is presumably further worsened by synergistic effects with other mitochondrial toxicant such as polycyclic aromatic hydrocarbons.[47]

Chemicals released from various agricultural, industrial, and pharmaceuticals activities are an environmentally important category of mitochondrial toxicants.[48] Documents of drug-induced mitochondrial toxicity (e.g., effects on mtDNA replication, mitochondrial transport pathways, electron transport chain, etc.) are quickly accumulating.[12] Due to the common evolutionary origins of mitochondria and bacteria, some antibacterial antibiotics are of considerable concern as they can induce ROS overproduction and mitochondrial dysfunction, potentially through disruption of mitochondrial protein synthesis and antibacterial antibiotic that are designed to target bacterial ribosomes which could also target mitochondrial ribosomes.[49] Pharmaceuticals and their metabolites are mostly detected in the environmental sources higher than prescribed for human consumption, particularly in aquatic environment.[50] Pharmaceuticals, particularly the persistent and lipophilic ones, have the potential to accumulate in the mitochondria and biomagnify in nontargeted aquatic species.[51] Certain pharmaceuticals, potentially including drugs that target mitochondria, that have insignificant biological outcomes at prescribed concentrations in human health contexts may have unintended physical and chemical interactions in the environment.[52] These chemicals may potentially lead to toxic effect and ecological impacts in the environment.[52]

Risk of mitochondrial toxicity can also elevate with changes in transport and fate of environmental contaminants in ecosystems due to anthropogenic and natural activities. For instance, the exposure with trace and toxic metals can alter quickly as a result of forecasted changes in transport and fate of compounds in the biotic and abiotic environment with weather alterations.[53] Several studies have explained mechanisms and forecasted outcomes of changed fate and transport of environmental toxicant due to alterations in pH, salinity, temperature, as well as alterations in frequency of climate events such as floods and sedimentation, forest fires, and cloud cover.[53] For example, metal speciation, distribution, and biotransformation will be affected with changes in chemical and geophysical properties, and bioavailability and accumulation of these environmental toxicant presumably alter as a result.[54] An indirect outcome of warming is enhanced release and bioavailability of sequestered mitochondrial toxicants in environment such as methyl mercury and polychlorinated biphenyls (PCBs) with melting snow and sea ice.[55] In addition, weather-induced changes in physiology and behavior of organisms especially in polar regions result in shifted biological toxicokinetics within a living thing as well as across trophic cascades with quick environmental alteration.[56] Altogether it is obvious that alteration in abiotic agents have considerable effects on transport and fate of mitochondrial toxicants in the abiotic and biotic environment.[57] Due to previously mentioned reasons and different sites in mitochondria, these organelles can be vulnerable to environmental toxicants (Fig. 1).

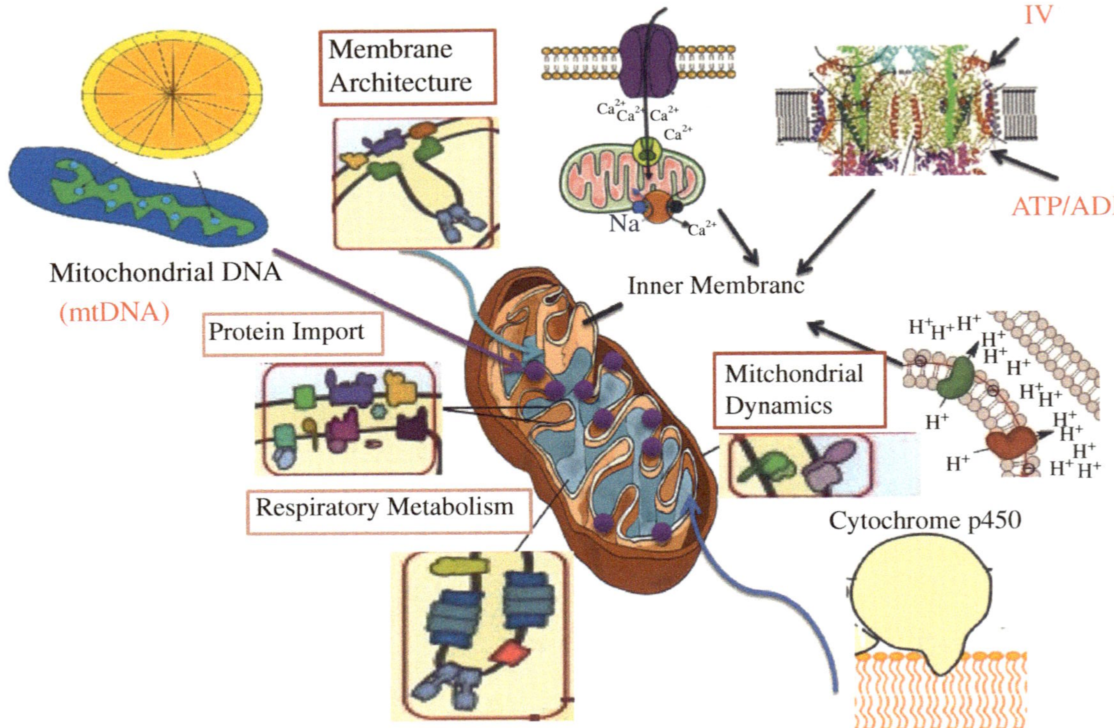

Fig. 1 Different sites in mitochondria and mitochondria's sensitivity to exposures to environmental toxicants.

4 Mitochondrial toxicity by environmental toxicants

With exposure to environmental contaminants, a series of toxic effects are induced on mitochondria, which originate from mitochondrial respiratory chain dysfunction, structure/membrane damage, and dynamic change in terms of fusion and fission.[12] Under slight exposure with environmental toxicants, fusion may apply an adaptive response to reduce cellular damage and enhance survival, through enhancing ATP production and increasing metabolic efficiency.[58] Mitophagy as a selective degradation by autophagy is induced when the separation of fragmented mitochondria is generated through fission.[59] Mitochondrial respiratory chain dysfunction can barricade ATP generation and increase ROS production, which can damage mtDNA more, and consequently lead to deficiency of mtDNA-coding mitochondrial proteins.[60] Moreover, oxidative stress and membrane damage of mitochondria increase the mitochondrial membrane permeability and formation of mitochondrial permeability transition (MPT) pores, allowing the release of cytochrome C into the cytosol to trigger apoptosis.[19] Therefore, environmental toxicants can induce dysfunction in the respiratory chain, mtDNA damages, mitodynamic alterations, bioenergetic imbalance, as well as oxidative stress and interaction with mitochondrial permeability transition pore (mPTP) components, which consequently result in toxicological effects and alteration of cell fates.[61] The previously mentioned toxic responses are described in more details later in the section.

5 Inhibition of the mitochondrial electron transport chain by environmental toxicants

The production of ATP in mitochondria needs a functional (ETC) and the oxidative phosphorylation system (OXPHOS). The proteins involved in oxidative phosphorylation are localized in the IMM and are organized into five multisubunit complexes (I-V). Complexes I to IV form a functional ETC, and complex V is the mitochondrial ATP synthase.[14] Exceptionally for complex II, the mitochondrial supercomplexes have subunits encoded both in the nuclear genes and the mitochondrial genomic DNA.[62] Furthermore, the production of the majority of ATP in the cells by oxidative phosphorylation, the respiratory chain (RC) can produce ROS as a byproduct.[63] The respiratory chain (RC) of mitochondria is responsible for the majority of mtROS generation. The main sources of ROS in mitochondria are complexes I and III.[64] Obviously, the respiratory chain of mitochondria is at the vanguard of attempts to figure out the physiopathology and physiology of mitochondria-related disorders. Therefore, environmental toxicants that target the respiratory chain could influence mitochondrial functions, containing diffusion of proteins, ROS generation, energy metabolism, and other vital pathological processes in mitochondria.[65] More studies have shown that the stress of different environmental toxicants such as cigarette smoke, fine particulate matter (PM2.5), heavy metals, bisphenol A, ionizing radiation, pesticides, and nanoparticles can induce mitochondrial dynamic morphology or changes to the structural integrity, DNA (mtDNA) damage, dysfunction of ATP production, and dysregulation of ROS generation which subsequently result in toxic effects and alteration of cell fates (Fig. 2).[66] Consistent exposure to environmental toxicants may affect mitochondrial electron transport chain and then damage the function and structure of mitochondria, leading to increased ROS production and decreased ATP

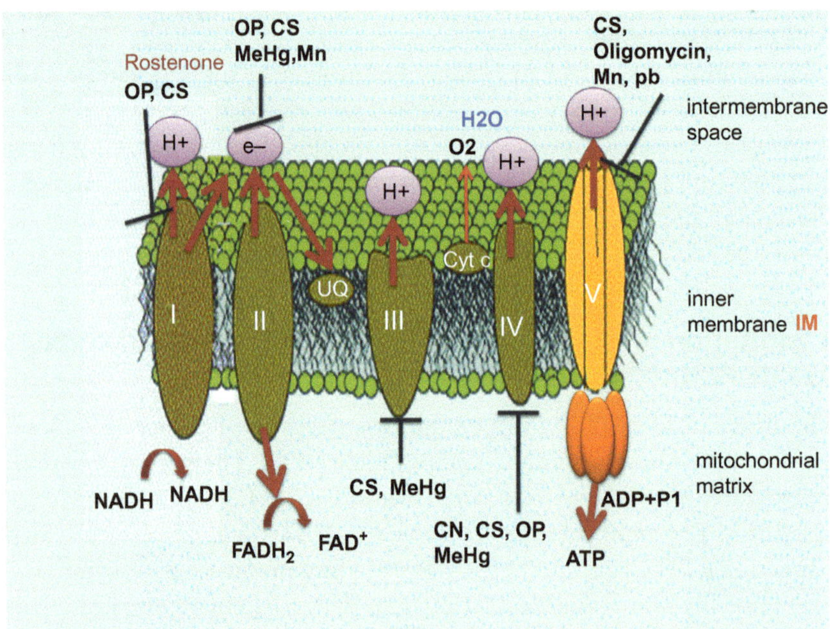

Fig. 2 Electron respiratory chain and toxicant sites of inhibition. Flow of electrons through the electron respiratory chain and pumping out of protons leading to the formation of H_2O and ATP are shown. Inhibition sites of the mitochondrial complexes by xenobiotic are shown by black blunted arrows. I, complex I; II, complex II; III, complex III; IV, complex IV; V, complex V; ETC, electron transport chain; Cyt c, cytochrome *c*; CN-, cyanide; UQ, ubiquinone.

yield, and consequently interrupt the intracellular homeostasis and finally alter the cell fate.[20] Previous studies have reported the effect of environmental toxicants on electron transport chain (ETC).[20]

5.1 Metalloids

Recently, Yu et al. and Fan et al. studied this topic by closely examining the mitochondrial toxicities such as thiourea gold (I) complexes (AuTuCl) and organic arsenical chemical 2-methoxy-4-(((4-(oxoarsanyl)phenyl) imino) methyl) phenol (MOPIMP) on respiratory chain function.[67, 68] The compound GSK932121A, a suspected complex III electron transport chain (ETC) inhibitor, displayed an adapted response state of increased basal oxygen consumption rate and calcium retention capacity in vivo study.[69] It has been reported that a short-term exposure of mitochondria with the organic arsenical MOPIMP increased the inhibition of RC complexes I, II, III, and IV, and with damage to the mitochondrial respiration function.[68] This study indicated that the terminal complex of the ETC, the activity of complex IV was significantly destroyed even at a low concentration level of MOPIMP.[68] Yu and coworkers reported that AuTuCl impaired the functioning of the mitochondrial respiratory chain and as well as changed the mitochondrial permeability transition.[67] They showed the underlying mechanism of mitochondrial respiratory chain dysfunction by AuTuCl exposure via inhibiting the activities of complexes II and IV, leading to a deficiency in ATP generation and release of cytochrome *c*.[67] Clearly, dysfunction of the mitochondrial respiratory chain directly resulted in the hindered ATP production and burst of ROS, which may trigger mitochondria-dependent apoptosis in the cells.[66]

5.2 Rotenone

Rotenone as a natural compound found in plants and as a potent insecticide in past is a potent complex I inhibitor that is fatal for humans and other mammals as well as for insects and fish.[70]

5.3 Cyanide

Cyanide as a bioproduct of normal human and plant metabolism as well as a byproduct of industrial metal and chemical processing, blocks complex IV of respiration.[71] This chemical competes with oxygen for binding to the Fe-Cu center in complex IV, inhibiting cytochrome *c* oxidase activity, and making it one of the most toxic compounds to humans.[71]

5.4 Cigarette smoke

Cigarette smoke extract (CSE) inhibits both complexes I and II enzymatic activity leading to decreased oxygen consumption in human primary bronchial epithelial cells.[72] The cigarette smoke component acrolein has been shown to inhibit enzymatic activity of complexes I, II, and III in primary and transformed human retinal cells.[72] Also, exposure of 4 consecutive days with the cigarette smoke component acrolein decreased in vivo complex I enzymatic activity in kidney and lung extracts, unchanged in the heart, but increased in the liver of BALB/c mice.[73] Complex IV was decreased in liver, heart, lung, and kidney.[73] Exposure with cigarette smoke for 4 or 8 weeks in mice led to increased expression of complexes II, III, IV, and V, while complex I decreased for both exposures.[74] It has been reported that the cigarette smoke induces mitochondrial membrane potential and the activation of autophagy.[75] It has been reported that cigarette smoke damages mitochondria leading to loss of membrane potential. Cigarette smoke stabilizes the accumulating PTEN-induced putative kinase 1 (PINK1) protein, which recruits the E3 ubiquitin ligase Parkin. Mitochondrial proteins are ubiquitinated by Parkin, allowing for mitochondria to be engulfed by isolation membranes that fuse with lysosomes (Fig. 3).[76]

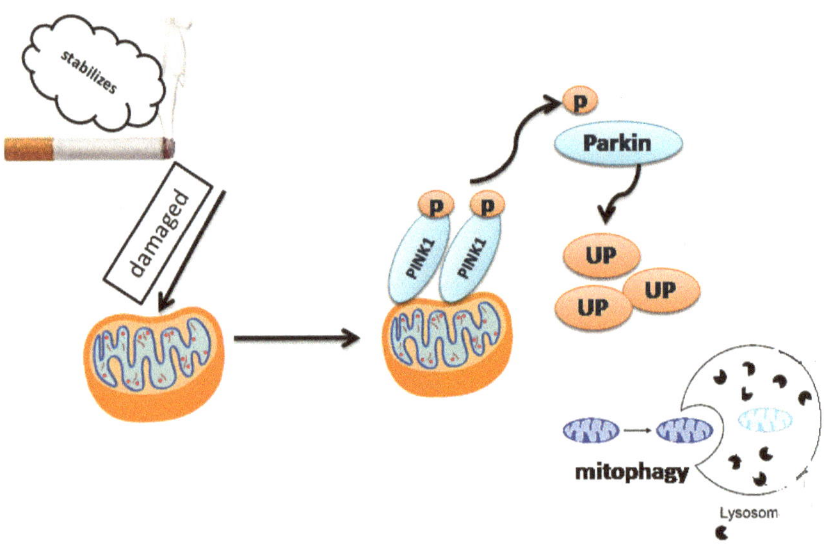

Fig. 3 Proposed mechanism of cigarette smoke-induced mitochondrial damage and mitophagy. Cigarette smoke damages mitochondria leading to loss of membrane potential. Cigarette smoke stabilizes the accumulating PINK1 protein, which recruits the E3 ubiquitin ligase Parkin. Mitochondrial proteins are ubiquitinated by Parkin, allowing for mitochondria to be engulfed by isolation membranes that fuse with lysosomes. PINK1, PTEN-induced putative kinase 1; Ub, ubiquitin.

5.5 Organophosphates

Previous studies have confirmed that organophosphates (OP) as a class of insecticides and environmental toxicants are inhibitors of mitochondrial respiration in multiple tissues.[77] For example, dichlorvos and monocrotophos decreased complexes I and II activities in brain stem, cerebellum, and cortex of rats.[78] Also, mevinphos inhibits the complexes I and IV in hearts of rats. Exposure to parathion or dichlorvos decreased the activities of complexes I, II, and IV in the liver and brain.[79]

5.6 Metals

ROS formation by methylmercury (MeHg) has also been shown to occur through inhibition of the electron transport chain of mitochondria.[80] In rats exposed to MeHg, there is enhanced mitochondrial oxygen consumption in the cerebrum and cerebellum accompanied with complex II inhibition in the cerebellum but not in the cerebrum.[80] Inhibition of complexes III and IV decreased ATP generation, increased mitochondrial-derived superoxide production, caused disruption of mitochondrial membrane potential, and also opening of mPTP was observed in rat primary cerebellar granule neurons treated with MeHg.[71]

6 mtDNA damage and mitochondrial dysfunction induced by environmental toxicants

Many environmental toxicants are known to react with DNA and induce damage and mutations, finally leading to cancer disease.[81] Assessment of the genotoxic effects of environmental toxicants has been developed and validated during the last decades.[81] Surface waters, drinking water, ground waters, industrial effluents, sediments, airborne particulate matter, and domestic discharges were evaluated using different endpoints or approaches, ultimately confirming the genotoxic effect induced by toxicants present in these different environments.[81, 82] Studies on mtDNA damage and mitochondrial dysfunction caused by environmental toxicants have been reported in many in vitro studies.[12] Here, we summarize the data on the known effects of environmental toxicant on mtDNA stability and mitochondrial function.

6.1 Pesticides

Several pesticides are known to inhibit complexes I, II and/or III and lead to ROS formation and mitochondrial membrane depolarization.[83] Rotenone, as a lipophilic pesticide, is able to freely cross biological membranes, and is a known mitochondrial toxicant. It was demonstrated to induce mtDNA damage in skeletal muscle and blood of rats.[84] Exposure to organophosphorus compounds such as parathion, triphenyl phosphite, and tri-ortho tolyl phosphate in cultured SH-SY5Y human neuroblastoma cells leads to changes in mitochondrial transmembrane potential.[85] Chronic exposure to organophosphorus compounds can produce ROS and impair the cellular and mitochondrial antioxidant defense system.[86]

6.2 Polychlorinated biphenyls (PCBs)

Polychlorinated biphenyls (PCBs) are persistent organic pollutants (POP) used in hundreds of commercial and industrial applications as capacitors and transformers, thermal insulation, plastics, oil-based paint, adhesives, and many others.[87] Rats treated with PCBs-contaminated milk showed mitochondrial alterations in liver and brain and several organ disruptions, with decreased activity of complexes II and III in brain tissue, but no differences in liver and brain mitochondria.[88] In addition, a study performed with the PCB metabolite (PCB29-pQ) showed decreased mitochondrial membrane potential ($\Delta\Psi m$), increased cellular ROS levels and apoptosis due to the release of cytochrome *c* from mitochondria into cytosol, and initiation of a caspase cascade in human hepatoma HepG2 cells.[89] It has been reported that treatment with Aroclor-1254 as a polychlorinated biphenyl caused a decline in ATP levels accompanied by a reduction in mitochondrial membrane potential ($\Delta\Psi m$) in human SH-SY5Y neuroblastoma cells.[90]

6.3 Dioxins

Dioxins are persistent organic pollutants (POPs) produced exclusively as byproducts of industrial processes such as production of certain pesticides, incineration, and chlorine bleaching of pulp and paper.[91] This group comprises dibenzofurans (DBF), biphenyls, polychlorinated dibenzo-p-dioxins, and other related compounds.[91] Among above chemicals, dibenzofuran is a relevant pollutant, as exposure to DBF may occur through dermal and inhalation contact, particularly at sites engaged in carbonization and combustion processes.[92] In vitro models, using rat liver mitochondria demonstrated that dibenzofurans might impair mitochondrial function, mainly due to a decrease in phosphorylative efficiency.[92] Also, TCDD (2,3,7,8-tetrachlorodibenzodioxin) exposure induces mitochondrial dysfunction by disruption of mitochondrial membrane potential and initiates mitochondrial stress signaling.[93] TCDD markedly decreased the $\Delta\Psi m$ in rat pheochromocytoma PC12 cells and induced mitochondrial dysfunction ROS production.[94] ATP-sensitive potassium channels activity in rat heart mitochondria was demonstrated to be higher than rat liver mitochondria after treatment with TCDD.[95] It was reported that this effect would counteract TCDD-induced oxidative stress in rat heart mitochondria, preventing additional damage.[95] TCDD as an endocrine disruptor disrupts hormone signaling, interferes with embryo development, reduces fertility, and causes spontaneous miscarriage in humans.[96] Treatment JAR cells, derived from a human trophoblastic tumor of the placenta with different concentrations of TCDD, showed a lower mitochondrial membrane potential and reduction in ATP content and mtDNA copy number.[96] Large-scale deletions in the mitochondrial genome and mtDNA rearrangements were observed in the treated JAR cells with TCDD.[96] A 7599-bp mtDNA deletion that encompasses the entire NADH dehydrogenase genes ND3, ND4, ND4L, ND4, ND5, and ND6 (C-I), the cytochrome *c* oxidase gene COIII (C-IV), and the cytochrome *b* gene Cytb (C-III), and part of the ATP synthase 6 (C-V) gene was identified after treatment with TCDD.[97] In addition, a significant raise in 8-hydroxy-2′-deoxyguanosine (8-OHdG) levels in mtDNA was observed 6 h after treatment with TCDD.[98]

6.4 Metalloids and metals

Heavy metals including vanadium, arsenic, beryllium, lead, cadmium, chromium, nickel, and cobalt are considered environmental toxicants, as they are known to damage cells via production of ROS and oxidative stress.[99] It has been reported that cadmium induces oxidative damage and cell death through ROS formation, mitochondrial dysfunction, deregulation of intracellular antioxidants, and apoptosis.[100] Treatment of rat cell-line PC12 with Hg^{2+}, Cu^{2+}, and Cd^{2+} was shown to impair respiratory function, mitochondrial membrane potential, intracellular ROS formation, and cell viability.[101] In addition, it was demonstrated that exposure with relatively low concentrations of arsenic trioxide (As_2O_3) results in low copy number of mtDNA in a dose dependence of environmental pollution levels in mouse oocytes both in vivo and in vitro.[102] Moreover, treatment with arsenic trioxide led to accumulation of a 3867 bp deletion in mtDNA that could affect the development of the oocytes.[103] Exposure to low-concentration methylmercury was associated with increased numbers of specific point mutations in the D-Loop of human embryonic neural progenitor cells mtDNA, leading to membrane potential loss, ATP depletion, and apoptosis (Fig. 4).[104]

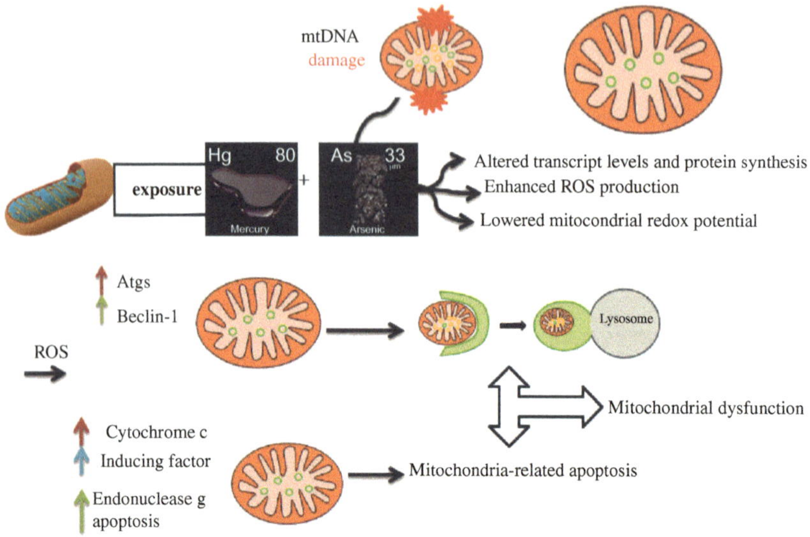

Fig. 4 Proposed roles of heavy metals (arsenic and mercury) in the development of mitochondrial dysfunction. Heavy metals induced mtDNA damage, excess Ca^{2+} uptake, changes in transcript levels and protein synthesis, as well as ROS production. Increased ROS further exacerbated the impairment of mitochondria and resulted in apoptosis and mitophagy.

6.5 Air contaminants

Polycyclic aromatic hydrocarbons (PAHs) cumulate in the aerosphere owing to incomplete combustion of organic material, coal burning, industrial production processes, and automobile exhaust.[105] The International Agency for Research on Cancer (IARC) has classified many of them as human carcinogens. Adverse health effects through the production of reactive radicals and oxidative stress can be an outcome of exposure to particulate air contaminants.[105] The adverse effects of ambient ultrafine particles and diesel exhaust were investigated in a murine macrophage cell line and a loss of mitochondrial mass, decrease in mitochondrial membrane potential, and induction of cell death through apoptosis were detected.[106] Moreover, isolated liver mitochondria obtained from treated rats with benzo[*a*]pyrene (BaP) showed higher Ca^{2+} uptake and enhancement of oxidative phosphorylation, suggesting that it stimulated mitochondrial respiration, which is associated with ROS generation.[107] It has been reported that in vivo treatment of zebrafish embryos to benzo[*a*]pyrene leads to changes in the proteome and an increase in mtDNA copy number.[108] Previous studies have suggested that zebrafish, specifically at the embryo stage, is a suitable in vivo model for monitoring polycyclic aromatic hydrocarbons exposure to hematopoietic tissue and other organs.[108] A comparison of mitochondrial and nuclear DNA damage in fishes living in a reference site and in a heavily contaminated site proved that samples living in the reference site showed low levels of basal mtDNA and nDNA damage compared with fish from the polluted waters.[109] On the other hand, when injected with benzo[*a*]pyrene, in comparison with the other animals, the level of damage was lower in fishes from the contaminated site.[110] The authors point out that multigenerational exposure in the wild may lead to adaptations that depress DNA damage arising from benzo[*a*]pyrene exposure.[110] Surprisingly, it has reported a controversial study about exposure to benzo[*a*]pyrene. The authors examined the potential effect of benzo[*a*]pyrene on induction of mtDNA mutations in bone marrow and liver in exposed mice.[111] They did not observe deletion and point mutations in mtDNA after 28 days of daily benzo[*a*]pyrene exposures despite the abundance of benzo[*a*]pyrene-induced DNA lesions.[111]

Nitrogen oxide (NOx) pollution is formed whenever fuel is combusted at high temperatures by a "fixing" of the nitrogen in the combustion chamber's dilution air into NOx. Also, nitrogen oxides originate from fuel burning.[112] Nitrogen dioxide (NO_2) exposure can cause respiratory damages, and previous studies show an increasing risk of neurological disorders and cardiovascular diseases.[113] Morphological changes in mitochondria of rat cortex were observed, with increased levels of ROS and decreased activity of the respiratory complexes and ATP production.[114]

A common environmental pollutant is sulfur dioxide (SO_2), formed during the combustion of sulfur-containing fuels by power plants, industrial facilities, and diesel engines; also, as emissions from volcanoes it can originate from natural sources.[115] It has been reported that exposure of brain cells of rats to SO_2 in gas chambers for 4 h/day during 30 days indicated an increase in inner mitochondrial membrane potential, ATP production, and mtDNA content.[116] Due to its coexistence in the air complex, the combination mixture SO_2 and benzo[*a*]pyrene was evaluated in mice liver cells and the results showed mRNA suppression of subunits of the mitochondrial oxidative phosphorylation complexes IV and V accompanied by a decrease in mitochondrial membrane potential.[117] Therefore, mitochondrial dysfunction and damages were suggested as an initial event following coexposure with combination benzo[*a*]pyrene and SO_2, leading to apoptotic as well as antiapoptotic signals at different post-exposure times.[118]

6.6 Cigarette smoke

Passive and active tobacco smoke decrease mitochondrial membrane potential and respiration leading to increased ROS/RNS production and decreased ATP content in a variety of tissues in a dose- and time-dependent manner.[119] Mitochondria are rich in enzymes with active sites easily inactivated and modified by oxidant from the inhalation of cigarette smoke, or from ROS/RNS production within the mitochondria.[120] For instance, it has been reported that activity of the mitochondrial isoform of superoxide dismutase was decreased after exposure with cigarette smoke.[121] Inactivation of oxidative phosphorylation complexes by tobacco smoke-induced RPOS/RNS directly impairs proton pumping efficiency leading to decreased MM and ATP generation.[122] Surprisingly, several evidences have demonstrated that cessation of both passive and active smoking improves the mitochondrial function.[123] It has been shown that exposure to cigarette smoke causes alterations in mitochondrial morphology, the formation of mtDNA adducts, increases mtDNA damage, alterations in mtDNA copy number, and promotes mitophagy in both cell culture and animal models.[119] Acrolein, as one of the primary reactive aldehydes within cigarette smoke, forms DNA adducts which are likely to persist in the mtDNA owing to the inefficiency of the DNA polymerase gamma of mitochondria.[124] Chemicals found within tobacco smoke such as benzene cause changes in mitochondrial DNA copy number, however, it is important to note that mitochondrial DNA copy number is influenced by several factors including health status and age.[125] Differences in mitochondrial DNA copy number may be masked by other diseases and may be difficult to attribute to toxicant exposure.[119] Altogether, mitochondrial DNA damage may be used as a more reliable indicator of cigarette smoke-induced injury than copy number based on the conflicting observations concerning copy number.[119] There are several toxicants in cigarette smoke which can induce mitochondrial dysfunction (Table 1).

6.7 Algal toxins

Anthropogenic activities lead to load of phosphorous and nitrogen in coastal marine systems and freshwaters, which may result in eutrophication of water bodies. Enhancement of the nutrient amount in aquatic environments often facilitates the occurrence of odorous, unsightly, and toxic cyanobacterial blooms.[126] Blue-green algae, known as cyanobacteria, have various species, some of them can produce and

Table 1 Common constituents of cigarette smoke.

Chemical component in cigarette	Commonly found in other sources
Arsenic	Common component of pesticides
Benzene	Used in the production of gasoline
Carbon monoxide	Product of vehicle exhaust, gas stoves, and coal heating processes
2-Ethylpyridine	Chemical syntheses
Acrolein	Product of burning organic matter
Acrylonitrile	Used in manufacture of plastics, resins, and nitriles
Cadmium	Major component of batteries
Chloroform	Found in pesticides, fumigants, and fire extinguishers
Chromium	Used in production of steel
Hexane	Used in glues
Hydrogen cyanide	Chemical weapons
Lead	Once used as a pigment in paint
Manganese	Used in production of steel and aluminum alloys
Mercury	Chemical manufacturing and fluorescent lamps
Naphthalene	Mothballs
Phenol	Used in production of plastics and detergents
Toluene	Used to produce paint thinners
Vinyl chloride	Used in production of PVC pipes, bottles, and some upholstery

release toxins into the aquatic environments.[127] Species of Anabaena, Microcystis, or Planktothrix produce microcystins; Aphanizomenon and Cylindrospermopsis produce anatoxin and cylindrospermopsin; and paralytic shellfish poison is released by various Aphanizomenon and Anabaena species.[127] Algal toxins increase the risk of human and animal intoxication and decrease water quality.[127]

A family of cyclic heptapeptide endotoxins, known as microcystins (MCs), have more than 80 different structural analogues, among which the hepatotoxic microcystin-LR (MC-LR) is the most common variant.[128] Several studies showed that microcystin-LR could affect mitochondrial function in human bronchial epithelial cells, mouse liver, and rat testis treated in vivo.[128] These studies reported changes in mitochondrial DNA copy number, increase in ROS levels, expression of mitochondrial genes, and apoptosis associated with the mitochondria-dependent apoptotic pathway.[129] All the previously mentioned studies suggest that damage to mitochondrial DNA is a common effect of exposure to microcystin-LR.[128]

7 Bioenergetic imbalance and mitochondrial dysfunction induced by environmental toxicants

The altered energy homeostasis is a direct consequence of mitochondrial dysfunction and it is a critical parameter in induction of toxicity through mitochondria-mediated pathways in cells.[130] For example, acute and developmental exposure to complex PAH mixtures and individual PAH compounds affects standard and maximum metabolic rates, leading to reduced aerobic scope in juvenile and adult teleost fishes.[131] It has been reported that dietary exposure to PAHs has transgenerational effects on metabolic rates in fishes. Exposure to polychlorinated biphenyls (PCB) resulted in altered energy metabolic enzymes in killifish and reduced standard metabolic rates in European eel.[132] Furthermore, exposure to organochlorine pesticides and metals can modify metabolic rates in mollusks, crustaceans, and fish.[133] It is conceivable that observed alterations are independent of mitochondrial toxicity and owing to modified metabolic demand and signaling or substrate composition shift as a result of physiological and behavioral alterations with chemical exposure.[12] Enhanced basal metabolic function after exposure to metals was observed in aquatic invertebrates, which may be due to increased ATP demand to maintain cellular integrity.[134]

8 Oxidative stress and mitochondrial dysfunction induced by environmental toxicants

ROS production, oxidative stress response, and oxidative damage derived from mitochondrial dysfunction may play an important role in the mitochondria-related diseases.[135] Direct inhibition of mitochondrial function (e.g., inhibition of complex I by rotenone, which is a main source of ROS) and indirect effects on mitochondria as well as environmental toxicant-specific detoxification and biotransformation processes can result in increased ROS production in the cells.[9] A number of evidences indicate that both physical (e.g., temperature and UV radiation) and chemical (e.g., metals and POPs) environmental stressors are ROS inducers, eliciting oxidative stress responses. Exposure to pH, temperature, and Cd stress indicated damage depending on the magnitude of the stressors and tissue-specific oxidative stress response in Antarctic scallop.[136] It has been reported that arsenic effects on antioxidant capacity are increased at higher salinities in oysters.[137] Exposure to pesticides and thermal stress leads to differential effects on oxidative stress response markers in coral (*Porites astreoides*) larvae.[138] It has been demonstrated that antioxidant response pathways are overwhelmed by combined exposure to temperature stress, ZnO nanoparticles, and organics pollutants in freshwater mussels (Unio tumidus).[139]

9 Environmental and occupational toxicants induce mitochondrial dysfunction in human

Experimental and epidemiological studies that figure out the effects of environmental contaminants on human mitochondrial DNA and function are scarcer and most of them are related to occupational exposure.[140] The few studies which directly evaluate the effects of environmental contaminants on human mitochondria are summarized later in the section.

Budnik and coworkers measured circulating mitochondrial DNA in serum individuals with confirmed exposure to off-gassing haloalkane pesticides and nonexposed controls, and this group found that exposure through storage facilities or in home environment was significantly associated with enhanced circulating mitochondrial DNA levels in serum.[141] The human exposure to haloalkane pesticides can cause alterations in ROS generation and apoptosis.[142] The relative quantities of mitochondrial DNA in serum were associated with the lag time after exposure to the contaminants, and suggested that mitochondrial DNA could be used as a biomarker to identify risk groups after exposure to environmental toxicants.[143]

There are experimental and epidemiological studies that environmental contaminants, especially pesticides, contribute to the risk of developing Parkinson's disease (PD) associated through impairment of mitochondrial function.[144] Oxidative stress and respiratory chain inhibition linked to pesticide exposure such as paraquat and rotenone were associated with Parkinson's disease in a case–control study with 110 Parkinson's disease cases and 358 controls.[145] Wang and Zhao in 2012 evaluated lung tissue from fruit growers exposed to different pesticides, such as organochlorines, pyrethroids, and carbamates, for at least 5 years, for mitochondrial DNA mutations.[146] These authors identified a high frequency of mitochondrial DNA mutations, and attributed this effect to ROS formation[147]; mitochondrial dysfunction and oxidative stress generated by the metabolism of the different pesticides.[147]

An investigation on benzene-exposed individuals was done with individuals from different Italian cities.[125] Occupational categories that contain exposure to low levels of benzene were 77 police officers, 153 bus drivers, 33 petrochemical plant workers, and 78 gas station attendants.[125] An increase in mitochondrial DNA copy number was determined in blood cells associated with airborne benzene exposure. The association of increased mtDNA copy number and mitochondrial mass in response to enhanced ROS formation and exposure of cell lines and blood cells to benzene was demonstrated by Eom and coworkers.[125]

The effects of chronic exposure to polycyclic aromatic hydrocarbons on mitochondrial DNA were analyzed in human peripheral blood lymphocytes from male coke-oven workers, and significant correlations between exposure to high levels of polycyclic aromatic hydrocarbons and higher mitochondrial DNA copy number as well as genetic alterations in nuclear DNA were found.[148] It has been reported that exposure of individuals to airborne polycyclic aromatic hydrocarbons by measuring eight urinary their metabolites, and proved that increased levels of four of PAHs metabolites was associated with decreased sperm mitochondrial DNA copy number, but not with mitochondrial DNA integrity nor sperm mitochondrial membrane potential.[148]

The influence of environmental contaminants in epigenetic mechanisms in mitochondrial DNA was investigated.[149] These studies reported mitochondrial DNA methylation in three DNA regions in human peripheral blood mononuclear cells (PBMC) of steel workers exposed to metal-rich particulate matter (PM), truck drivers exposed to traffic-derived elemental carbon, and gas station attendants exposed to air benzene.[149] The steel workers with high metal-rich PM exposure showed higher methylation at the 12S ribosomal RNA loci and phenylalanine tRNA than low-exposed controls, as well as higher mitochondrial DNA copy number; no alterations were seen on mitochondrial DNA from the other two exposed groups.[149] The above study showed that locus-specific mitochondrial DNA methylation is correlated with mitochondrial DNA damage and selected exposures.[149] However, a controversial study found that methylation of mitochondrially encoded tRNA phenylalanine (MT-TF) and mitochondrially encoded 12S RNA (MT-RNR1) was significantly lower in chrome plating workers compared with controls, and no link existed between mitochondrial DNA copy number and chrome exposure.[150] The authors suggest that the differences between the two studies may be associated with different exposure conditions that might have various effects on mitochondrial DNA methylation.[150] However, another study demonstrated a reduction in blood mitochondrial DNA copy number in association with elevated exposure to ambient PM10 and elemental carbon in 60 truck drivers during work hours.[151]

Previous studies have shown that occupational and environmental exposure to particle-containing welding fumes was associated with an increase in mitochondrial DNA copy number and decreased mitochondrial DNA methylation in worker, as well as with increased welder's blood pressure (BP).[152] This study suggests that mitochondrial DNA could be involved in the etiology of particle-related cardiovascular disorders.[152]

The influence of ambient air pollution on mitochondrial function of the fetus in the exposed mothers in late pregnancy was evaluated.[153] The authors found an association with lower mitochondrial DNA copy number in cord blood with exposure to PM2.5.[153] No such association was observed in Gent, Belgium, with exposure to PM10, where the lower mitochondrial DNA copy number in the placentas was correlated with maternal exposure.[153]

10 Environmental toxicants and diseases with mitochondrial dysfunction

Environmental toxicants–induced mitochondrial dysfunction causes complex and diverse diseases in humans. The defect of mitochondrial function has been reported in major environment-related multifactorial diseases, such as viral infections, respiratory disease, cardiovascular diseases, cancer, and neurological disorders.[20] However, exposure to harmful environmental contaminants is ubiquitous among all populations, but environment-related diseases occur in only a subset of the population.[20] Interaction between genetic predisposition and environmental factors is therefore presumably a necessary requisite to disease manifestation. In the following, we listed several environment-related diseases with focus on mitochondrial dysfunction induced by environmental toxicants.[20]

10.1 Alzheimer's disease

Alzheimer's disease (AD) is the most common cause of dementia and the third leading cause of death in the United States after cardiovascular disease and cancer.[154] Previous studies suggest that epigenetic factors and/or environment play an important role in influencing and initiating the cascade of events that contribute to the appearance of Alzheimer's disease.[154] A link between environmental exposure and Alzheimer's disease has been described for absorbed, inhaled, and ingested toxins.[155] Key neuropathologic features of Alzheimer's disease are neurofibrillary tangles and p-amyloid plaques.[156] The probable relationship between Alzheimer's disease and oxidative stress derived from the free radical hypothesis of aging suggests that age-related accumulation of ROS could be responsible for damage in major cell components.[156] The role of mitochondrial damage and elevated oxidative stress has been suggested in the pathogenesis of Alzheimer's disease. Exposure to environmental toxicant presumably contributes to Alzheimer's disease pathogenesis through mitochondrial oxidative stress.[157]

As mentioned earlier, most organophosphates, such as chlorpyrifos (CPF), dichlorodiphenyltrichloroethane (DDT), and paraquat (PQ), evaluated have been shown to induce oxidative stress, mitofusin 1-mediated mitochondrial fragmentation, reduction in mitochondrial ETC activity, and activation of the mitochondrial apoptosis pathway[158]. Degeneration of hippocampal cholinergic neurons happens in Alzheimer's disease and results in memory deficits due to loss of cholinergic modulation of hippocampal synaptic circuits.[159] Chlorpyrifos was shown to induce concentration-dependent cell death in SN56 cells, as a cholinergic murine neuroblastoma cell line.[160] In rats, exposure to chlorpyrifos and subsequent intracerebroventricular β-amyloid (Aβ) resulted in reduced microtubule-associated proteins and altered memory spatial task in the prefrontal cortex.[161] Treatment of mice with paraquat produced a significant increase in Aβ levels that was associated with increased 4-HNE and nitrotyrosine levels consistent with mitochondrial oxidative damage in cerebral cortex, leading to the impairment of memory and learning.[162]

It has been reported that lead (Pb) affects intelligence, cognitive abilities, speed processing, motor functions, and memory in children. Pb ions have been shown to alter mitochondrial respiration.[163] Aluminum (Al) has a high affinity for the large pyramidal neurons of the human brain hippocampus, and has been shown to promote amyloid aggregation and accumulation, a key feature of AD neuropathology.[164] Al has been reported to generate free radicals, reduced ETC activity, and decreased ATP synthesis in rat brains.[162] It has been reported that Al increases lipid peroxidation, protein and DNA oxidation, and decreased polyunsaturated fatty acids in AD brain.[165] Methylmercury (MeHg) crosses the blood–brain barrier due to affinity to cysteine disrupt neurodevelopment, and is associated with altered cognition and memory loss in adults.[166] MeHg is reported to induce mitochondrial dysfunction and ROS formation, and decrease glutathione levels.[166] Exposure to Pb, As, and Cd and their combination in the AD-amyloid pathway in male rats activated the synthesis of Ab in the frontal cortex and/or hippocampus, mediated by an increase in amyloid precursor protein (APP) and in APP-processing enzymes such as BACE1.[167] All metals were found to increase the APP production, increase levels of malondialdehyde (MDA), reduce activity of antioxidant enzymes glutathione-S-transferase, catalase, and superoxide dismutase, and the induction of 1L-1α and IL-1β in the frontal cortex.[168] Inorganic Hg has been related to increased oxidative stress and cell damage, and secretion of Aβ-40 and Aβ-42 in SHSY5Y neuroblastoma cells.[169]

10.2 Parkinson's disease

Parkinson's disease (PD) as a chronic progressive neurodegenerative disease is the second most common neurodegenerative disorder after Alzheimer's disease.[170] There is ample evidence of mitochondrial dysfunction in Parkinson's disease that includes oxidative stress, complex I inhibition, DJI and PINK1 dysfunction, and an interaction between PINK1 and Parkin that influences mitochondrial dynamics.[170] Considerable studies suggest a multifactorial etiology for Parkinson's disease, involving environmental factors and genetic.[170] Epidemiological studies suggest that exposure to environmental contaminants such as metals, flame retardants, pesticides, and solvents could increase the risk of developing Parkinson's disease.[170] The role of mitochondrial dysfunction and environmental factors in the pathophysiology of Parkinson's disease was first suggested by the discovery of 1-methyl-4-phenyl-1,2,3,6-tetrahydropyridine (MPTP). 1-Methyl-4-phenylpyridinium (MPP) as active metabolite of MPTP is selective for dopaminergic neurons and acts by inhibiting mitochondrial complex I of the ETC, and induces increased ROS production and selective degeneration of nigrostriatal neurons.[171]

The main mechanisms suggested for the relationship between Parkinson's disease and pesticide exposure are mitochondrial complex I inhibition and oxidative stress by paraquat.[172] Paraquat is taken up into dopaminergic neurons and causes oxidative stress, mitochondrial dysfunction, and cellular toxicity.[172] Administrations of paraquat to adult rats and mice elevate ROS formation in the striatal homogenate, induce a dose-dependent decrease in dopaminergic neurons from the substantia nigra, a neurobehavioral syndrome characterized by reduced ambulatory activity, and a decline in striatal dopamine nerve terminal density.[172] It has been reported that combined exposure to paraquat and the fungicide Maneb (MB) in rural workers increased the risk of developing Parkinson's disease by 75% in agricultural areas of California.[173] Animal studies have reported that concurrent exposure to paraquat and MB results in a significant dopamine (DA) fiber loss, altered DA turnover, and decreased locomotor activity in adult mice.[173]

Rotenone treatment in animal models produces postural instability, rigidity, and bradykinesia, induces a loss of striatal dopamine, reduces the tyrosine hydroxylase-positive neurons in the substance nigra, and the accumulation of α-synuclein- and poly-ubiquitin-positive aggregates in remaining dopaminergic neurons.[174] Dieldrin as a lipophilic organochlorine has shown to be relatively selective dopaminergic neurotoxin in mesencephalic cultures, induces a conformational change in α-syn, depletes intracellular DA levels, generates ROS, depolarizes mitochondrial membrane potential, and induces apoptosis.[175] Exposure of mice to low levels of dieldrin resulted in a long-term enhancement of protein and mRNA levels of the vesicular monoamine transporter 2 (VMAT2) and dopamine transporters.[176]

It has been reported that occupational exposure to metals, such as iron, manganese (Mn), mercury, aluminum, zinc, lead, and copper, may be a risk factor for Parkinson's disease-like disorder.[177] The underlying mechanism is unknown, although evidence of mitochondrial dysfunction and oxidative stress is well documented in brains of Parkinson's disease patients.[178] Experimental studies have reported that iron may play a significant role in Parkinson's disease neurodegeneration.[178] The administration of iron into rat brains results in behavioral changes and parkinsonism.[178] The 8-hydroxyquinoline as a metal ion chelator prevented neurodegeneration in Parkinson's disease models. Ferritin transgenic mice have been shown to be resistant to paraquat-induced lipid peroxidation and neuronal loss, possibly by avoiding the formation of ROS.[20] It has been reported that manganese exposure causes overexpression of synuclein in neuroblastoma cells and implicated to be involved in neuronal apoptosis.[20] PCI 2 cells treated with manganese demonstrated increased apoptosis related to mitochondrial dysfunction and increased ROS.[179]

11 Summary

In this chapter, we discuss that mitochondria are main targets in environmental toxicant-induced toxicity and diseases, where direct impact of environmental toxins leads to diverse aspects of mitochondrial metabolism, mtDNA genetics, oxidative stress, respiratory chain dysfunction, bioenergetic disbalance, and cell death. Previous studies showed that mitochondria are susceptible to certain environmental toxicants due to the negative charge of the mitochondrial matrix facilitates accumulation of cationic metals and the high lipid content in membrane facilitates accumulation of lipophilic compounds. Available experimental studies increasingly support a link between exposure to environmental toxins and several diseases that share a common feature of mitochondrial dysfunction. Future studies specifically focusing on the impact of environmental toxins in mitochondrial-related diseases are needed, especially association of cigarette smoke, air pollutions, metals, and pesticides.

References

1. Osellame LD, Blacker TS, Duchen MR. Cellular and molecular mechanisms of mitochondrial function. *Best Pract Res Clin Endocrinol Metab.* 2012;26(6):711–723.
2. Bolisetty S, Jaimes EA. Mitochondria and reactive oxygen species: physiology and pathophysiology. *Int J Mol Sci.* 2013;14(3):6306–6344.
3. Seungyoon BY, Pekkurnaz G. Mechanisms orchestrating mitochondrial dynamics for energy homeostasis. *J Mol Biol.* 2018;430(21):3922–3941.
4. Hoffmann A, Spengler D. The mitochondrion as potential interface in early-life stress brain programming. *Front Behav Neurosci.* 2018;12:306.
5. Meyer JN, Leung MC, Rooney JP, et al. Mitochondria as a target of environmental toxicants. *Toxicol Sci.* 2013;134(1):1–17.
6. Newman JR, Schreiber RK, Novakova E. Air pollution effects on terrestrial and aquatic animals. In: *Air pollution effects on biodiversity.* Springer; 1992:177–233.
7. Keunen E, Remans T, Bohler S, Vangronsveld J, Cuypers A. Metal-induced oxidative stress and plant mitochondria. *Int J Mol Sci.* 2011;12(10):6894–6918.
8. Scatena R, Bottoni P, Botta G, Martorana GE, Giardina B. The role of mitochondria in pharmacotoxicology: a reevaluation of an old, newly emerging topic. *Am J Physiol Cell Physiol.* 2007;293(1):C12–C21.
9. Wang L, Duan Q, Wang T, et al. Mitochondrial respiratory chain inhibitors involved in ROS production induced by acute high concentrations of iodide and the effects of SOD as a protective factor. *Oxid Med Cell Longev.* 2015;2015.
10. Bonneh-Barkay D, Reaney SH, Langston WJ, Di Monte DA. Redox cycling of the herbicide paraquat in microglial cultures. *Mol Brain Res.* 2005;134(1):52–56.
11. Leavesley HB, Li L, Prabhakaran K, Borowitz JL, Isom GE. Interaction of cyanide and nitric oxide with cytochrome c oxidase: implications for acute cyanide toxicity. *Toxicol Sci.* 2008;101(1):101–111.
12. Meyer JN, Hartman JH, Mello DF. Mitochondrial toxicity. *Toxicol Sci.* 2018;162(1):15–23.
13. Mejia EM, Hatch GM. Mitochondrial phospholipids: role in mitochondrial function. *J Bioenerg Biomembr.* 2016;48(2):99–112.
14. Tang JX, Thompson K, Taylor RW, Oláhová M. Mitochondrial OXPHOS biogenesis: co-regulation of protein synthesis, import, and assembly pathways. *Int J Mol Sci.* 2020;21(11):3820.
15. Zsurka G, Peeva V, Kotlyar A, Kunz WS. Is there still any role for oxidative stress in mitochondrial DNA-dependent aging? *Genes (Basel).* 2018;9(4):175.
16. Guerra F, Arbini AA, Moro L. Mitochondria and cancer chemoresistance. *Biochim Biophys Acta Biomembr.* 2017;1858(8):686–699.
17. Hertweck KL, Dasgupta S. The landscape of mtDNA modifications in cancer: a tale of two cities. *Front Oncol.* 2017;7:262.
18. Kalani K, Yan SF, Yan SS. Mitochondrial permeability transition pore: a potential drug target for neurodegeneration. *Drug Discov Today.* 2018;23(12):1983–1989.
19. Kroemer G, Galluzzi L, Brenner C. Mitochondrial membrane permeabilization in cell death. *Physiol Rev.* 2007;87(1):99–163.

20. Zolkipli-Cunningham Z, Falk MJ. Clinical effects of chemical exposures on mitochondrial function. *Toxicology.* 2017;391:90–99.
21. Wasilewski M, Chojnacka K, Chacinska A. Protein trafficking at the crossroads to mitochondria. *Biochim Biophys Acta (BBA) - Mol Basis Dis.* 2017;1864(1):125–137.
22. Zorov DB, Juhaszova M, Sollott SJ. Mitochondrial reactive oxygen species (ROS) and ROS-induced ROS release. *Physiol Rev.* 2014;94(3):909–950.
23. Alberts B, Johnson A, Lewis J, Raff M, Roberts K, Walter P. The mitochondrion. In: *Molecular biology of the cell.* 4th ed. Garland Science; 2002.
24. Lenaz G, Fato R, Formiggini G, Genova ML. The role of coenzyme Q in mitochondrial electron transport. *Mitochondrion.* 2007;7:S8–S33.
25. Coricovac DE, Dehelean CA. Pathological aspects with global impact induced by toxicants at cellular level. In: *Toxicology studies-cells, drugs and environment.* IntechOpen; 2015.
26. van der Bliek AM, Sedensky MM, Morgan PG. Cell biology of the mitochondrion. *Genetics.* 2017;207(3):843–871.
27. McStay GP. Contributions of plasma protein binding and membrane transporters to drug-induced mitochondrial toxicity. In: *Mitochondrial dysfunction caused by drugs and environmental toxicants.* Wiley; 2018:1–14.
28. Kühlbrandt W. Structure and function of mitochondrial membrane protein complexes. *BMC Biol.* 2015;13(1):1–11.
29. Paradies G, Paradies V, Ruggiero FM, Petrosillo G. Role of cardiolipin in mitochondrial function and dynamics in health and disease: molecular and pharmacological aspects. *Cell.* 2019;8(7):728.
30. Levings III C, Brown G. Molecular biology of plant mitochondria. *Cell.* 1989;56(2):171–179.
31. Piel III RB, Dailey Jr HA, Medlock AE. The mitochondrial heme metabolon: insights into the complex (ity) of heme synthesis and distribution. *Mol Genet Metab.* 2019;128(3):198–203.
32. Szabo I, Zoratti M. Mitochondrial channels: ion fluxes and more. *Physiol Rev.* 2014;94(2):519–608.
33. Poburko D, Santo-Domingo J, Demaurex N. Dynamic regulation of the mitochondrial proton gradient during cytosolic calcium elevations. *J Biol Chem.* 2011;286(13):11672–11684.
34. Zielonka J, Joseph J, Sikora A, et al. Mitochondria-targeted triphenylphosphonium-based compounds: syntheses, mechanisms of action, and therapeutic and diagnostic applications. *Chem Rev.* 2017;117(15):10043–10120.
35. Simpson DP. Mitochondrial transport functions and renal metabolism. *Kidney Int.* 1983;23(6):785–793.
36. Will Y, Dykens JA. *Mitochondrial dysfunction caused by drugs and environmental toxicants.* Wiley Online Library; 2018.
37. Chen LB. Mitochondrial membrane potential in living cells. *Annu Rev Cell Biol.* 1988;4(1):155–181.
38. Lodish H, Berk A, Zipursky SL, Matsudaira P, Baltimore D, Darnell J. Electron transport and oxidative phosphorylation. In: *Molecular cell biology.* 4th ed. WH Freeman; 2000.
39. Schönfeld P, Wojtczak L. Short-and medium-chain fatty acids in energy metabolism: the cellular perspective. *J Lipid Res.* 2016;57(6):943–954.
40. Cohen BH. Pharmacologic effects on mitochondrial function. *Dev Disabil Res Rev.* 2010;16(2):189–199.
41. Phillips L, Johnson M, Deener K, Bonanni C. EPA's exposure assessment toolbox (EPA-Expo-Box). *J Environ Inf.* 2015;25(2).
42. Mattingly CJ, Rosenstein MC, Davis AP, Colby GT, Forrest Jr JN, Boyer JL. The comparative toxicogenomics database: a cross-species resource for building chemical-gene interaction networks. *Toxicol Sci.* 2006;92(2):587–595.

43. Y-J H, Liao W, Yan Z-F, et al. Progress in the research of the toxicity effect mechanisms of heavy metals on freshwater organisms and their water quality criteria in China. *J Chem*. 2020;2020.
44. Selvaraj V, Tomblin J, Armistead MY, Murray E. Selenium (sodium selenite) causes cytotoxicity and apoptotic mediated cell death in PLHC-1 fish cell line through DNA and mitochondrial membrane potential damage. *Ecotoxicol Environ Saf*. 2013;87:80–88.
45. Rahman M, Yanful E, Jasim S. Endocrine disrupting compounds (EDCs) and pharmaceuticals and personal care products (PPCPs) in the aquatic environment: implications for the drinking water industry and global environmental health. *J Water Health*. 2009;7(2):224–243.
46. Drakvik E, Altenburger R, Aoki Y, et al. Statement on advancing the assessment of chemical mixtures and their risks for human health and the environment. *Environ Int*. 2020;134:105267.
47. Gauthier P. *Mixtures of metals and polycyclic aromatic hydrocarbons lead to complex toxic outcomes in the freshwater amphipod, Hyalella azteca*; 2015.
48. Griffith CM, Baig N, Seiber JN. Contamination from industrial toxicants 22. In: *Handbook of food chemistry*. vol. 719; 2015.
49. Kalghatgi S, Spina CS, Costello JC, et al. Bactericidal antibiotics induce mitochondrial dysfunction and oxidative damage in mammalian cells. *Sci Transl Med*. 2013;5(192):192ra85.
50. World Health Organization. *Pharmaceuticals in drinking-water*. World Health Organization; 2012.
51. Zenker A, Cicero MR, Prestinaci F, Bottoni P, Carere M. Bioaccumulation and biomagnification potential of pharmaceuticals with a focus to the aquatic environment. *J Environ Manage*. 2014;133:378–387.
52. Hargreaves IP, Al Shahrani M, Wainwright L, Heales SJ. Drug-induced mitochondrial toxicity. *Drug Saf*. 2016;39(7):661–674.
53. Biswas B, Qi F, Biswas JK, Wijayawardena A, Khan MAI, Naidu R. The fate of chemical pollutants with soil properties and processes in the climate change paradigm—a review. *Soil Syst*. 2018;2(3):51.
54. Reeder RJ, Schoonen MA, Lanzirotti A. Metal speciation and its role in bioaccessibility and bioavailability. *Rev Mineral Geochem*. 2006;64(1):59–113.
55. Carlsson P, Breivik K, Brorström-Lundén E, et al. Polychlorinated biphenyls (PCBs) as sentinels for the elucidation of Arctic environmental change processes: a comprehensive review combined with ArcRisk project results. *Environ Sci Pollut Res*. 2018;25(23):22499–22528.
56. Krause JS, Pérez JH, Chmura HE, et al. Weathering the storm: do arctic blizzards cause repeatable changes in stress physiology and body condition in breeding songbirds? *Gen Comp Endocrinol*. 2018;267:183–192.
57. Moore B, Allard G. Abiotic disturbances and their influence on forest health: a review. In: *Forest health and biosecurity working paper*. 35E; 2011.
58. Grünewald A, Kumar KR, Sue CM. New insights into the complex role of mitochondria in Parkinson's disease. *Prog Neurobiol*. 2019;177:73–93.
59. Twig G, Shirihai OS. The interplay between mitochondrial dynamics and mitophagy. *Antioxid Redox Signal*. 2011;14(10):1939–1951.
60. Guo C, Sun L, Chen X, Zhang D. Oxidative stress, mitochondrial damage and neurodegenerative diseases. *Neural Regen Res*. 2013;8(21):2003.
61. Ma L, Liu J-Y, Dong J-X, Xiao Q, Zhao J, Jiang F-L. Toxicity of Pb2+ on rat liver mitochondria induced by oxidative stress and mitochondrial permeability transition. *Toxicol Res*. 2017;6(6):822–830.
62. Burger G, Lang BF, Reith M, Gray MW. Genes encoding the same three subunits of respiratory complex II are present in the mitochondrial DNA of two phylogenetically distant eukaryotes. *Proc Natl Acad Sci*. 1996;93(6):2328–2332.

63. Stowe DF, Camara AK. Mitochondrial reactive oxygen species production in excitable cells: modulators of mitochondrial and cell function. *Antioxid Redox Signal.* 2009;11(6):1373–1414.
64. Murphy MP. How mitochondria produce reactive oxygen species. *Biochem J.* 2009;417(1):1–13.
65. Kowaltowski AJ, Vercesi AE. Mitochondrial damage induced by conditions of oxidative stress. *Free Radic Biol Med.* 1999;26(3–4):463–471.
66. Zhou P, Huang R. Targeting of the respiratory chain by toxicants: beyond the toxicities to mitochondrial morphology. *Toxicol Res.* 2018;7(6):1008–1011.
67. Yu B, Ma L, Jin J, et al. Mitochondrial toxicity induced by a thiourea gold (i) complex: mitochondrial permeability transition and respiratory deficit. *Toxicol Res.* 2018;7(6):1081–1090.
68. Fan X-Y, Yuan L, Wu C, et al. Mitochondrial toxicity of organic arsenicals: membrane permeability transition pore opening and respiratory dysfunction. *Toxicol Res.* 2018;7(2):191–200.
69. Broom AJ, Ambroso J, Brunori G, et al. Effects of mid-respiratory chain inhibition on mitochondrial function in vitro and in vivo. *Toxicol Res.* 2016;5(1):136–150.
70. Sherer TB, Betarbet R, Testa CM, et al. Mechanism of toxicity in rotenone models of Parkinson's disease. *J Neurosci.* 2003;23(34):10756–10764.
71. Caito SW, Aschner M. Mitochondrial redox dysfunction and environmental exposures. *Antioxid Redox Signal.* 2015;23(6):578–595.
72. van der Toorn M, Slebos D-J, de Bruin HG, et al. Cigarette smoke-induced blockade of the mitochondrial respiratory chain switches lung epithelial cell apoptosis into necrosis. *Am J Physiol Lung Cell Mol Physiol.* 2007;292(5), L1211-L8.
73. Raza H, John A, Nemmar A. Short-term effects of nose-only cigarette smoke exposure on glutathione redox homeostasis, cytochrome P450 1A1/2 and respiratory enzyme activities in mice tissues. *Cell Physiol Biochem.* 2013;31(4–5):683–692.
74. Agarwal AR, Zhao L, Sancheti H, Sundar IK, Rahman I, Cadenas E. Short-term cigarette smoke exposure induces reversible changes in energy metabolism and cellular redox status independent of inflammatory responses in mouse lungs. *Am J Physiol Lung Cell Mol Physiol.* 2012;303(10), L889-L98.
75. Miao Q, Xu Y, Zhang H, Xu P, Ye J. Cigarette smoke induces ROS mediated autophagy impairment in human corneal epithelial cells. *Environ Pollut.* 2019;245:389–397.
76. Jin SM, Youle RJ. PINK1-and Parkin-mediated mitophagy at a glance. *J Cell Sci.* 2012;125(4):795–799.
77. Terry Jr AV. Functional consequences of repeated organophosphate exposure: potential non-cholinergic mechanisms. *Pharmacol Ther.* 2012;134(3):355–365.
78. Imam A, Ogunniyi A, Ibrahim A, et al. Dichlorvos induced oxidative and neuronal responses in rats: mitigative efficacy of Nigella sativa (black cumin). *Niger J Physiol Sci.* 2018;33(1):83–88.
79. Karami-Mohajeri S, Abdollahi M. Mitochondrial dysfunction and organophosphorus compounds. *Toxicol Appl Pharmacol.* 2013;270(1):39–44.
80. Mori N, Yasutake A, Marumoto M, Hirayama K. Methylmercury inhibits electron transport chain activity and induces cytochrome c release in cerebellum mitochondria. *J Toxicol Sci.* 2011;36(3):253–259.
81. Basu AK. DNA damage, mutagenesis and cancer. *Int J Mol Sci.* 2018;19(4):970.
82. Guan Y, Wang X, Wong M, et al. Evaluation of genotoxic and mutagenic activity of organic extracts from drinking water sources. *PLoS One.* 2017;12(1), e0170454.
83. Gomez C, Bandez MJ, Navarro A. Pesticides and impairment of mitochondrial function in relation with the parkinsonian syndrome. *Front Biosci.* 2007;12(1079):93.
84. Heinz S, Freyberger A, Lawrenz B, Schladt L, Schmuck G, Ellinger-Ziegelbauer H. Mechanistic investigations of the mitochondrial complex I inhibitor rotenone in the context of pharmacological and safety evaluation. *Sci Rep.* 2017;7(1):1–13.

85. Carlson K, Ehrich M. Organophosphorus compound-induced modification of SH-SY5Y human neuroblastoma mitochondrial transmembrane potential. *Toxicol Appl Pharmacol.* 1999;160(1):33–42.
86. Pearson JN, Patel M. The role of oxidative stress in organophosphate and nerve agent toxicity. *Ann NY Acad Sci.* 2016;1378(1):17.
87. Falck Jr F, Ricci Jr A, Wolff MS, Godbold J, Deckers P. Pesticides and polychlorinated biphenyl residues in human breast lipids and their relation to breast cancer. *Arch Environ Health.* 1992;47(2):143–146.
88. Ounnas F, Privé F, Lamarche F, et al. A relevant exposure to a food matrix contaminated environmentally by polychlorinated biphenyls induces liver and brain disruption in rats. *Chemosphere.* 2016;161:80–88.
89. Xu D, Li L, Liu L, et al. Polychlorinated biphenyl quinone induces mitochondrial-mediated and caspase-dependent apoptosis in HepG2 cells. *Environ Toxicol.* 2015;30(9):1063–1072.
90. Cocco S, Secondo A, Del Viscovo A, et al. Polychlorinated biphenyls induce mitochondrial dysfunction in SH-SY5Y neuroblastoma cells. *PLoS One.* 2015;10(6), e0129481.
91. World Health Organization. *Dioxins and their effects on human health*; 1999.
92. Duarte F, Gomes A, Teodoro J, et al. Dibenzofuran-induced mitochondrial dysfunction: interaction with ANT carrier. *Toxicol In Vitro.* 2013;27(8):2160–2168.
93. Stohs SJ. Oxidative stress induced by 2, 3, 7, 8-tetrachlorodibenzo-p-dioxin (TCDD). *Free Radic Biol Med.* 1990;9(1):79–90.
94. Sánchez-Martín FJ, Fernández-Salguero PM, Merino JM. 2, 3, 7, 8-Tetrachlorodibenzo-p-dioxin induces apoptosis in neural growth factor (NGF)-differentiated pheochromocytoma PC12 cells. *Neurotoxicology.* 2010;31(3):267–276.
95. Pereira SP, Pereira GC, Pereira CV, et al. Dioxin-induced acute cardiac mitochondrial oxidative damage and increased activity of ATP-sensitive potassium channels in Wistar rats. *Environ Pollut.* 2013;180:281–290.
96. Chen S-C, Liao T-L, Wei Y-H, Tzeng C-R, Kao S-H. Endocrine disruptor, dioxin (TCDD)-induced mitochondrial dysfunction and apoptosis in human trophoblast-like JAR cells. *Mol Hum Reprod.* 2010;16(5):361–372.
97. Mustafa MF, Fakurazi S, Abdullah MA, Maniam S. Pathogenic mitochondria DNA mutations: current detection tools and interventions. *Gene.* 2020;11(2):192.
98. Türkez H, Geyikoglu F, Yousef MI. Ameliorative effect of docosahexaenoic acid on 2, 3, 7, 8-tetrachlorodibenzo-p-dioxin-induced histological changes, oxidative stress, and DNA damage in rat liver. *Toxicol Ind Health.* 2012;28(8):687–696.
99. Tchounwou PB, Yedjou CG, Patlolla AK, Sutton DJ. Heavy metal toxicity and the environment. *Experientia Suppl.* 2012;133–164.
100. Ghasemi H, Rostampour F, Ranjbar A. *The role of oxidative stress in metals toxicity/mitochondrial dysfunction as a key player*; 2014.
101. Belyaeva EA, Sokolova TV, Emelyanova LV, Zakharova IO. Mitochondrial electron transport chain in heavy metal-induced neurotoxicity: effects of cadmium, mercury, and copper. *Scientific World Journal.* 2012;2012.
102. Ameer SS, Xu Y, Engström K, et al. Exposure to inorganic arsenic is associated with increased mitochondrial DNA copy number and longer telomere length in peripheral blood. *Front Cell Dev Biol.* 2016;4:87.
103. Zhang W, Liu Y, An Z, Huang D, Qi Y, Zhang Y. Mediating effect of ROS on mtDNA damage and low ATP content induced by arsenic trioxide in mouse oocytes. *Toxicol In Vitro.* 2011;25(4):979–984.
104. Wang X, Yan M, Zhao L, et al. Low-dose methylmercury-induced apoptosis and mitochondrial DNA mutation in human embryonic neural progenitor cells. *Oxid Med Cell Longev.* 2016;2016.
105. Llamas A, Al-Lal A-M, García-Martínez M-J, et al. Polycyclic aromatic hydrocarbons (PAHs) produced in the combustion of fatty acid alkyl esters from different feedstocks:

quantification, statistical analysis and mechanisms of formation. *Sci Total Environ.* 2017;586:446–456.
106. Sotty J, Kluza J, De Sousa C, et al. Mitochondrial alterations triggered by repeated exposure to fine (PM2. 5–0.18) and quasi-ultrafine (PM0. 18) fractions of ambient particulate matter. *Environ Int.* 2020;142, 105830.
107. Salazar I, Pavani M, Aranda W, Maya JD, Morello A, Ferreira J. Alterations of rat liver mitochondrial oxidative phosphorylation and calcium uptake by benzo [a] pyrene. *Toxicol Appl Pharmacol.* 2004;198(1):1–10.
108. Kim H-Y, Kim H-R, Kang M-G, et al. Profiling of biomarkers for the exposure of polycyclic aromatic hydrocarbons: Lamin-A/C isoform 3, poly [ADP-ribose] polymerase 1, and mitochondria copy number are identified as universal biomarkers. *Biomed Res Int.* 2014;2014.
109. Mulvey M, Newman MC, Vogelbein W, Unger MA. Genetic structure of Fundulus heteroclitus from PAH-contaminated and neighboring sites in the Elizabeth and York Rivers. *Aquat Toxicol.* 2002;61(3–4):195–209.
110. Honda M, Suzuki N. Toxicities of polycyclic aromatic hydrocarbons for aquatic animals. *Int J Environ Res Public Health.* 2020;17(4):1363.
111. Valente WJ, Ericson NG, Long AS, White PA, Marchetti F, Bielas JH. Mitochondrial DNA exhibits resistance to induced point and deletion mutations. *Nucleic Acids Res.* 2016;44(18):8513–8524.
112. Hachenberg N. *Predictions of NOx emissions in pulverized coal combustion*; 2014.
113. Lee B-J, Kim B, Lee K. Air pollution exposure and cardiovascular disease. *Toxicol Res.* 2014;30(2):71–75.
114. Stefanatos R, Sanz A. The role of mitochondrial ROS in the aging brain. *FEBS Lett.* 2018;592(5):743–758.
115. Joskow PL, Schmalensee R, Bailey EM. The market for sulfur dioxide emissions. *Am Econ Rev.* 1998;669–685.
116. Qin G, Wu M, Wang J, Xu Z, Xia J, Sang N. Sulfur dioxide contributes to the cardiac and mitochondrial dysfunction in rats. *Toxicol Sci.* 2016;151(2):334–346.
117. Dandawate PR, Vyas AC, Padhye SB, Singh MW, Baruah JB. Perspectives on medicinal properties of benzoquinone compounds. *Mini Rev Med Chem.* 2010;10(5):436–454.
118. Qin G, Wu M, Sang N. Sulfur dioxide and benzo (a) pyrene trigger apoptotic and anti-apoptotic signals at different post-exposure times in mouse liver. *Chemosphere.* 2015;139:318–325.
119. Fetterman JL, Sammy MJ, Ballinger SW. Mitochondrial toxicity of tobacco smoke and air pollution. *Toxicology.* 2017;391:18.
120. Phaniendra A, Jestadi DB, Periyasamy L. Free radicals: properties, sources, targets, and their implication in various diseases. *Indian J Clin Biochem.* 2015;30(1):11–26.
121. Yang Z, Harrison CM, Chuang GC, Ballinger SW. The role of tobacco smoke induced mitochondrial damage in vascular dysfunction and atherosclerosis. *Mutat Res Fundam Mol Mech Mutagen.* 2007;621(1–2):61–74.
122. Hoffmann RF, Zarrintan S, Brandenburg SM, et al. Prolonged cigarette smoke exposure alters mitochondrial structure and function in airway epithelial cells. *Respir Res.* 2013;14(1):1–13.
123. Kaplan A, Abidi E, Ghali R, Booz GW, Kobeissy F, Zouein FA. Functional, cellular, and molecular remodeling of the heart under influence of oxidative cigarette tobacco smoke. *Oxid Med Cell Longev.* 2017;2017.
124. Zhang S, Balbo S, Wang M, Hecht SS. Analysis of acrolein-derived 1, N 2-propanodeoxyguanosine adducts in human leukocyte DNA from smokers and nonsmokers. *Chem Res Toxicol.* 2011;24(1):119–124.
125. Carugno M, Pesatori AC, Dioni L, et al. Increased mitochondrial DNA copy number in occupations associated with low-dose benzene exposure. *Environ Health Perspect.* 2012;120(2):210–215.

126. Wurtsbaugh WA, Paerl HW, Dodds WK. Nutrients, eutrophication and harmful algal blooms along the freshwater to marine continuum. *Wiley Interdiscip Rev Water.* 2019;6(5), e1373.
127. Zanchett G, Oliveira-Filho EC. Cyanobacteria and cyanotoxins: from impacts on aquatic ecosystems and human health to anticarcinogenic effects. *Toxins.* 2013;5(10):1896–1917.
128. Yi X, Xu S, Huang F, et al. Effects of chronic exposure to microcystin-LR on kidney in mice. *Int J Environ Res Public Health.* 2019;16(24):5030.
129. Bouaïcha N, Miles CO, Beach DG, et al. Structural diversity, characterization and toxicology of microcystins. *Toxins.* 2019;11(12):714.
130. Malińska D, Więckowski MR, Michalska B, et al. Mitochondria as a possible target for nicotine action. *J Bioenerg Biomembr.* 2019;51(4):259–276.
131. Brown D, Bailey J, Oliveri A, Levin E, Di Giulio R. Developmental exposure to a complex PAH mixture causes persistent behavioral effects in naive Fundulus heteroclitus (killifish) but not in a population of PAH-adapted killifish. *Neurotoxicol Teratol.* 2016;53:55–63.
132. van Ginneken V, Palstra A, Leonards P, et al. PCBs and the energy cost of migration in the European eel (Anguilla anguilla L.). *Aquat Toxicol.* 2009;92(4):213–220.
133. Gheorghe S, Stoica C, Vasile GG, Nita-Lazar M, Stanescu E, Lucaciu IE. Metals toxic effects in aquatic ecosystems: modulators of water quality. *Water Qual.* 2017;60–89.
134. Sokolova I. Mitochondrial adaptations to variable environments and their role in animals' stress tolerance. *Integr Comp Biol.* 2018;58(3):519–531.
135. Belhadj Slimen I, Najar T, Ghram A, Dabbebi H, Ben Mrad M, Abdrabbah M. Reactive oxygen species, heat stress and oxidative-induced mitochondrial damage. A review. *Int J Hyperthermia.* 2014;30(7):513–523.
136. Xie X, He Z, Chen N, Tang Z, Wang Q, Cai Y. The roles of environmental factors in regulation of oxidative stress in plant. *Biomed Res Int.* 2019;2019.
137. Moreira A, Freitas R, Figueira E, et al. Combined effects of arsenic, salinity and temperature on Crassostrea gigas embryotoxicity. *Ecotoxicol Environ Saf.* 2018;147:251–259.
138. Ross C, Olsen K, Henry M, Pierce R. Mosquito control pesticides and sea surface temperatures have differential effects on the survival and oxidative stress response of coral larvae. *Ecotoxicology.* 2015;24(3):540–552.
139. Falfushynska H, Gnatyshyna L, Yurchak I, Sokolova I, Stoliar O. The effects of zinc nanooxide on cellular stress responses of the freshwater mussels Unio tumidus are modulated by elevated temperature and organic pollutants. *Aquat Toxicol.* 2015;162:82–93.
140. Xu Y, Li H, Hedmer M, et al. Occupational exposure to particles and mitochondrial DNA-relevance for blood pressure. *Environ Health.* 2017;16(1):1–10.
141. Budnik LT, Kloth S, Baur X, Preisser AM, Schwarzenbach H. Circulating mitochondrial DNA as biomarker linking environmental chemical exposure to early preclinical lesions elevation of mtDNA in human serum after exposure to carcinogenic halo-alkane-based pesticides. *PLoS One.* 2013;8(5), e64413.
142. Ahmadi N, Mandegary A, Jamshidzadeh A, et al. Hematological abnormality, oxidative stress, and genotoxicity induction in the greenhouse pesticide sprayers; investigating the role of NQO1 gene polymorphism. *Toxics.* 2018;6(1):13.
143. Bordoni L, Smerilli V, Nasuti C, Gabbianelli R. Mitochondrial DNA methylation and copy number predict body composition in a young female population. *J Transl Med.* 2019;17(1):1–11.
144. Chin-Chan M, Navarro-Yepes J, Quintanilla-Vega B. Environmental pollutants as risk factors for neurodegenerative disorders: Alzheimer and Parkinson diseases. *Front Cell Neurosci.* 2015;9:124.
145. Tanner CM, Kamel F, Ross GW, et al. Rotenone, paraquat, and Parkinson's disease. *Environ Health Perspect.* 2011;119(6):866–872.

146. Wang C-Y, Zhao Z-B. Somatic mtDNA mutations in lung tissues of pesticide-exposed fruit growers. *Toxicology.* 2012;291(1–3):51–55.
147. Cortés-Iza SC, Rodríguez AI. Oxidative stress and pesticide disease: a challenge for toxicology. *Rev Fac Med.* 2018;66(2):261–267.
148. Pavanello S, Dioni L, Hoxha M, Fedeli U, Mielzynska-Švach D, Baccarelli AA. Mitochondrial DNA copy number and exposure to polycyclic aromatic hydrocarbons. *Cancer Epidemiol Biomarkers Prev.* 2013;22(10):1722–1729.
149. Byun H-M, Baccarelli AA. Environmental exposure and mitochondrial epigenetics: study design and analytical challenges. *Hum Genet.* 2014;133(3):247–257.
150. Linqing Y, Bo X, Xueqin Y, et al. Mitochondrial DNA hypomethylation in chrome plating workers. *Toxicol Lett.* 2016;243:1–6.
151. Hou L, Zhang X, Dioni L, et al. Inhalable particulate matter and mitochondrial DNA copy number in highly exposed individuals in Beijing, China: a repeated-measure study. *Part Fibre Toxicol.* 2013;10(1):1–9.
152. Leso V, Vetrani I, Della Volpe I, Nocera C, Iavicoli I. Welding fume exposure and epigenetic alterations: a systematic review. *Int J Environ Res Public Health.* 2019;16(10):1745.
153. Brunst KJ, Sanchez-Guerra M, Chiu Y-HM, et al. Prenatal particulate matter exposure and mitochondrial dysfunction at the maternal-fetal interface: effect modification by maternal lifetime trauma and child sex. *Environ Int.* 2018;112:49–58.
154. Korolev IO. Alzheimer's disease: a clinical and basic science review. *Med Stud Res J.* 2014;4(1):24–33.
155. Yegambaram M, Manivannan B, Beach GT, Halden UR. Role of environmental contaminants in the etiology of Alzheimer's disease: a review. *Curr Alzheimer Res.* 2015;12(2):116–146.
156. Zhao Y, Zhao B. Oxidative stress and the pathogenesis of Alzheimer's disease. *Oxid Med Cell Longev.* 2013;2013.
157. Wang W, Zhao F, Ma X, Perry G, Zhu X. Mitochondria dysfunction in the pathogenesis of Alzheimer's disease: recent advances. *Mol Neurodegener.* 2020;15:1–22.
158. Yamada S, Kubo Y, Yamazaki D, Sekino Y, Kanda Y. Chlorpyrifos inhibits neural induction via Mfn1-mediated mitochondrial dysfunction in human induced pluripotent stem cells. *Sci Rep.* 2017;7(1):1–12.
159. Ferreira-Vieira HT, Guimaraes MI, Silva RF, Ribeiro MF. Alzheimer's disease: targeting the cholinergic system. *Curr Neuropharmacol.* 2016;14(1):101–115.
160. del Pino J, Moyano P, Anadon MJ, et al. Acute and long-term exposure to chlorpyrifos induces cell death of basal forebrain cholinergic neurons through AChE variants alteration. *Toxicology.* 2015;336:1–9.
161. Ruiz-Muñoz AM, Nieto-Escamez FA, Aznar S, Colomina MT, Sanchez-Santed F. Cognitive and histological disturbances after chlorpyrifos exposure and chronic Aβ (1–42) infusions in Wistar rats. *Neurotoxicology.* 2011;32(6):836–844.
162. Chen L, Yoo S-E, Na R, Liu Y, Ran Q. Cognitive impairment and increased Aβ levels induced by paraquat exposure are attenuated by enhanced removal of mitochondrial H2O2. *Neurobiol Aging.* 2012;33(2), 432.e15-e26.
163. Sanders T, Liu Y, Buchner V, Tchounwou PB. Neurotoxic effects and biomarkers of lead exposure: a review. *Rev Environ Health.* 2009;24(1):15.
164. Bhattacharjee S, Zhao Y, Hill JM, et al. Selective accumulation of aluminum in cerebral arteries in Alzheimer's disease (AD). *J Inorg Biochem.* 2013;126:35–37.
165. Markesbery WR. The role of oxidative stress in Alzheimer disease. *Arch Neurol.* 1999;56(12):1449–1452.
166. Dos Santos AA, Hort MA, Culbreth M, et al. Methylmercury and brain development: a review of recent literature. *J Trace Elem Med Biol.* 2016;38:99–107.
167. Ashok A, Rai NK, Tripathi S, Bandyopadhyay S. Exposure to As-, Cd-, and Pb-mixture induces Aβ, amyloidogenic APP processing and cognitive impairments via oxidative stress-dependent neuroinflammation in young rats. *Toxicol Sci.* 2015;143(1):64–80.

168. Aliahmat NS, Noor MRM, Yusof WJW, Makpol S, Ngah WZW, Yusof YAM. Antioxidant enzyme activity and malondialdehyde levels can be modulated by Piper betle, tocotrienol rich fraction and Chlorella vulgaris in aging C57BL/6 mice. *Clinics.* 2012;67(12):1447–1454.
169. Olivieri G, Brack C, Müller-Spahn F, et al. Mercury induces cell cytotoxicity and oxidative stress and increases β-amyloid secretion and tau phosphorylation in SHSY5Y neuroblastoma cells. *J Neurochem.* 2000;74(1):231–236.
170. De Lau LM, Breteler MM. Epidemiology of Parkinson's disease. *Lancet Neurol.* 2006;5(6):525–535.
171. Mapa MS, Le VQ, Wimalasena K. Characteristics of the mitochondrial and cellular uptake of MPP+, as probed by the fluorescent mimic, 4'I-MPP+. *PLoS One.* 2018;13(8), e0197946.
172. Spivey A. *Rotenone and paraquat linked to Parkinson's disease: human exposure study supports years of animal studies.* National Institute of Environmental Health Sciences; 2011.
173. Desplats P, Patel P, Kosberg K, et al. Combined exposure to Maneb and Paraquat alters transcriptional regulation of neurogenesis-related genes in mice models of Parkinson's disease. *Mol Neurodegener.* 2012;7(1):1–10.
174. Norazit A, Meedeniya AC, Nguyen MN, Mackay-Sim A. Progressive loss of dopaminergic neurons induced by unilateral rotenone infusion into the medial forebrain bundle. *Brain Res.* 2010;1360:119–129.
175. Kitazawa M, Anantharam V, Kanthasamy AG. Dieldrin-induced oxidative stress and neurochemical changes contribute to apoptopic cell death in dopaminergic cells. *Free Radic Biol Med.* 2001;31(11):1473–1485.
176. Richardson JR, Caudle WM, Wang M, et al. Developmental exposure to the pesticide dieldrin alters the dopamine system and increases neurotoxicity in an animal model of Parkinson's disease. *FASEB J.* 2006;20(10):1695–1697.
177. Caudle WM, Guillot TS, Lazo CR, Miller GW. Industrial toxicants and Parkinson's disease. *Neurotoxicology.* 2012;33(2):178–188.
178. Subramaniam SR, Chesselet M-F. Mitochondrial dysfunction and oxidative stress in Parkinson's disease. *Prog Neurobiol.* 2013;106:17–32.
179. Zeng J, Wang L, Xia W, et al. Mechanism of reactive oxygen species in manganese chloride-induced apoptosis in PC12 cells. *Zhonghua Lao Dong Wei Sheng Zhi Ye Bing Za Zhi.* 2006;24(3):157–160.

CHAPTER 5

Updates on mitochondria, calorie restriction, and aging

Bahareh Sadat Yousefsani[a,b], Majid Dadmehr[a,b], and Jalal Pourahmad[c,*]
[a]Research Institute for Islamic and Complementary Medicine, Iran University of Medical Sciences, Tehran, Iran
[b]School of Persian Medicine, Iran University of Medical Sciences, Tehran, Iran
[c]Department of Pharmacology & Toxicology, School of Pharmacy, Shahid Beheshti University of Medical Sciences, Tehran, Iran

Contents

Abbreviations

AMPK	AMP-activated protein kinase
Atg	autophagy-related
ATP	adenosine triphosphate
BMI	body mass index
CR	calorie restriction
CVD	cardiovascular disease
DM	diabetes mellitus
FGF2	fibroblast growth factor 2
IF	intermittent fasting
LC3	microtubule-associated protein light chain 3
mTOR	mammalian target of rapamycin
OXPHOS	oxidative phosphorylation
ROS	reactive oxygen species
SIRT1	silent mating type information regulation-1
SOD	superoxide dismutase
AD	alzheimer's disease
HD	huntington's disease

*Main contributor.

Mitochondrial Metabolism
https://doi.org/10.1016/B978-0-12-822416-8.00005-1

PD parkinson's disease
AL ad libitum
mt-mKeima mitochondria-targeted mKeima plasmid

1 Introduction

Aging has been described as a reducing ability to maintain homeostasis and raising the risk of death. Malfunctioning happens due to the aging of the quality control systems. At an older age, the accumulation of ruined macromolecules and membranes may damage cell functioning; the accumulation of unhealthy mitochondria and peroxisomes, which may enhance reactive oxygen species (ROS) per unit of produced energy and accelerate the aging action. In recent years, nutritional approaches have been considered as possible alternatives to the currently existing medications to increase the response to therapy. For example, the Mediterranean diet, ketogenic diet, fasting, etc., are among these that are recommended to patients along with the main treatment. In addition, a diet full of antioxidants exerts protective effects on all people, especially on the elderly people. The bioactive food components have a potential preventive action on different types of chronic diseases like various types of cancer. Another example is consuming a Mediterranean diet with rich plant foods that decreased the risk of developing multiple chronic disorders and increased life expectancy. Nowadays, there has been a great deal of interest in the effects of intermittent fasting (IF) and calorie restriction (CR) on different diseases. The effects of CR have been investigated on many diseases such as nonalcoholic fatty liver disease (NAFLD), alzheimer's disease (AD), vascular diseases, depression, and even different types of cancer. Studies revealed that IF or CR has beneficial effects on aging and aging-related diseases.[1–6] This chapter provides an overview of the mitochondrial mechanisms of aging and the effects of diet on it.

2 Aging and mitochondria

Harman et al. in 1956 proposed that aging is happening due to progressive macromolecules damage by ROS, which leads to reduced cellular function.[7] Mitochondria are double membrane–bound organelles responsible for many processes in eukaryotic cells, such as the production of adenosine triphosphate (ATP). Mitochondria are the main site of ROS production. They are responsible for cellular death, regulation of calcium hemostasis, and immunity. Due to the production of the energy for the cell and in the electron transport chain, ROS is produced as a byproduct.[8] Oxidative phosphorylation (OXPHOS) in the mitochondrial respiratory chain produces ATP. This process can result in the generation of different types of ROS, such as peroxides, superoxide, hydroxyl radical, and singlet oxygen.[9] The constituents of mitochondria, like components of the electron transport chain and the genome of mitochondria, are damaged in aging, proportional to adjacency to the site of ROS production. Besides, mitochondrial genetic diseases lead to shortening life span and premature aging.[10, 11] Point mutations and deletions of mitochondrial DNA accumulate in different tissues during aging in humans, monkeys, and rodents. These mutations are distributed and even can accumulate in certain cells, causing a mosaic pattern of respiratory chain deficiency in tissues such as the heart, skeletal muscle, and brain. Therefore, impaired respiratory chain function will increase ROS production and thereby increment the rate of mitochondrial DNA mutation accumulation, which will compromise respiratory chain functions.[12] Thus, mitochondrial metabolism and mitophagy/autophagy process are the two most active metabolic cellular processes that play a key role in regulating the life span of different types of organism.[13] In the mitophagy process, damaged or dysfunctional mitochondria are destroyed. The human mitochondrial half-life is 10–25 days, and the mitophagy process is to help the balance between mitochondrial biosynthesizing and degrading.[14] Mitophagy removes mitochondria with mutated DNA.[15] We have two types of mitophagy in the human body: micromitophagy and macromitophagy. In micromitophagy, vacuolar/lysosomal membrane directly attacks the mitochondria. But, in macromitophagy, the removal of targeted mitochondria happens by hydrolases in which autophagosomes and fusion of autophagosomes to vacuolar/lysosomal lead to degradation of dysfunctional mitochondria.[16, 17] Macromitophagy is only in mammalian cells and plays a key role in several biologic processes like the development of different illnesses. For instance, dysfunctional mitochondria are degraded by PINK1/Parkin-dependent process in the substantia nigra's dopaminergic neurons, leading to the pathology of parkinson's disease (PD).[18] In 2005, mitophagy was nominated when Lemasters et al. introduced Uth1p. It is a specific outer-membrane protein in yeast mitochondria, which is responsible for mitochondrial autophagy.[19] Tal et al. demonstrated that Aup1p exists in the intermembrane space of yeast mitochondria. Their studies showed that the localization of Aup1p is a key point for efficient mitophagy in stationary phase of cells.[20] Thus, mitochondria appear to play an important role in aging and age-related diseases. Accordingly, maintaining mitochondrial normal homeostasis is important for individual health.

Nowikovsky et al. revealed that evacuation of Mdm38, a mitochondrial inner-membrane protein, caused the loss of mitochondrial K^+/H^+ exchange, which leads to mitochondrial swelling, and then the mitochondrial reticulum fragments.[21] Atg32, the intramitochondrial domain, is a mitochondrial receptor for selective mitophagy in the yeast.[22] Aoki et al. concluded that the phosphorylation of serine on Atg32 by nitrogen is an important step for occurring mitophagy.[23] Two mitogen-activated protein kinases, Slt2, for degradation of both mitochondria and peroxisomes (via pexophagy), and Hog1, a mitogen-activated protein kinase, are required for mitophagy in *Saccharomyces cerevisiae*. Hog1 phosphorylates Atg11. Moreover, Hog1 is required to phosphorylate Pbs2, the downstream mitogen-activated protein kinase.[24] Besides, mitophagy plays an important role in human disease as well as adaptive responses to conditions such as starvation, hypoxia, or developmental signals.[25] There were two main pathways for the induction of mitophagy in mammalian cells: damage-induced mitophagy and developmental-induced mitophagy.[26] Two important proteins agitate damage-induced mitophagy in human: (1) Parkin is an E3 ubiquitin protein ligase and (2) serine/threonine-protein kinase PINK1 codifies the PTEN-induced putative kinase.[27] In polarized conditions, PINK1 is preserved at basal levels and is located in the inner mitochondrial membrane. In the damaged mitochondria, depolarization occurs because the mitochondrial membrane potential is not sufficient to pass PINK1 from the outer mitochondrial membrane to the inner mitochondrial membrane.[28] PINK1, on the outer membrane of mitochondria, leads to phosphorylation of Mitofusin-1, Mitofusin-2, ubiquitin, and ubiquitin-like domain of Parkin and activates the E3 ligase of parkin1.[29] Parkin raises the ubiquitination of voltage-dependent anion channels. These are required to keep the LC3-interacting region comprising p62 that acts as an adaptor molecule to recruit autophagosomal membranes in the mitochondria.[30] Proapoptotic proteins are responsible for developmentally induced mitophagy by the Nix or Bnip3L that are part of the Bcl-2 family.[31] Nix protein induces mitophagy, which interacts with LC3. After that, LC3 interacts with gamma-aminobutyric acid receptor-associated protein to generate the LC3/gamma-aminobutyric acid receptor-associated protein complex, which transfers the autophagosome to the mitochondria for elimination.[29] In addition, the beclin1/Bcl-2 interaction is broken by the binding of Nix and Bnip3 to Bcl-2. Therefore, beclin1 starts mitochondrial autophagy. Hypoxia induces FUNDC1, a mitochondrial outer-membrane receptor of mitophagy. FUNDC1 is dephosphorylated because of hypoxia and interacts with LC3-II to initiate the mitophagy process.[32]

3 Calorie restriction, intermittent fasting, and age-related diseases

Currently, the positive effect of fasting on health and the prevention of many diseases has been confirmed based on many different research studies. There are different types of fasting, alternate-day fasting, periodic fasting, etc. In alternate-day fasting, people receive very little or no energy over a period of time (e.g., 16–48 h). In periodic fasting or "fasting-mimicking diets," IF extended from 2 to 21 or more days. Time-restricted feeding is another type of fasting in which food access is forbidden to a time window (e.g., 8–9 h) daily.[23, 33, 34] In a study on metabolic diseases, it has been found that fasting has many positive effects on these complications. Fasting reduces the amount of leptin and insulin in the blood and increases the sensitivity of cells to them. Fasting also lowers blood pressure and ameliorates heart rate at rest condition. Reducing inflammation and total body fat is another beneficial effect of fasting. Besides, fasting ameliorates the resistance of the brain and heart to stressful situations by increasing tissue resistance and functional improvement in different models of myocardial infarction as well as stroke.[34, 35] Accordingly, different animal models revealed the affirmative effects of different types of above-mentioned fasting. All of these causes the onset of age-related diseases to be delayed or slowed down. Age-related diseases include different types of cancer, stroke, neurological disorders such as AD, PD, huntington's disease (HD) as well as cardiovascular complications and DM.[35] Recent researches revealed that CR is one of the strongest nongenetic triggers for initiating the autophagy and mitophagy processes, which increased mitochondrial activity. One of the dietary interventions is CR that reduces calorie intake without causing malnutrition.[36] CR modifies physiological activities, such as reducing cellular proliferation rates, decreasing ROS production, and reducing body temperature.[37–39] CR, which includes modifications in calorie intake dietary composition or timing of food intake, e.g., IF, has a fundamental role in chronic diseases.[40] Furthermore, CR has advantageous effects on increasing life span, presumably through a lag in the onset and reduction in the intensity of chronic aging-related disorders, such as CVD and different types of cancers.[40] The main role of CR on life span is regulated through overlapping following pathways: if nutrient supply is sufficient, nutrient-sensing signaling cascades will be activated, so cellular growth will be ameliorated. These lead to organismal growth and proliferation. On the other hand, the signaling pathways are downregulated in the low food availability or the lack of specific macromolecules, resulting in activation of protective metabolic pathways to improve the agglomeration of cellular damage, block cellular proliferation, and activate stress resistance transcription factors that negatively adjusted proaging pathways.[40–45] CR inhibited the insulin-like growth factor-1 receptor-dependent pathways and target of rapamycin (TOR)-dependent operations. These processes are involved in cellular proliferation and glycolysis. CR also has antiinflammatory efficacy; it decreases nuclear factor-κB activity and reduces the proinflammatory process of aging.[46] Recent studies demonstrated that CR induces mitochondrial function through activation of AMPK and sirtuins. At the same time, stronger acting of Forkhead box proteins happens that initiates the activation of the intracellular antioxidant system and autophagy/mitophagy process. Thus, both the inhibition of TOR and activation of AMPK can activate Forkhead box protein. Furthermore, CR activates nuclear factor erythroid 2-related factor 2, which leads to the expression of antioxidant proteins.[47] CR reduces the accumulation of demolished proteins via the following mechanisms: CR produces short-term starvation that leads to increment in proteolysis and reduces the accumulation of damaged protein. Furthermore, CR causes a decrease in protein turnover, reducing the need for protein replacement, CR increments glucose metabolism, reduces insulin levels, and has protective effects on neurodegeneration in animal models of HD, AD, PD, and stroke. Besides, CR can decrease the incidence of inflammatory and chronic disorders, such as DM different types of cancers, as well as CVD.[48–50]

4 Calorie restriction, intermittent fasting, and mitochondria

Due to the positive effects of fasting on various diseases, this field has attracted the attention of many researchers to conduct sequence further research. Because CR/IF is an accessible inexpensive way that the affirmative effects were wonderful.

Brandhorst and his colleagues proposed that IF is one of the best approaches for preventing different types of malignancies and betterment of the efficacy of cancer therapies.[51] The increment of DNA repair, mitochondrial health, and mitophagy/autophagy are the outcomes of the activation of the signaling pathways of the adaptive cellular stress response. IF inhibits disease progression and modifies health conditions through these cellular/molecular mechanisms. In addition, IF affects many signaling pathways like increment the function of mitochondria, keep the balance between mitochondrial biogenesis and mitophagy, decline in mTOR signaling, as well as upregulation of cyclic AMP response element-binding protein and neurotrophic factors such as brain-derived neurotrophic factor and fibroblast growth factor 2 (FGF2).[34] Scientists demonstrated that IF causes improvement in antiaging processes leading to prolonged life span.[52–54] Recent studies revealed that autophagy is the most important molecular mechanism, which has a key role in prolonging life span by eliminating cellular toxic composites and cleaning the cells.[55–57] Researchers screened the mitophagy-deficient mutants to find out which genes are essential for mitophagy. For this purpose, yeasts were grown on a nutrient-rich medium that contains a low glucose concentration (0.2%), which is considered a CR condition. The outcomes of this study revealed that the atg32Δ mutant is responsible for both reductions in the mean and maximum chronological life span of yeast. It must be noted that Atg32Δ damage is not related to nonselective macroautophagy and just belongs to the selective macroautophagy pathway.[22, 58]

Nevertheless, the exact impact of CR and IF on mitophagy/autophagy induction has not been assessed till now.

Price et al. evaluated the effect of long-term CR on *in vivo* model (male C57/BL6 mice) of hepatic mitophagy. For this purpose, two groups of CR mice and AL mice (with free access to water and food [ad libitum]) were evaluated. CR mice received 3.0 g of NIH-31/NIA fortified diet[a] once a day but AL mice have freely accessed that diet. The body weight of mice in each group was measured weekly. 0.5, 1, 4, 8, 15, or 32 days after beginning the experiment, mice were sacrificed for further evaluation. The results of this study demonstrated that protein synthesis and breakdown rates were significantly decreased in the hepatocytes in CR group. Besides, the mitochondrial protein half-life was significantly increased in the CR group in comparison with the AL mice. Furthermore, there was a significant decrease in hepatocyte mitochondria biogenesis in the CR group. Besides, a considerable decrease was observable in the CR group in the rate of whole protein metabolic and mitophagy.[36] The kidney effects of CR were evaluated in research by Cui and his colleagues.[59] For evaluating the effect of CR on NAFLD and DM, Linden et al. designed a study on diabetic Otsuka Long-Evans Tokushima fatty rats. For this purpose, rats were randomly divided into four groups: (1) AL diet, (2) metformin (300 mg/kg/day) with AL diet, (3) CR (~30% less than AL diet; 21 g/day), and (4) combination of metformin + CR for 12 weeks. Then, liver and blood samples were collected to recognize biochemical markers that belong to liver function and insulin resistance. The result of this study demonstrated that BNip3 concentrations

[a]This is a standard reference diet formulated used in biomedical and biological researches.

had a significant rise in the CR + metformin group compared with the AL diet. Furthermore, fasting glucose and insulin levels were elevated in the CR group. The combination of CR + metformin therapy causes increment glucose tolerance. Besides, serum alanine aminotransferases were significantly decreased in the CR + metformin group. A significant reduction in the markers of de novo lipogenesis in the liver like fatty acid synthase, acetyl-CoA carboxylase, and stearoyl-CoA desaturase-1 was shown in the CR group. But, hepatic mitochondrial activity such as palmitate oxidation and β-hydroxy acyl CoA dehydrogenase activity showed a significant increase in this group.[60]

Kidney mitochondrial function as well as the markers of protein expression in autophagy/mitophagy were also evaluated in this research. For this purpose, the rats were divided into three groups: (1) control group with normal rat chow (3.42 kcal/g), (2) high-calorie diet group (4.51 kcal/g), and (3) CR diet (30% less calorie in comparison with the control group). Blood samples and renal tissue of rats in each group were taken after 20 months of intervention. The results of this study revealed that in the high-caloric diet group, cholesterol and triglyceride levels, the concentration of creatinine, as well as kidney weight were significantly increased in comparison with the control rats. Furthermore, the expression of PINK1 in the high-calorie group showed a considerable increase indicating an increment in kidney mitochondrial damages in the high calorie receiving rats. On the other hand, in the rats which received the CR diet, the expression of PINK1 in the kidneys was significantly reduced indicating that CR diet causes a considerable decrease in kidney mitochondrial damages. It must be noted that there were no significant changes in the expression of Parkin between groups.[59] After examining the effect of CR on liver and kidney mitochondria, Zhao et al. evaluated the effect of exercise and CR on cardiac mitochondria.[61] In this study, mice were divided into four groups: (1) control group (C), (2) exercise-trained group (T) (swim training 5 days/week for 10 weeks), (3) CR group (in this group mice were fed 40% less calories than normal for 10 weeks), and (4) exercise training + CR (T + CR) group. The food administered in the research has 23% protein, 9% water, 4% crude fiber, and 6% fat. At the beginning of the study, the average food intake of the mice in the control group was measured weekly; the food consumption by the mice in the control group was 2.5–4.5 g. After that, 60% of calculated calories were given to the mice in the CR group. Results of the study demonstrated that there is a significant increase in mitophagy in CR and CR + T groups. Furthermore, PINK1 protein and mRNA expression have a considerable increase in these two groups. It must be considered that the expression of Parkin mRNA has a significant decrease in the mice of the T + CR group in comparison with the control group, indicating more sensitivity of PINK1 than Parkin in the PINK1/Parkin pathway. Besides, Binp3 was found out to have a considerable decrease in the mice in the T and T + CR groups than in the control group. The NIX protein expression was significantly decreased in all intervention groups in comparison with the control group. The last step of mitochondrial fission, which controls with Drp1 showed a significant increment in the mice in the CR and T + CR groups. All these findings revealed that CR and exercise training stimulate mitophagy in the heart because of the upregulation of Drp1 and PINK1, respectively. On the other hand, Nix and Bnip3 have no effect on mitophagy activation in the mice's cardiac cells.[61] In CR conditions, macroautophagy plays an important role in prolonging life span in yeast aging by sustaining the function of mitochondria and maintaining cellular lipid homeostasis.[18]

In a crossover study by Schwalm et al., the effects of exercise and diet on mitophagy/autophagy in skeletal muscle were evaluated.[62] Therefore, healthy athletes were chosen for the research. Participants were randomly divided into two groups: (1) fed with a high-carbohydrate meal and (2) fasted group. Group 1 was fed with a 150 mL drink containing 6% carbohydrate every 15 min during the exercise; but group 2 with 150 mL water every 15 min during the 2 h of the exercise session. After the washout period (2 weeks), participants in each group received another intervention. The vastus lateral in muscle was chosen for biopsies at three special time points: (i) before initiation of exercise, (ii) immediately after, and (iii) 1 h after exercise to evaluate the mRNA level and protein expression, which are relevant mitophagy biomarkers. The outcomes of this study demonstrated that Mfn2 mRNA level was decreased in the fed state in comparison with baseline values. On the other hand, mitochondrial DRP1 and Fis1 protein expression were significantly elevated postexercise in fed state. The results of this research demonstrated that mitochondrial Parkin was similar among the two nutritional states, before and after exercise. But, the protein expression and mRNA levels of Bnip3 showed a considerable elevation in the fed with a high-carbohydrate diet group immediately after exercise in comparison with the fasted condition. Phospho-DRP1, which involved in mitochondrial fission showed a considerable rise instantly after exercise and 1 h after exercise in the fed group. The LC3bII/LC3bI ratio was reduced in the fasted state after exercise in comparison with the fed group, demonstrating decreased mitophagy in this situation.[62] In the same way, to examine the extent of mitophagy/autophagy in the muscles of athletes, in a study by Tarpey et al., a comparison was done between healthy individuals with sedentary lifestyles or endurance runners.[63] For this purpose, they choose two groups of participants in this research: (1) healthy sedentary males and (2) males with endurance training. The men in both groups refused to exercise 36 h before starting the study. After that, they were exposed to fasting conditions or a high-fat meal (52 g carbohydrate, 24 g protein, and 58 g fat). Then, to evaluate mitophagy biomarkers, muscle biopsies were taken at special time points (after 12 h of fasting and 4 h after a high-fat meal). Vastus lateralis muscle of individuals was chosen for taking muscle biopsies to determine autophagy/mitophagy protein markers. The results revealed that there were no differences in total P1NK1 levels between sedentary and endurance-trained males before or after a high-fat meal. On the other hand, the amount of phospho-P1NK1 was significantly higher in endurance-trained subjects in comparison with sedentary males in both fasted and high-fat meal states. Furthermore, total Parkin protein was considerably higher in sedentary men, especially after consumption of a high-fat meal, but skeletal muscle of endurance-trained men had a higher amount of phospho-Parkin after both fasted and fed intervention. Mitochondrial fusion protein was considerably raised in runners in both fasted and fed situations. The amount of mitochondrial fusion protein is more in the fasting state but significantly increased after the high-fat meal in the endurance-trained males. The higher amount of phospho-PINK1 and phospho-Parkin in runners demonstrated that endurance training enhances mitophagy and mitochondrial dynamics in skeletal muscle.[63]

In a study by Jamart et al., for evaluating the effects of CR and exercise on autophagy/mitophagy, mice were divided into four following groups: (1) fed and rested group, (2) fed and exercised group (90 min running, 10 m/min), (3) CR (9.5 h of food withdrawal) and rested (food was prohibited during the dark cycle), (4) CR and exercised (running started 8 h after food prohibition). After ending the trial, gastrocnemius muscle of mice was chosen for biopsy for finding out the mitophagy biomarkers. Based on the outcome of this experiment, Drp1 mRNA was significantly reduced in the fasted groups, but it was unchanged in the exercise state. There was a significant raise in the level of Binp3 protein in fasted plus rest group. However, it was reduced after exercise in the fasted mice group. BNip3 mRNA was significantly raised in the fasted state in both run and rest groups. Parkin protein levels showed a considerable rise in the fasted rested mice, despite Parkin mRNA did not change in all groups. It was interesting that mitochondrial fusion was not affected by any intervention, neither Mfn1 nor Mfn2 levels in all four groups.[64] AMPK activates autophagy; but its role in the function of muscles, especially during fasting periods, has not yet been known. Therefore, to identify the relationship among AMPK, autophagy, and fasting, Bujak et al. established a study on mice lacking skeletal muscle AMPK (AMPK MKO) in the following two groups: (1) young mice (1-month-old), (2) aged mice (18-month-old). The mice were fasted for 12, 24, or 48 h or fed AL before sacrificing. Then, their muscles were used for evaluating autophagy/mitophagy biomarker. The results of this research demonstrated that in the fasted state the expression of mRNA of the autophagy adaptor protein, p62, was significantly raised. Furthermore, Parkin protein level showed a significant elevation by more than twofold in AMPK-MKO group in comparison with wild mice. This elevation was not related to their diet. These findings demonstrated that AMPK in skeletal muscle is necessary for the induction of mitophagy/autophagy.[65] In a study by Oost et al., FGF21 was measured for evaluating mitophagy and muscle function.[66] For this purpose, a knockout muscle-derived FGF21 model was used. Different groups of mice fasted for 24 h and AL mice were examined. For evaluating mitophagy, mitochondria-targeted mKeima plasmid (mt-mKeima) was used. The levels of FGF21 mRNA showed a significant rise in the fasted wild-type group. This finding demonstrated that they were under physiological stress because FGF21 has low expression in healthy muscles. Like the wild-type fed group, the mice in the fasting FGF21 knockout group had normal muscle histology, which showed that this deletion protects the muscles against fasting-induced damages.[66]

Glic et al. aimed to determine the effect of Binp3 protein in the adjustment of the function of mitochondria. For this purpose, they used 3- to 5-month-old BNip3 heterozygous and BNip3 null mice. Mice were fasted for 16 h before glucose tolerance test and glucagon stimulation test or 4 h fasted for evaluation of the insulin tolerance tests, and 6 and 24 h fasted for mitophagy protein marker analysis. The outcomes of this research demonstrated the high levels of expression of Binp3 in the liver and heart of BNip3 heterozygous mice. Furthermore, the BNip3 protein was increased after the fasted state than basal levels. Besides, the induction of BNip3 protein levels raised more after 24 h of fasting in comparison with 6 h of fasting. On the other hand, nix protein levels did not change in the fasted state compared with their initial condition.[67]

5 Diet, calorie restriction, and life span

Aging leads to a progressive worsening in physiological functions, metabolic processes, and homeostasis through a noxious change at all levels of the biological organization in the organism. Moreover, it can also increase vulnerability to disease. Aging is a complex process, which increases the chance of degenerative diseases and eventually death. Many theories have been developed to describe the biology of aging; however, all aspects of the aging process have not been fully elucidated. Among them, the mitochondrial free radical theory of aging is considered one of the strongest theories of aging.[42, 68, 69] In the aging process, some changes occur in mitochondrial metabolism, which includes decreased respiratory capacity and increased oxidative stress. These age-related changes have been observed in a range of body tissues, including muscle and different areas of the brain.[70] Various aspects of aging in cell and organism are elucidated using mitochondrial/ free radical aging theory. According to this theory, decreased mitochondrial energy production and damage, which are caused by mitochondrial ROS are the most important reasons of aging.[7] Recent studies in invertebrate model organisms, rodents, primates, and humans have highlighted the role of proper food intake and diet in the improvement of the aging and aging-related diseases, and also extending healthy life span. Epidemiological and experimental evidence suggests that diet composition, meal timing, IF daily feeding rhythms, and reduced intake of certain nutrients play a prominent role in modulating the mechanisms of aging. It is now believed that certain dietary interventions can also lead to increased longevity and healthy aging.[71, 72] In simple model organisms and rodents, the role of diet in slowing the aging process and increasing life span has been well recognized. A uniform linear relationship between CR and longevity has been observed in many species of mice and rats. Although a reduction of about 10%–50% in calorie intake less than normal intake can lead to a proportional increase in maximum life expectancy, CR of more than 50% usually leads to starvation and increased mortality.[72] The findings of several human studies show that long-term CR, along with adequate nutrients intake, is widely known to increase life span, which can cause numerous metabolic adaptations. As a result, the risk of some chronic diseases such as DM, hypertension, CVD, and cancer is reduced.[71–73] A definite degree of CR, which is considered ideal for some people, may be excessive and harmful for others. Therefore, to guide dietary prescriptions, biomarkers of optimal calorie intake are needed to properly define the threshold of optimal health and starvation/malnutrition.[72] Colman et al. evaluated the effect of CR in the prevention of diseases and mortality in a 20-year longitudinal adult-onset CR study in rhesus monkeys. This study revealed that moderate CR could reduce the amount of aging-related deaths. Moreover, CR delayed the onset of age-related pathologies such as DM, cancer, CVD, and brain atrophy.[74] In contrast, Mattison et al. in their study did not confirm the effects of long-term CR on increasing the life span of rhesus monkeys. This 23-year study showed that survival curves did not improve and CR could not increase survival compared with the control group. However, age-related diseases were significantly reduced in CR monkeys.[75] Although aging has a complex etiology, simple genetic alterations in laboratory model organisms have been able to show a significant increase in healthy life span. A lot of these life span mutations are able to create a state similar to the physiological experience of malnutrition periods by reducing the activity of nutrient-signaling pathways. According to the findings of various studies, dietary restriction, reducing food intake without causing malnutrition, can increase the average and/or maximum life span of a wide range of organisms including yeasts, flies, worms, fish, rodents, and rhesus monkeys. They have also shown other health benefits such as delays in reducing body function and reducing the incidence of major diseases in monkeys and rodents.[76] Also in human subjects, CR causes beneficial health effects through reducing metabolic markers of a number of diseases.[77]

6 Diet, calorie restriction, and mitochondria

There is evidence that CR induces mitochondrial biogenesis. Nisoli et al. demonstrated that after 30% CR in mice for either 3 or 12 months, a large increase in mitochondria was observed in brain, heart, liver, and adipose tissue. This was accompanied by increases in mitochondrial DNA, the proteins cytochrome *c* (cyt c), and cytochrome oxidase subunit IV (COXIV).[78] Another study in healthy humans showed that CR could increase the biogenesis of mitochondria in skeletal muscles.[79] There is a controversy around this idea. Hancock et al. have reevaluated the concept of whether calorie restriction can lead to mitochondria. In this study, rats were divided into two groups of freely eating and CR rats. They were evaluated for 2 weeks. After anesthesia, the tissues of the heart, brain, liver, epididymal fat pads, and triceps muscles were analyzed. Finally, they concluded that 30% CR in rats could not increase mitochondria in the heart, brain, liver, white adipose tissue, and skeletal muscle. Thus, this finding contradicts what Nisoli et al.

reported that 30% CR significantly increased mitochondrial biogenesis in laboratory rodents.[80] Miller et al. performed a 6-week study on young (6 months), middle (12 months), and old (24 months) male mice that were on lifelong CR, as well as the AL feeding control group. They studied the synthesis of cumulative mitochondrial protein and the regulation of its transcription and translation in the liver, heart and skeletal muscle. The results of this study showed that after 6 weeks, there were no difference between two groups on mitochondrial protein synthesis. However, in CR group, a reduced cellular proliferation was seen in the liver and heart. Finally, they concluded that CR was able to maintain mitochondrial protein synthesis while, during periods of energetic stress, it had the ability to reduce cellular proliferation, which leads to increased somatic maintenance.[81] It is suggested that CR reduces the production of ROS and mitochondrial proton leakage by induction of a hypometabolic state. This condition can delay a variety of diseases and increase life span in a wide range of animal species. Bevilacqua et al. evaluated the effects of long-term CR on total energy, muscle mitochondrial energy, ROS production, and its damage in 6-month-old male rats for 12–18 months.[82] A number of hypotheses have been proposed for antiaging mechanisms of CR, including CR can cause changes in mitochondrial function. Lambert et al. evaluated this hypothesis on mitochondria isolated from different tissues of male rats in both fully fed and CR groups. In this study, it was shown that CR can lead to specific tissue changes in mitochondrial respiration rate.[83] There are many factors that can increase mitochondrial enzyme content and its activity in skeletal muscle, one of them is exercise. In addition, exercise is involved in the mitochondrial biogenesis in a wide variety of body tissues, which are not generally related to the metabolic needs of exercise.[84] Lauren et al. evaluated effects of 12 months of CR on muscle mitochondrial function and associated transcriptional profiles in 51 healthy nonobese humans in two groups of 25% CR or AL (control) diet. At the end of the study, weight, body mass index (BMI), fat, and fat-free mass were significantly reduced in the CR group compared with control group. In CR group, there were no changes in maximal ATP synthesis rate and ATPflux/O2. There were also no effects in the pathways associated with the mitochondrial function, its biogenesis, and oxidative stress. In healthy nonobese individuals, CR appears to have no effect on muscle mitochondrial function. On the other hand, having a "more coupled" phenotype may make this effect possible.[85] Mitochondria have been considered as a major contributor in pathophysiology of aging. Based on genetic studies, sirtuins are antiaging genes in lower organisms.

They have seven homologs in mammals, among them, the function of four genes dependent on mitochondria CR seems to be an important intervention to increase the life span of mammals. It has been shown that after CR without malnutrition, there is an increase in SIRT1 levels in several tissues and a corresponding increase in mitochondrial components. Among the several models that describe how mitochondrial augmentation is effective in reducing the aging process and related diseases, some of them have highlighted the significant role of ROS.[86] Heilbronn et al. evaluated the effects of short-term CR (for 6–12 months) on human health. They found that subjects who were moderately overweight, after undergoing a CR period, experienced other benefits in addition to weight loss, including increased glucose tolerance and insulin sensitivity, and diminished metabolic rate.[87] Short-term CR of 6 months (as much as 10% or 30%) can result in elevation of plasma glutathione peroxidase activity and decreased the levels of plasma carbonyl protein. However, significant change was not detected in other antioxidants (e.g., superoxide dismutase (SOD) and catalase). Based on these findings, it can be said that CR can have modulatory effects on antioxidant defense and oxidative stress in humans.[88] DR-induced changes in mammalian mitochondrial function include: increased respiratory capacity, increased OXPHOS efficiency, decreased ROS emission, increased ROS scavengers, increased resistance to apoptotic challenge, maintenance of mitochondrial dynamics, maintenance of mtDNA content, and increased mitophagy.[89]

7 Conclusion

Based on the studies that have been reviewed in this section, it can be concluded that paying attention to the food consumed has a great role in an individual's health. In addition, varieties of studies conducted in the field of CR have different results depending on the type of study. But in general, CR increases mitophagy/autophagy by increasing the factors that lead to mitophagy. Therefore, it can be said that CR is a nonexpensive and low-adverse effect way to deal with diseases and in many cases increases the effects of treatment. CR improves a wide range of age-related diseases, from neurological diseases such as AD and PD to CVD and various types of cancers. However, more studies are needed to find the exact mechanism of the relationship between mitophagy and CR. In addition, a careful examination should be done on the different types of CR, for example, restricting carbohydrate, fat, or protein, or different types of fasting such as intermittent, time-restricted feeding, and alternative day fasting, to find out the best type for each disease.

References

1. Kim DH, et al. Anti-aging effects of calorie restriction (CR) and CR mimetics based on the senoinflammation concept. *Nutrients.* 2020;12(2).
2. Li J, et al. The role of vitamins in the pathogenesis of non-alcoholic fatty liver disease. *Integr Med Insights.* 2016;11:19–25.
3. Negri R, et al. Randomised clinical trial: calorie restriction regimen with tomato juice supplementation ameliorates oxidative stress and preserves a proper immune surveillance modulating mitochondrial bioenergetics of T-lymphocytes in obese children affected by non-alcoholic fatty liver disease (NAFLD). *J Clin Med.* 2020;9(1).
4. Bhatti GK, et al. Lifestyle modifications and nutritional interventions in aging-associated cognitive decline and Alzheimer's disease. *Front Aging Neurosci.* 2019;11:369.
5. Zhang Y, et al. The effects of calorie restriction in depression and potential mechanisms. *Curr Neuropharmacol.* 2015;13(4):536–542.
6. Klempel MC, et al. Intermittent fasting combined with calorie restriction is effective for weight loss and cardio-protection in obese women. *Nutr J.* 2012;11:98.
7. Harman D. Aging: a theory based on free radical and radiation chemistry. *J Gerontol.* 1956;11(3):298–300.
8. Salimi A, et al. Mitochondrial and lysosomal protective agents ameliorate cytotoxicity and oxidative stress induced by cyclophosphamide and methotrexate in human blood lymphocytes. *Hum Exp Toxicol.* 2019;38(11):1266–1274.
9. Zorov DB, Juhaszova M, Sollott SJ. Mitochondrial reactive oxygen species (ROS) and ROS-induced ROS release. *Physiol Rev.* 2014;94(3):909–950.
10. Kujoth GC, et al. Mitochondrial DNA mutations, oxidative stress, and apoptosis in mammalian aging. *Science.* 2005;309(5733):481–484.
11. Trifunovic A, et al. Premature ageing in mice expressing defective mitochondrial DNA polymerase. *Nature.* 2004;429(6990):417–423.
12. Trifunovic A, et al. Somatic mtDNA mutations cause aging phenotypes without affecting reactive oxygen species production. *Proc Natl Acad Sci U S A.* 2005;102(50):17993–17998.
13. Gabandé-Rodríguez E, de Las Heras MMG, Mittelbrunn M. Control of inflammation by calorie restriction mimetics: on the crossroad of autophagy and mitochondria. *Cell.* 2019;9(1):82.
14. Pfeifer U. Inhibition by insulin of the formation of autophagic vacuoles in rat liver. A morphometric approach to the kinetics of intracellular degradation by autophagy. *J Cell Biol.* 1978;78(1):152–167.
15. Suen DF, et al. Parkin overexpression selects against a deleterious mtDNA mutation in heteroplasmic cybrid cells. *Proc Natl Acad Sci U S A.* 2010;107(26):11835–11840.
16. Bhatia-Kiššová I, Camougrand N. Mitophagy in yeast: actors and physiological roles. *FEMS Yeast Res.* 2010;10(8):1023–1034.
17. Kanki T, Klionsky DJ. The molecular mechanism of mitochondria autophagy in yeast. *Mol Microbiol.* 2010;75(4):795–800.
18. Richard VR, et al. Macromitophagy is a longevity assurance process that in chronologically aging yeast limited in calorie supply sustains functional mitochondria and maintains cellular lipid homeostasis. *Aging (Albany NY).* 2013;5(4):234–269.
19. Lemasters JJ. Selective mitochondrial autophagy, or mitophagy, as a targeted defense against oxidative stress, mitochondrial dysfunction, and aging. *Rejuvenation Res.* 2005;8(1):3–5.
20. Tal R, et al. Aup1p, a yeast mitochondrial protein phosphatase homolog, is required for efficient stationary phase mitophagy and cell survival. *J Biol Chem.* 2007;282(8):5617–5624.
21. Nowikovsky K, et al. Mdm38 protein depletion causes loss of mitochondrial K+/H+ exchange activity, osmotic swelling and mitophagy. *Cell Death Differ.* 2007;14(9):1647–1656.
22. Okamoto K, Kondo-Okamoto N, Ohsumi Y. Mitochondria-anchored receptor Atg32 mediates degradation of mitochondria via selective autophagy. *Dev Cell.* 2009;17(1):87–97.

23. Aoki Y, et al. Phosphorylation of serine 114 on Atg32 mediates mitophagy. *Mol Biol Cell.* 2011;22(17):3206–3217.
24. Mao K, et al. Two MAPK-signaling pathways are required for mitophagy in Saccharomyces cerevisiae. *J Cell Biol.* 2011;193(4):755–767.
25. Springer MZ, Macleod KF. In brief: Mitophagy: mechanisms and role in human disease. *J Pathol.* 2016;240(3):253–255.
26. Novak I. Mitophagy: a complex mechanism of mitochondrial removal. *Antioxid Redox Signal.* 2012;17(5):794–802.
27. Gouspillou G, et al. The relationship between muscle fiber type-specific PGC-1α content and mitochondrial content varies between rodent models and humans. *PLoS One.* 2014;9(8), e103044.
28. Ding WX, Yin XM. Mitophagy: mechanisms, pathophysiological roles, and analysis. *Biol Chem.* 2012;393(7):547–564.
29. Hamacher-Brady A, Brady NR. Mitophagy programs: mechanisms and physiological implications of mitochondrial targeting by autophagy. *Cell Mol Life Sci.* 2016;73(4):775–795.
30. Palikaras K, Tavernarakis N. Mitochondrial homeostasis: the interplay between mitophagy and mitochondrial biogenesis. *Exp Gerontol.* 2014;56:182–188.
31. Ney PA. Mitochondrial autophagy: origins, significance, and role of BNIP3 and NIX. *Biochim Biophys Acta.* 2015;1853(10 Pt B):2775–2783.
32. Liu L, et al. Mitochondrial outer-membrane protein FUNDC1 mediates hypoxia-induced mitophagy in mammalian cells. *Nat Cell Biol.* 2012;14(2):177–185.
33. Chaix A, et al. Time-restricted feeding is a preventative and therapeutic intervention against diverse nutritional challenges. *Cell Metab.* 2014;20(6):991–1005.
34. Mattson MP, Longo VD, Harvie M. Impact of intermittent fasting on health and disease processes. *Ageing Res Rev.* 2017;39:46–58.
35. Cherif A, et al. Effects of intermittent fasting, caloric restriction, and Ramadan intermittent fasting on cognitive performance at rest and during exercise in adults. *Sports Med.* 2016;46(1):35–47.
36. Price JC, et al. The effect of long term calorie restriction on in vivo hepatic proteostatis: a novel combination of dynamic and quantitative proteomics. *Mol Cell Proteomics.* 2012;11(12):1801–1814.
37. Caro P, et al. Forty percent and eighty percent methionine restriction decrease mitochondrial ROS generation and oxidative stress in rat liver. *Biogerontology.* 2008;9(3):183–196.
38. Soare A, et al. Long-term calorie restriction, but not endurance exercise, lowers core body temperature in humans. *Aging (Albany NY).* 2011;3(4):374.
39. Shaddock J, Chou M, Casciano D. Effects of age and caloric restriction on cell proliferation in hepatocyte cultures from control and hepatectomized Fischer 344 rats. *Mutagenesis.* 1996;11(3):281–284.
40. Brandhorst S, et al. Protective effects of short-term dietary restriction in surgical stress and chemotherapy. *Ageing Res Rev.* 2017;39:68–77.
41. Madia F, et al. Oncogene homologue Sch9 promotes age-dependent mutations by a superoxide and Rev1/Polζ-dependent mechanism. *J Cell Biol.* 2009;186(4):509–523.
42. Fontana L, Partridge L, Longo VD. Extending healthy life span—from yeast to humans. *Science.* 2010;328(5976):321–326.
43. Longo VD, Finch CE. Evolutionary medicine: from dwarf model systems to healthy centenarians? *Science.* 2003;299(5611):1342–1346.
44. Kenyon C. A conserved regulatory system for aging. *Cell.* 2001;105(2):165–168.
45. Guarente L, Kenyon C. Genetic pathways that regulate ageing in model organisms. *Nature.* 2000;408(6809):255–262.
46. Picca A, Pesce V, Lezza AMS. Does eating less make you live longer and better? An update on calorie restriction. *Clin Interv Aging.* 2017;12:1887.

47. López-Lluch G, Navas P. Calorie restriction as an intervention in ageing. *J Physiol.* 2016;594(8):2043–2060.
48. Boya P, et al. Inhibition of macroautophagy triggers apoptosis. *Mol Cell Biol.* 2005;25(3):1025–1040.
49. Bordone L, Guarente L. Calorie restriction, SIRT1 and metabolism: understanding longevity. *Nat Rev Mol Cell Biol.* 2005;6(4):298–305.
50. Fontana L, et al. Long-term calorie restriction is highly effective in reducing the risk for atherosclerosis in humans. *Proc Natl Acad Sci U S A.* 2004;101(17):6659–6663.
51. Brandhorst S, Longo VD. Fasting and caloric restriction in cancer prevention and treatment. In: *Metabolism in cancer.* Springer; 2016:241–266.
52. Longo VD, Mattson MP. Fasting: molecular mechanisms and clinical applications. *Cell Metab.* 2014;19(2):181–192.
53. Castrogiovanni P, et al. Fasting and fast food diet play an opposite role in mice brain aging. *Mol Neurobiol.* 2018;55(8):6881–6893.
54. Longo VD, et al. Interventions to slow aging in humans: are we ready? *Aging Cell.* 2015;14(4):497–510.
55. Madeo F, et al. Essential role for autophagy in life span extension. *J Clin Invest.* 2015;125(1):85–93.
56. Madeo F, Tavernarakis N, Kroemer G. Can autophagy promote longevity? *Nat Cell Biol.* 2010;12(9):842–846.
57. Petrovski G, Das DK. Does autophagy take a front seat in lifespan extension? *J Cell Mol Med.* 2010;14(11):2543–2551.
58. Kanki T, et al. Atg32 is a mitochondrial protein that confers selectivity during mitophagy. *Dev Cell.* 2009;17(1):98–109.
59. Cui J, et al. Mitochondrial autophagy involving renal injury and aging is modulated by caloric intake in aged rat kidneys. *PLoS One.* 2013;8(7), e69720.
60. Linden MA, et al. Combining metformin therapy with caloric restriction for the management of type 2 diabetes and nonalcoholic fatty liver disease in obese rats. *Appl Physiol Nutr Metab.* 2015;40(10):1038–1047.
61. Zhao Y, et al. Exercise training and dietary restriction affect PINK1/Parkin and Bnip3/nix-mediated cardiac mitophagy in mice. *Gen Physiol Biophys.* 2018;37(6):657–666.
62. Schwalm C, Deldicque L, Francaux M. Lack of activation of mitophagy during endurance exercise in human. *Med Sci Sports Exerc.* 2017;49(8):1552–1561.
63. Tarpey MD, et al. Skeletal muscle autophagy and mitophagy in endurance-trained runners before and after a high-fat meal. *Mol Metab.* 2017;6(12):1597–1609.
64. Jamart C, et al. Higher activation of autophagy in skeletal muscle of mice during endurance exercise in the fasted state. *Am J Physiol Endocrinol Metab.* 2013;305:E964–E974.
65. Bujak AL, et al. AMPK activation of muscle autophagy prevents fasting-induced hypoglycemia and myopathy during aging. *Cell Metab.* 2015;21(6):883–890.
66. Oost LJ, et al. Fibroblast growth factor 21 controls mitophagy and muscle mass. *J Cachexia Sarcopenia Muscle.* 2019;10(3):630–642.
67. Glick D, et al. BNip3 regulates mitochondrial function and lipid metabolism in the liver. *Mol Cell Biol.* 2012;32(13):2570–2584.
68. Pamplona R, Barja G. Mitochondrial oxidative stress, aging and caloric restriction: the protein and methionine connection. *Biochim Biophys Acta Biomembr.* 2006;1757(5–6):496–508.
69. Shinmura K. Effects of caloric restriction on cardiac oxidative stress and mitochondrial bioenergetics: potential role of cardiac sirtuins. *Oxid Med Cell Longev.* 2013;2013.
70. Lin MT, Beal MF. Mitochondrial dysfunction and oxidative stress in neurodegenerative diseases. *Nature.* 2006;443(7113):787–795.
71. Fontana L, Partridge L. Promoting health and longevity through diet: from model organisms to humans. *Cell.* 2015;161(1):106–118.

72. Rizza W, Veronese N, Fontana L. What are the roles of calorie restriction and diet quality in promoting healthy longevity? *Ageing Res Rev.* 2014;13:38–45.
73. Serna JDC, da Silva CCC, Kowaltowski AJ. Functional changes induced by caloric restriction in cardiac and skeletal muscle mitochondria. *J Bioenerg Biomembr.* 2020;52:269–277.
74. Colman RJ, et al. Caloric restriction delays disease onset and mortality in rhesus monkeys. *Science.* 2009;325(5937):201–204.
75. Mattison JA, et al. Impact of caloric restriction on health and survival in rhesus monkeys from the NIA study. *Nature.* 2012;489(7415):318–321.
76. Anderson RM, Shanmuganayagam D, Weindruch R. Caloric restriction and aging: studies in mice and monkeys. *Toxicol Pathol.* 2009;37(1):47–51.
77. Fontana L, Klein S. Aging, adiposity, and calorie restriction. *JAMA.* 2007;297(9):986–994.
78. Nisoli E, et al. Calorie restriction promotes mitochondrial biogenesis by inducing the expression of eNOS. *Science.* 2005;310(5746):314–317.
79. Civitarese AE, et al. Calorie restriction increases muscle mitochondrial biogenesis in healthy humans. *PLoS Med.* 2007;4(3), e76.
80. Hancock CR, et al. Does calorie restriction induce mitochondrial biogenesis? A reevaluation. *FASEB J.* 2011;25(2):785–791.
81. Miller BF, et al. A comprehensive assessment of mitochondrial protein synthesis and cellular proliferation with age and caloric restriction. *Aging Cell.* 2012;11(1):150–161.
82. Bevilacqua L, et al. Long-term caloric restriction increases UCP3 content but decreases proton leak and reactive oxygen species production in rat skeletal muscle mitochondria. *Am J Physiol Endocrinol Metab.* 2005;289(3):E429–E438.
83. Lambert A, et al. The effect of aging and caloric restriction on mitochondrial protein density and oxygen consumption. *Exp Gerontol.* 2004;39(3):289–295.
84. Little JP, et al. Skeletal muscle and beyond: the role of exercise as a mediator of systemic mitochondrial biogenesis. *Appl Physiol Nutr Metab.* 2011;36(5):598–607.
85. Sparks LM, et al. Effects of 12 months of caloric restriction on muscle mitochondrial function in healthy individuals. *J Clin Endocrinol Metabol.* 2017;102(1):111–121.
86. Guarente L. Mitochondria—a nexus for aging, calorie restriction, and sirtuins? *Cell.* 2008;132(2):171–176.
87. Heilbronn LK, et al. Effect of 6-month calorie restriction on biomarkers of longevity, metabolic adaptation, and oxidative stress in overweight individuals: a randomized controlled trial. *JAMA.* 2006;295(13):1539–1548.
88. Meydani M, et al. The effect of caloric restriction and glycemic load on measures of oxidative stress and antioxidants in humans: results from the CALERIE trial of human caloric restriction. *J Nutr Health Aging.* 2011;15(6):456–460.
89. Ruetenik A, Barrientos A. Dietary restriction, mitochondrial function and aging: from yeast to humans. *Biochim Biophys Acta Biomembr.* 2015;1847(11):1434–1447.

CHAPTER 6

Mitochondrial metabolism and carcinogenesis

Siddhesh Aras[a,b], Neeraja Purandare[a,b], and Lawrence I. Grossman[a,b]
[a]Center for Molecular Medicine and Genetics, Wayne State University School of Medicine, Detroit, MI, United States
[b]Perinatology Research Branch, Division of Obstetrics and Maternal-Fetal Medicine, Division of Intramural Research, Eunice Kennedy Shriver National Institute of Child Health and Human Development, National Institutes of Health, U.S. Department of Health and Human Services, Bethesda, MD and Detroit, MI, United States

Contents

1 Introduction

Mitochondria are semiautonomous, endosymbiotic organelles maintaining their own DNA and best known for generating energy in the form of ATP for a multitude of cellular processes.[1] However, studies in recent years have uncovered a much broader role of mitochondria in cell physiology. In addition to substrate metabolism, mitochondria have been shown to play a vital role in regulating cell signaling, apoptosis, and immune responses.[2] Although mitochondria harbor over 1000 proteins, the mitochondrial genome encodes for only 13 of these[3] with the majority encoded by the nuclear genome.[4] The mitochondria have evolved to coordinate ATP production in response to substrate availability (nutrients and oxygen).[5] Functionally, the mitochondria undergo

Mitochondrial Metabolism
https://doi.org/10.1016/B978-0-12-822416-8.00001-4

the process of oxidative phosphorylation (OxPhos) by utilizing substrates from the cellular cytoplasm, such as pyruvate and fatty acids, to generate reducing equivalents in the form of NADH and FADH.[6] Sequential passage of these through a series of protein complexes called the electron transport chain (ETC) generates potential energy via a proton gradient[7] that is used to produce ATP. Mitochondria also generate reactive oxygen species (ROS) as a byproduct of OxPhos. Most physiologically generated ROS are maintained at low levels by ROS scavengers.[8] The normally low level of ROS function as a signaling molecule[9]; however, when the levels increase beyond the scavenging capacity, ROS induce DNA,[10] lipid,[11] and protein damage.[12] Another critical function regulated by the mitochondria is apoptosis.[13] In response to stress, the ratio of proapoptotic to antiapoptotic proteins on the mitochondria regulate mitochondrial permeabilization and therefore activation of the apoptotic cascade.[14]

Cancer is a multifaceted syndrome that results in abnormal and uncontrolled cell proliferation with or without spread to other parts of the body.[15] In addition to significant morbidity, cancer mortality is a major hurdle to increasing global life expectancy.[16] In very simple terms, carcinogenesis can be defined as the conversion of a normal cell to a cancerous one. This process can be subdivided into three stages: (a) initiation, (b) promotion, and (c) progression.[17] Initiation involves the changes in normal cells in response to a cancerous stimulus, both external such as ultraviolet radiation,[18] and internal, such as genetic mutations.[19] The next stage is the promotion stage wherein an "initiated" cell undergoes clonal expansion brought about by promoting agents such as chemicals and epigenetic regulators that alter gene expression.[20] These two steps lead to abnormal appearance that on histopathology are labeled as preneoplastic or benign neoplasms.[21] The final step is progression, which is irreversible and is characterized by uncontrolled growth even in the absence of the promoting stimulus.[22]

Uncontrolled growth and proliferation of cells requires energy to support vital cellular functions. Almost a century ago, Warburg observed that glucose uptake in cancer cells was enhanced with a secretion of lactate despite the presence of oxygen,[23] widely known now as the Warburg effect or aerobic glycolysis. This use of glycolysis for aerobic growth initially led to the belief that impaired mitochondrial function underlies carcinogenesis, with glycolysis as the major ATP source in rapidly proliferating cells. The dependence of tumor cells on mitochondrial function was uncovered by studies involving mitochondrial DNA depletion.[24–26] Additionally, studies have shown that the amount of ATP generated via glycolysis in cancers is much less compared with that produced by OxPhos, like any other cell,[27] suggesting that aerobic glycolysis is more of an adaptive response observed in cancer cells. That is, the generation of energy by glycolysis is a choice the cancer cell makes for its benefit. In the last decade, the term "Reverse Warburg" has been used to highlight the importance of tumor microenvironment on cancer cell growth.[28] According to this effect, it is the cancer-associated fibroblasts that undergo glycolysis to provide pyruvate and lactate to the cancer cells for mitochondrial function.

In this review, we will focus on mitochondrial metabolism in carcinogenesis, specifically the regulators and mechanisms of regulation of mitochondrial function.

2 Tumor initiation and mitochondrial function

Initiation is the first step that predisposes normal cells to undergo a malignant transformation. Tumor initiation occurs as a result of genetic changes that induce the cell to undergo uncontrolled growth and proliferation.[29] The role of the cellular genome in carcinogenesis was uncovered by the seminal discoveries in the early 1980s showing that transfection of DNA from tumors and tumor cell lines resulted in transformation of NIH 3T3 cells.[30, 31] Later, the identification of mutant *HRAS* opened the novel concept of human DNA mutations as a causative agent for cancers.[32, 33] In addition to the concept of oncogenes and proto-oncogenes to promote cancer,[34] the existence of genes that could suppress tumors was also suggested.[35] Cancer research since then has been focused on identifying and elucidating the mechanisms by which oncogenes and tumor suppressors arise and result in initiation of cancers.

Oncogenes and mitochondrial metabolism. Oncogenes, as the name suggests, encode proteins that function in oncogenesis. In general, they represent mutations that are gain of function. The products of oncogenes, oncoproteins, can be broadly classified into categories based on their role in cell physiology.[36] The effect of these is to activate cell proliferation and inhibit apoptosis. The most well-studied oncogenes fall in one of these categories.

Transcription factors. Oncogenic transcription factors have been found to be drivers of a selective cancer type. These factors increase transcription of genes that in turn control multiple aspects of carcinogenesis.

(a) Maintenance of cancer stem cells: Cancer stem cells are a population of cells within a tumor. These stem cells have the capacity of self-renewal and carcinogenesis.[37] Two well-studied leukemic transcription factors that function toward the maintenance and self-renewal properties of cancer stem cells are fusion proteins; one involves Mixed Lineage Leukemia fusion protein resulting from a translocation of MLL gene to Chromosome 3 Protein (*MLLT3*), called *MLL-AF9*, and the other is from a chimera of Core binding Factor β (CBFβ) and Smooth Muscle Myosin Heavy chain 11 (*SMMHC*) (CBFβ-SMMHC). Both these result from gene rearrangements leading to translation of a fusion protein. MLL-AF9 transactivates genes such as *Hoxa9* and *MEIS1*[38]; Hoxa9 regulates Bcl-2 to affect survival of stem cells[39] whereas MEIS1 reduces mitochondrial ROS via inhibition of mitochondrial oxygen consumption.[40] CBFβ-SMMHC, on the other hand, functions to inhibit apoptosis by regulating expression of *Myc*.[41]

(b) Induction of unlimited proliferative capacity: One of the key features of a cancer cell is its ability to replicate indefinitely. Telomerase reverse transcriptase (TERT) activation has been shown to contribute to immortalization via multiple mechanisms in addition to stabilizing telomere length.[42] GA-binding protein (GABP) has been shown to function as an oncogenic transcription factor activating TERT in glioblastomas.[43] GABP, a transcription factor required for cell cycle progression,[44] is also an inducer of mitochondrial biogenesis. It does so by transcriptionally activating TFAM to stimulate mitochondrial gene expression.[45] Additionally, GABP also transcriptionally induces genes for nuclear-encoded mitochondrial ETC subunit proteins.[46,47] GABP is required for optimal mitochondrial function as $GABP^{-/-}$ cells are defective in their ability to consume oxygen, the substrate for mitochondrial function.[48]

(c) Regulation of cellular metabolism: *c-Myc* oncogene is a classic example in the regulatory category. The role of *c-Myc* as an oncogene was confirmed by studies in Burkitt's lymphoma cells.[49] MYC is a transcription factor that activates key genes in metabolic pathways responsible for rapid cell growth.[50] MYC also activates ROS, inducing DNA damage that contributes to genomic instability in carcinogenesis.[51] MYC has been shown to enhance the expression of glutaminase 1 via transcriptional inhibition of microRNAs to allow utilization of glutamine by cancer cells as an alternative energy source.[52,53] MYC is a transcriptional activator of genes to induce mitochondrial biogenesis.[54] MYC via its effect on OxPhos in the mitochondria expands the cancer stem cell pool, thereby conferring resistance of cancer cells to chemotherapy.[55]

Growth factors. One of the first pieces of evidence that growth factors could lead to oncogenesis was the observation that Platelet-derived growth factor 2 (PDGF2) could transform NIH3T3 cells.[56] As carcinogenesis is associated with uncontrolled cell proliferation, growth factors indeed could be potential oncogenes. Growth factors are peptides that, upon binding specific receptors, activate signaling cascades and therefore can be conducive for carcinogenesis.

(a) Platelet-derived growth factor (PDGF): There are four characterized growth factors that belong to the PDGF family (PDGF-A, PDGF-B, PDGF-C, and PDGF-D) and can activate platelet-derived growth factor receptor (PDGFR) alpha (PDGFR-α) and beta (PDGFR-β)[57] PDGF-B is homologous to the *v-sis* oncogene of simian sarcoma virus.[58, 59] The role of PDGF in regulating mitochondrial function has been studied in vascular smooth muscle cells, where it has been shown to induce mitochondrial fragmentation (fission) via signaling that results in phosphorylation of DRP1, the mitochondrial fission protein, at Ser-616. Additionally, PDGF treatment of cells enhanced proliferation and mitochondrial oxygen consumption.[60] From a metabolic perspective, PDGF enhances the cellular energy dependence on fatty acids,[61] a critical pathway for cancer cells.[62]

(b) Wnt-1-induced secreted protein 1 (WISP-1): WISP-1 belongs to the CCN family of growth factors and has been shown to function as an oncogene.[63] Via the PI3K/AKT pathway, WISP-1 functions as an antiapoptotic in neuronal[64] and cardiac[65] cells by upregulation of the antiapoptotic protein Bcl-xL. The antiapoptotic role of WISP-1 contributes to carcinogenesis by preventing cancer cell death.

Growth factor receptors. Growth factor receptors play a key role in execution of the signaling events that result in cell growth and proliferation, both hallmarks of cancer. Growth factor signaling not only supports carcinogenesis but also supports the maintenance, angiogenesis, and metastatic potential of tumors.[66]

(a) Human Epidermal Growth Factor Receptor family (HER): This family has four members. Human EGFR (HER1), upon ligand-induced dimerization and autophosphorylation, activates intracellular signaling pathways.[67] Multiple studies have characterized the oncogenic potential of EGFR to transform cells.[68–71] Ligand-dependent activation of EGFR causes its translocation to the mitochondria, where it induces a fission phenotype. However, in contrast to PDGF, EGFR inhibits the fusion protein mitofusin 1 to induce fission.[72] EGFR signaling in breast cancer cells induces a glycolytic phenotype via phosphorylation of PKM2. Additionally, the glycolytic metabolite fructose 1,6 bis-phosphate interacts with and induces EGFR activity.[73]

(b) HER2 is another well-studied member of this receptor oncogene family implicated in breast cancers. The transformation potential of HER2 was evident upon observations that *HER2* expression in NIH3T3 cells induced malignant transformation,[74] also in the absence of EGFR in NR6 cells.[75] HER2 has been shown to translocate to mitochondria in cancer cells and induce a switch in metabolism. It caused a reduction in ETC complex activity with a reduction in the mitochondrial membrane potential and enhanced lactate production, giving rise to a glycolytic phenotype.[76] HER2 also plays a key role in activating the mitochondrial phosphocreatine shuttle as a source of energy for cancer cells.[77] *HER2* expressing breast cancer tissues were shown to display an increase in the levels of markers for the mitochondrial unfolded protein response (UPR^{mt}) and could underlie the antiapoptotic phenotype observed in these cancers.[78] UPR^{mt} is a stress response that induces pathways to achieve metabolic homeostasis.[79]

Intracellular signaling regulators. Many oncogenes encode proteins that function as critical regulators of intracellular signaling pathways that allow cells to grow and transform into a cancerous state. These include (a) GTP-binding proteins (G-proteins)[80] and (b) nonreceptor kinases.[81] G-proteins are structurally distinct and can be classified into five groups. Of these, contributions by two groups of proteins like Ras[82] and Rho[83] proteins to carcinogenesis have been well studied. With respect to nonreceptor kinases, there are two categories (a) Serine–Threonine kinases such as AKT and (b) Tyrosine kinases such as Abl.[84] These proteins function toward oncogenesis upon gain-of-function mutations.

(a) Ras: Ha-Ras, N-Ras, and Ki-Ras are proto-oncogenes that, upon suffering activating mutations, encode proteins capable of transforming cells.[85–87] Pancreatic, lung, and colorectal cancers have been shown to harbor mutations in Ras,[88–90] which have many functions in cancer cells ranging from oncogenic transactivation and cell cycle regulation to regulation of metabolic signaling. With respect to the latter, Ras promotes glycolysis in tumor cells.[91] However, contrasting studies at the mitochondrial level have shown that K-Ras has to induce glutamine metabolism to support ATP production as well as maintenance of cellular redox.[92] In lung cancers, K-Ras induces beta-oxidation to fuel energy production.[93] In breast cancer MCF-7 cells, H-Ras has an additive effect with c-Myc toward maintenance of stemness and increasing mitochondrial function. The most upregulated protein identified on a mass spectrometric screen was MNRR1 (CHCHD2) that we and others have shown to be upregulated in cancers.[94–96] MNRR1 is a bi-organellar regulator that functions both in the mitochondria to enhance OxPhos and inhibit apoptosis[96,97] and in the nucleus to transcriptionally activate a number of genes.[97,98] In light of recent findings regarding the increased levels of MNRR1 in cancers, it remains to be evaluated whether MNRR1 is a bona fide oncogene.

(b) Rho: Although initially thought to function as cytoskeletal regulators, Rho proteins also regulate cell proliferation and survival. One of the first reports demonstrated that constitutively active Rac1, a Rho protein, induced transformation of cells[99] and that it cooperates with Ras to affect a transformative phenotype.[100] Mitochondria harbor a Rho-GTPase, Miro,[101] that function in the transport of mitochondria.[102] Recent evidence also suggests that Miro functions as a sensor of Ca^{2+} with a key role in the process of mitophagy.[103] Mitophagy is the selective autophagy of mitochondria and is required to prevent the accumulation of damaged mitochondria in cells. Mitophagy is critical for carcinogenesis,[104,105] being required for mitochondrial fission,[106] a phenotype observed in cancer cells as a result of growth factor signaling. $Miro^{-/-}$ cells are deficient in their migratory capacity[107] and therefore may reduce cancer metastasis.

(c) Nonreceptor tyrosine kinases: As the name suggests, these kinases phosphorylate tyrosine residues on substrates to regulate signaling communication within cells and maintain homeostasis. The two most well-studied in this category are Abl kinase and Src kinase.

(i) The Abl family includes Abl1 and Abl2. Both these kinases play a vital role in cell growth and survival, DNA-damage response, and cell migration.[108] Chromosomal translocation t(9;22)(q34.1;q11.2), resulting in a fusion protein, BCR-Abl, confirmed the role of Abl kinase as an oncogene in cancers such as acute lymphoid leukemia (ALL), acute myeloid leukemia (AML), and chronic myeloid leukemia (CML).[109] Similarly, TEL-Abl2 and RCSD1-Abl2 fusions have been reported in leukemia and B-cell ALL, respectively.[110,111] BCR-Abl functions as an antiapoptotic by preventing mitochondrial membrane potential changes and therefore inhibiting proapoptotic Bax from accumulating on the mitochondrial outer membrane to induce permeabilization.[112]

We have previously shown in K562 cells (BCR-Abl-positive) that treatment with bosutinib reduces mitochondrial oxygen consumption by reducing tyrosine phosphorylation of MNRR1.[113] Abl2, on the other hand, has also been shown to be upregulated in nonsmall-cell lung cancer (NSCLC)[114] and breast cancer cells[115] and is associated with increased invasive capacity of breast cancer cells.[116] Abl2 also phosphorylates MNRR1 to induce OxPhos, suggesting the importance of mitochondrial function in carcinogenesis.[113]

(ii) The oncogenic Src kinase, upregulated for expression and activity in solid tumors, is associated with cell growth and proliferation, motility, and reduction of apoptosis.[117, 118] Src in the mitochondria reduces levels of mtDNA, with a reduction in oxygen consumption rate resulting in a glycolytic phenotype in breast cancers.[118] In contrast, Src maintained OxPhos in osteosarcoma and prostate cancer cells by phosphorylating NDUFB10, a subunit of complex I of the ETC.[119] Additionally, Src kinase also targets other subunits in the ETC such as COXII,[120] NDUFV2, and SDHA.[121] Studies on Src kinases have shown contrasting results.

(d) Nonreceptor serine–threonine kinases: This category includes kinases such as AKT, PIM1, and RAF. These kinases phosphorylate serine and threonine residues on target proteins to facilitate cell proliferation.

(i) *AKT* was identified as an oncogene in viral studies on mouse T-cell lymphoma.[122, 123] *AKT2* expression and knockdown has been shown to induce and inhibit transformation of cells.[124, 125] Dominant negative mutants of AKT inhibit transformation.[126] AKT translocation to the mitochondria has been identified[127] and shown to induce a glycolytic phenotype by phosphorylating PDK1.[128] Another important target of AKT in the mitochondria is the calcium uniporter. In cancers, AKT phosphorylates MICU1 to increase mitochondrial calcium levels and generate ROS to facilitate cancer cell growth.[129] ROS, on the other hand, has been shown to inactivate PTEN and to activate AKT to support tumor cell growth in uterine leiomyomas.[130] AKT also functions as an antiapoptotic by phosphorylating and inhibiting the proapoptotic protein BAD.[131]

(ii) PIM1 and PIM2 have been shown to be oncogenic in lymphocytes[132, 133] and overexpression of PIM1 was found in leukemias and oral cancers.[134, 135] Activating mutations in PIM1 have also been found in B-cell lymphomas.[133, 136] Functionally, PIM1 acts in synergy with c-Myc to induce transformation.[137] PIM1, via phosphorylation and alteration of the subcellular localization of p21, regulates the cell cycle.[138] PIM1 also phosphorylates and inactivates the proapoptotic protein BAD, thereby contributing to the phenotype of cancer cells.[139, 140] In ovarian cancer cells, PIM1 inhibits mitochondrial respiration and increases lactate production to impart the glycolytic phenotype.[141]

(iii) RAF: The RAF oncogene functions at least in part to induce cell proliferation via activation of the MEK/ERK pathway.[142] The role of RAF as an oncogene was discovered in retroviral studies of murine leukemia virus.[143] There are three RAF proteins, A, B, and C. Of these, B-RAF is the most active and is activated by up-stream RAS.[144–146] Many cancers have an underlying mutation in B-RAF.[147] RAF kinases localize to the mitochondria. For instance, the BRAFV600E mutant observed in thyroid cancers localized to the mitochondria to function as an anti-apoptotic and is resistant to conventional RAF inhibitors.[148] Similarly, A-RAF and C-RAF localize to the mitochondria.[149, 150] Functionally, A-RAF has been shown to interact with PKM2 to induce a glycolytic phenotype underlying cancer.[151]

Bcl-2 apoptosis regulators. Studies have shown these proteins to function as potential oncogenes since apoptosis inhibition is one of the hallmarks of cancers.[152] Bcl-2 proteins function to prevent Bax and Bak accumulation on the mitochondrial outer membrane and thus prevent cytochrome *c* release and apoptosis.[153] Bcl-2 and Bcl-xL are thus upregulated in many cancers.[154] For instance, Bcl-2 has been implicated in the carcinogenesis of follicular lymphoma.[155] Bcl-2 and Myc function synergistically to induce a transformative phenotype.[156]

Tumor suppressors and mitochondrial metabolism. Similar to the concept of oncogenes, tumor suppressing functions were identified using somatic cell fusion experiments.[35, 157] In contrast to oncogenes, a loss of function mutation in tumor suppressors generally results in cancer.[158] The most common tumor suppressors are p53 and Rb.[159, 160] In addition, studies have also shown BRCA1, BRCA2, and PTEN along with others to also function as tumor suppressors.[161]

(a) p53: This gene holds the position as the most mutated gene found in cancers.[162] Although initially its role in carcinogenesis was thought to be a result of mutation-dependent loss of function as an inducer of apoptosis and cell cycle arrest,[163] recent studies have brought to light its role in regulating two critical features of cancer cells: redox and metabolism.[164] p53 induces OxPhos by transactivating SCO2.[165] Additionally, p53 activates glutaminase 2 to generate more energy via OxPhos and stimulates GSH to act as a ROS scavenger, thereby maintaining cellular redox.[166] Wild-type p53, in addition to increasing OxPhos, also inhibits glycolysis.[167, 168] Mutations in p53, as seen in many cancers, therefore are drivers of the glycolytic phenotype.[169] p53 is also a negative regulator of lipid synthesis.[170] This feature is important as rapidly proliferating cancer cells require lipids for (a) synthesis of membrane components, and (b) as alternative energy sources.[170] In breast tumors, mutant p53 via its effect on SREBPs regulates the mevalonate metabolic pathway to induce lipid biosynthesis by the cancer cells.[171] p53 also plays a key role in DNA damage response pathways by facilitating repair proteins to restore stability of the genome.[172] Similarly, p53 functions in the maintenance of the mitochondrial genome either directly in coordination with SSBP1 via its 3′-5′ exonuclease activity,[173] or indirectly via transcriptionally regulating mitophagy.[174] p53 also regulates mitochondrial dynamics. A balance between fusion and fission is required to produce homeostatic physiology[175, 176] and Mitofusin-2, a fusion regulator, is a direct transcriptional target of p53.[177] Finally, mitochondrial p53 also affects mitochondrially encoded proteins at a posttranslational level and abnormal functioning can cause a defect in OxPhos.[178]

(b) Rb: Rb proteins function as negative regulators of the cell cycle by either (a) transcriptionally inhibiting genes required for cell cycle progression by interaction with transcription factor E2F[179–181] or (b) inhibiting Cyclin E/cdk2 and A/cdk2 kinases.[182] In the mitochondria, Rb directly interacts with Bax and induces apoptosis, suggestive of its tumor suppressive role.[183] In contrast, colorectal carcinoma cells display an antiapoptotic role for Rb by its ability to enhance the BAG-1/NF-κB axis.[184] Rb loss has been shown to both activate and inhibit mitochondrial function. Loss of Rb was associated with reduction in the levels of mtDNA and of OxPhos subunit proteins, leading to a reduction in mitochondrial oxygen consumption.[185] In contrast, Rb loss has been shown to induce mitochondrial protein translation in breast cancers to support a metastatic phenotype.[186] Rb-deficient cells were also observed to express ASCT2, the glutamine transporter, at high levels, promoting uptake of glutamine, thereby fueling the TCA cycle[187] and regulating cellular redox.[188] Rb depletion increased fatty acid levels and transcriptionally upregulated *Elov16* and *Scd1,* suggestive of a tumor suppressor role involving regulation of fatty acid metabolism.[189] Rb has also been associated with a phenotypic feature of enhanced ROS levels.[190]

(c) Mitochondrial tumor suppressors: Multiple studies have uncovered a potential oncogenic role for nuclear-encoded mitochondrial gene products. Two enzymes of the mitochondrial tricarboxylic acid cycle (TCA), fumarate hydratase (FH) and succinate dehydrogenase (SDH), have been linked to carcinogenesis. SDH catalyzes the conversion of succinate to fumarate and FH catalyzes fumarate to malate. Loss of function mutations in these two are associated with HIF1 activation as a result of accumulation of metabolites.[191] Additionally, defective FH in tumors also inhibits AMP kinase to result in a glycolytic phenotype.[192, 193]

3 Mitochondrial function during tumor promotion

In simple terms, tumor promotion is the stage wherein an "initiated" cell upon chronic exposure to an agent expresses a transformed phenotype.[194] Inflammation and ROS contribute maximally toward tumor promotion. Here, we discuss the regulatory crosstalk between inflammation, ROS, and mitochondrial metabolism. The role of inflammation in tumorigenesis was evident from early studies showing infiltration of tumors by immune cells.[195] Inflammation is indeed regarded as a hallmark of cancer,[196] promoting carcinogenesis rather than inhibiting it.[197–199] Inflammation associated with cancers can be broadly classified on the basis of stimulus, intensity, and mechanism into distinct categories, all of which contribute to tumor development.[200] ROS can be viewed as a mediator of inflammatory response.[201] Indeed, animal studies have shown that bona fide oncogenes fail to induce cancers in the absence of an inflammatory response.[202, 203] Inflammation induces signaling pathways to enhance tumor proliferation[204] and also enhance vascularization to support tumor growth.[205]

Inflammation due to environmental and dietary factors. The National Toxicology Program's 14th report lists the environmental toxins that could affect carcinogenesis[206] (Table 1). Tobacco smoke, for example, induces inflammatory mediators and functions as a promoter of lung cancer.[207] Another example is asbestos, which promotes lung cancer via activation of

Table 1 List of environmental toxins that could affect carcinogenesis as per the National Toxicology Program's 14th report.[206]

Category	Substance
Chemical	Aflatoxins, Alcoholic Beverage Consumption, 4-Aminobiphenyl, Analgesic Mixtures Containing Phenacetin, Aristolochic Acids, Arsenic and Inorganic Arsenic Compounds, Asbestos, Azathioprine, Benzene, Benzidine and Dyes Metabolized to Benzidine, Beryllium and Beryllium Compounds, Bis(chloromethyl) Ether and Technical-Grade Chloromethyl Methyl Ether, 1,3-Butadiene, 1,4-Butanediol Dimethanesulfonate, Cadmium and Cadmium Compounds, Chlorambucil1-(2-Chloroethyl)-3-(4-methylcyclohexyl)-1-nitrosourea, Chromium Hexavalent Compounds, Cyclophosphamide, Cyclosporin A, Diethylstilbestrol, Dyes Metabolized to Benzidine, Erionite, Estrogens, Ethylene Oxide, Formaldehyde, Melphalan, Methoxsalen with Ultraviolet A Therapy, Mineral Oils: Untreated and Mildly Treated, Mustard Gas, 2-Naphthylamine, Nickel Compounds, Tamoxifen, 2,3,7,8-Tetrachlorodibenzo-*p*-dioxin, Thiotepa, o-Toluidine, Trichloroethylene, Vinyl Chloride
Physical	Coal Tars and Coal-Tar Pitches, Coke-Oven Emissions, Neutrons, Radon, Crystalline Silica (Respirable Size), Solar Radiation, Soots Strong Inorganic Acid Mists Containing Sulfuric Acid, Exposure to Sunlamps or Sunbeds, Tobacco (Environmental, Smoking, Smokeless), Thorium Dioxide (ionizing radiation), Broad-Spectrum Ultraviolet Radiation, Wood Dust, X-Radiation and Gamma Radiation
Biological	Epstein–Barr Virus, Hepatitis B Virus, Hepatitis C Virus, Human Immunodeficiency Virus Type 1, Human Papillomaviruses: Some Genital-Mucosal Types, Human T-Cell Lymphotrophic Virus Type 1, Kaposi Sarcoma–Associated Herpesvirus, Merkel Cell Polyomavirus

chronic inflammation.[208] Diets rich in lipids are associated with inflammatory promotion of hepatocellular carcinoma[209] and pancreatic intraepithelial neoplasia.[210]

Intrinsic inflammation. This includes an inflammatory response in the tumor microenvironment that contributes toward cancer growth. Tumor cells release proinflammatory cytokines such as HMGB1,[211] IL-1, and ASC[212] as a result of necrosis due to limiting oxygen availability[213] that, in turn, activates signaling pathways supporting tumor growth.[214,215] Animal studies have also characterized proinflammatory matrix proteoglycans released by tumor cells that function toward tumor progression and metastasis.[216]

Inflammation secondary to infection. Infection-induced inflammatory signaling also contributes to the promotion of cancer cell growth. Animal studies have shown *Listeria monocytogenes* to induce inflammatory signaling via TLR2 to promote carcinogenesis.[217] Similarly, *Helicobacter pylori* activates immune signaling to promote cancer cell growth[218] via activation of the cyclooxygenase pathway.[219] Furthermore, normal colonic flora activates immune signaling to drive the growth of colonic tumors.[220] The role of Hepatitis C-induced inflammation in liver cancers is well known.[221]

Inflammation due to autoimmunity or cancer therapy. Autoimmunity has been reviewed to have a diametrically opposite role in tumor promotion.[222] Chronic inflammation as a result of autoimmunity, such as celiac disease and its activation of the IFN-γ, is suggested to have a procancer effect.[223] Another example is psoriasis, wherein keratinocyte proliferation is induced as a result of Wnt signaling.[224] Chemotherapy[225] and radiation[226, 227] also induce inflammation as a result of cell death, which in turn promotes tumor growth.

Inflammatory mediators in cancer and mitochondrial metabolism. Most inflammation associated tumor growth is mediated by immune cells in the tumor milieu.[228] However, the inflammatory cells mediate effects that are pro- as well as antitumor. Studies in a mouse model suggest that the tumor environment dictates the balance between protumor and antitumor effects.[229] By and large, the cancer promoting functioning of immune cells is the result of cytokine induction that stimulates the cell to undergo proliferation.[200] Cytokines induce transcription factors, such as AP-1, STAT3, and NF-κB to enhance cancer cell proliferation, survival, and invasion. For example, the role of AP-1 activation via the TRAF2-Jnk pathway downstream of TNF-α has been documented for skin cancers.[230] TNF-α also has been shown to promote tumor growth via inducing angiogenesis.[231] Across multiple organs, TNF-α has been shown to induce mitochondrial dysfunction and enhance oxidative stress by generation of ROS.[232–234] A second function of TNF-α with respect to the mitochondria is its ability to induce mitophagy.[235] NIX (also known as BNIP3L), a non-canonical mitophagy inducer, has been shown to promote glioblastoma tumor growth.[236] On the other hand, PINK1/PARKIN, the classical mitophagy effectors, function as tumor suppressors.[237]

Two other inflammatory mediators are IL-1 and IL-6. IL-1α has been shown to drive the induction of NF-κB-regulated genes.[238] The importance of IL-1 has been uncovered for almost all tumors.[239] IL-1β has been proposed to be a master cytokine driving tumor progression. Blocking of this cytokine with anti-PD-1 inhibited tumor progression in mice.[240] Very recently, IL-1 has been shown in adipocyte cells to inhibit OxPhos with a reduction in the functioning of Complex I and III of the ETC.[241] Additionally, IL-1 reduced supercomplex formation, a phenomenon that enhances the efficiency of the ETC.[242] In fibroblasts, IL-1 treatment increased Ca^{2+} uptake by the mitochondria from ER stores.[243] Complex II inhibition, in addition to inhibition of complexes I and III, was also observed in cardiomyocytes treated with IL-1.[244] IL-6, a proinflammatory cytokine, signals through the gp130 receptor to activate the Jak/STAT pathway.[245] Studies have shown a poor prognostic outcome in patients with high serum levels of IL-6.[246, 247] Lung adenocarcinomas harboring an EGFR mutation transcriptionally induce *IL-6*; that, in turn, via autocrine and paracrine effects, stimulates cellular growth via phosphorylation of STAT3.[248] Additionally, studies in breast cancer mammosphere models have suggested IL-6 to be a procancer factor that activates NOTCH signaling to support cell growth and survival.[249] IL-6 thus affects multiple aspects of carcinogenesis from proliferation, apoptosis, and metastasis.[250] IL-6 is prooxidant in cells[251]: cells treated with IL-6 show an increase in mitochondrial membrane potential and ROS.[252] ROS levels, in turn, affect NRF2-dependent redox balance,[253] suggesting ROS to be the underlying player in cancer cell growth and proliferation. In strong contrast, a study using pancreatic cells has shown IL-6 to function as an antioxidant by regulating autophagy.[254] Taken together, these studies uncover a procarcinogenic role of inflammatory mediators.

4 Tumor progression and mitochondrial function

As the name indicates, tumor progression is a phenomenon wherein tumors become more aggressive or "malignant."[255] Histopathologically, the first two stages, namely, initiation and promotion, are labeled as "benign" whereas the stage of progression constitutes "malignancy."[256] An important characteristic of the progression stage is that cell growth continues independent of any stimulus.[21] This stage is characterized by an enhanced ability of the cell to grow with a more aggressive phenotype and biochemical changes that may lead to metastasis. Another important characteristic of this stage, in contrast to initiation and promotion, is that this stage is irreversible.[17] The primary tumor may receive an additional alteration to its genome, for example in the form of gain of copy number[257] or a chemical carcinogen-induced mutation[258] that allows the cell to acquire a more invasive and metastatic phenotype. Additionally, metabolic rewiring accompanying the genetic changes allows the cells to survive in diverse metabolic environments and under various stresses found in the metastatic sites. For metastasis to occur, the cells have to detach from the primary tumor and migrate through the vasculature to lodge at a distinct site. All these confer great stress on the cancer cell ranging from anoikis[259] to hypoxia and other microenvironmental stresses such as substrate availability.[260, 261] Metabolic rewiring of cells is thus vital for tumor progression. In strong contrast to the Warburg phenotype, recent studies have shown that mitochondrial function is required for the aggressive and metastatic phenotype of cancer cells.[95, 262–264] Studies in a glioblastoma model suggest the presence of an oxygen gradient within the primary tumor. Cells farther away from blood vessels are more hypoxic compared with those in the vicinity.[265] The cells closer to the blood vessels display increased mitochondrial function that has been correlated to their enhanced aggressiveness and chemoresistant phenotype. Unbiased proteomic approaches in mammosphere models have also identified a significant induction of mitochondrial proteins that then initiate organellar function.[266] Elegant experiments in mice show that mitochondrial DNA-deficient cancer cells acquire host cellular mitochondria, restoring organellar function in the cancer cells[267] and suggesting that mitochondrial function is required and is regulated to sustain tumor progression.

By and large, tumor cells that are at the stage of progression are under conditions of oxidative stress that are detrimental for metastasis.[268] Studies in multiple cancer models have shown antioxidants to facilitate metastasis of cancer cells.[269–271] Mitochondrial one-carbon metabolism via the enzyme SHMT2 required for mitochondrial function[272] regulates cellular redox under stress to facilitate progression.[273] Cellular redox is maintained by generation of NADPH, which enhances GSH and TRX levels to build up the antioxidant capacity of cells. One carbon metabolism generates one carbon units (methyl groups) that are utilized for generation of anabolic precursors required for the enhanced growth phenotype of cancer cells.[274] For metastasis, the cancer cells induce a genetic program called the Epithelial to Mesenchymal transition (EMT) that allows the cancer cell to metastasize and then revert back to the epithelial phenotype in the new MET niche.[275] The EMT-MET program is dictated by the oxygen tensions prevalent at the primary and metastatic sites.[276] EMT is not a requirement for metastasis[277] but does facilitate the process.[278]

During progression, cancer cells also undergo a metabolic adaptation to support mitochondrial function depending on the availability of specific metabolites and the oxygen tension in the microenvironment.[279] Depending on the organ, the median oxygen tension for primary tumor and normal tissue ranges from 0.3% to 4.2% and 3.4% to 6.8%, respectively.[280] For example, a breast cancer cell exposed at the primary site to a median oxygen tension of 1.3%, upon metastasis to liver or lungs, will be in an environment of approximately threefold higher oxygen. This is important with respect to the mitochondrial function because oxygen is the key substrate for mitochondrial metabolism. Similarly, chemical substrate flexibility in distinct sites also governs the success of a metastatic event. Brain utilizes ~20% of body glucose, where it is the most abundant substrate.[281] However, the levels of glutamine, the precursor of glutamate in the brain, are also high[282] and can be used as substrate by cells metastasized to the brain.[283] Glutamate funnels into the TCA cycle to support mitochondrial function.[284] Similarly, breast cancer cells were shown to utilize glutamine to fuel OxPhos upon metastasis to lungs.[285] The microenvironment of lymph nodes, a frequent metastatic site, is enriched in fatty acids[286] and cancer cells, upon metastasis, have been shown to utilize fatty acid oxidation as a source of energy via induction of the transcriptional activator YAP.[287] A fine-tuned interplay between the tumor cell microenvironmental oxygen tension and substrate availability dictates the adaptive regulation of cancer cells toward progression and metastasis.

5 Mitochondrial defects and cancer

The possibility of mitochondrial DNA mutations in carcinogenesis has been known for over 30 years.[288] Elegant experiments in the 1990s demonstrated that cells devoid of mtDNA (ρ^0) have reduced ability to form tumors in nude mice.[24, 26] mtDNA variations have also been associated with cancer risk. Specifically, mutations that enhance ROS levels such as T8993G on mtDNA encoding ATP6[289] and G13997A and 13885insC mutations in ND6,[290] have been shown to be procarcinogenic. Mutations in nuclear-encoded mitochondrial proteins have also been associated with cancers.

(1) Succinate Dehydrogenase (SDH): SDH is a vital enzyme of the TCA cycle that catalyzes the step of succinate oxidation to fumarate. It is a multi-subunit enzyme and mutations in all subunits have been found in adrenal tumors such as pheochromocytomas and paragangliomas.[291] Loss of function mutations cause an accumulation of succinate that results in a pseudo hypoxic state that inhibits prolyl hydroxylases and stabilizes HIF.[292] SDH-B mutations have also been shown to generate ROS[293] that can contribute toward carcinogenesis.

(2) Isocitrate Dehydrogenase 2 (IDH2): IDH2 is a mitochondrial enzyme that catalyzes the decarboxylation of isocitrate to α-ketoglutarate (α-KG), generating NADPH. IDH2 mutations have been found in cases of gliomas,[294] chondrosarcomas,[295] cholangiocarcinomas,[296] and Acute Myeloid Leukemias (AML).[297] The R140 mutation of IDH2, which is most common in AML, is a neomorphic mutant that generates *R*(−)-2-hydroxyglutarate (2HG) from α-KG. 2HG is a potential oncometabolite.[298] NADPH is used for its reducing equivalents to maintain antioxidant scavenging in cells.[299] With generation of 2HG, NADPH is converted to $NADP^+$, thereby increasing oxidative stress on cells[300] that may contribute toward cancer development.

(3) Fumarate Hydratase (FH): FH enzyme converts fumarate to malate in the TCA cycle. Mutations in this enzyme have been associated with uterine leiomyomas and renal cell carcinoma.[301] Loss of function results in fumarate accumulation, which is prooncogenic. At least three mechanisms have been shown to play a role. (1) Fumarate induces succination of at least 10 proteins involved in redox signaling, resulting in an imbalance and oxidative stress,[302] (2) Induction of histone and DNA methylation resulting in cell growth and proliferation,[303] and (3) Similar to SDH, stabilization of HIF to induce glycolytic genes, a part of the oncogenic phenotype.[304]

6 Mitochondria and interorganellar communication in cancers

Mitochondrial function is required for cancer cell growth and proliferation. However, recent evidence indicates that mitochondrial metabolism is regulated via its finely controlled interplay with other cellular organelles such as the nucleus, the endoplasmic reticulum, and the peroxisomes.[305] Metabolic regulation in cancer cells also occurs as a result of specific communications between mitochondria and other organelles.

Mitochondria and Peroxisomes: Peroxisomes are interconnected organelles within a cell that have diverse functions ranging from regulation of ROS levels to metabolism of fatty acids[306] and including a role in immune signaling and aging.[307] Mitochondria and peroxisomes physically interact[308] with the interaction facilitated by the tethering proteins Fzo1 and Pex34.[309] Experimental evidence using fibroblasts devoid of peroxisomes suggests that peroxisomal membrane proteins such as Pex3 and Pex14 target mitochondria[310] and that mitochondria are important in peroxisomal biogenesis. Mitochondria-peroxisomes cooperate with each other with respect to the above functions of beta-oxidation[311] and redox regulation.[312] For example, inhibition of peroxisomal catalase affects mitochondrial redox,[313, 314] and at the same time, expression of peroxisomal ROS scavenger PRDX5 offers protection against mitochondrial ROS.[315] In addition to mitochondria, the peroxisomes also are well known to generate ROS.[316] Peroxisomes also harbor a catalase to detoxify the peroxide generated.[317, 318] The role of peroxisomes in cancer was suggested by studies documenting increased levels of lipid metabolic enzymes in these organelles.[319] Inhibition of lipid processing in peroxisomes resulted in reduced growth and a less aggressive phenotype of cancer cells.[320] A vesicular transport system is postulated to play a role in the mitochondria-peroxisomal communication that delivers heme and mistargeted long chain fatty acids required for catalase and beta-oxidation.[321] Beta-oxidation as well as ROS regulation are needed for cancer cell growth. Fatty acid oxidation provides additional energy and is also a source of NADPH, a vital requirement for key anabolic reactions[322] for cancer cell growth.[62] Although ROS are a stimulus for DNA damage and carcinogenesis, maintaining their level is required for growth and proliferation of cancer cells.[323, 324]

Mitochondria and Endoplasmic Reticulum: Interactions between the mitochondria and the ER are the most well studied to date. ER-mitochondria interaction either physically or functionally regulates mitochondrial metabolism and is vital for growth and proliferation of cancer cells. The lipid raft between these two organelles, initially called membrane fraction "X,"[325] is now commonly called mitochondria-associated membrane (MAM)[326] and is a site for phospholipid synthesis.[325] Both organelles are connected via specific proteins, many of which have been identified.[327] Functionally, the MAMs are associated with regulation of lipid metabolism,[328] calcium homeostasis,[329] mitochondrial dynamics of fusion[330] and fission,[331] and mitophagy.[332] Mitochondrial function has been shown to be essential for survival of cancer cells[333]; for example, using ρ^0 143B cells, MAM-dependent calcium transfer into the mitochondria was shown to be required for survival of these cells.[334] Multiple proteins localized to the MAMs are oncoproteins or tumor suppressors and are involved in carcinogenesis and cancer progression.

Cardiolipin is a phospholipid in the inner mitochondrial membrane that plays a key role in the maintenance of mitochondrial ETC complexes. Phosphatidic acid, a precursor of cardiolipin,[335] is transported from ER to mitochondria via MAMs for cardiolipin biosynthesis.[336] ER-stress is a deviation from normal physiology of ER resulting in the accumulation of unfolded/misfolded proteins, triggering the homeostatic unfolded protein response (UPR).[337] The UPR is overactive in cancers[338] owing to the metabolic alterations that a cancer cell is exposed to. Indeed, UPR is required for carcinogenesis. Studies with XBP1 (a regulator of UPR)-deficient cells have demonstrated a significant reduction in the capacity of tumor cells to grow.[339] During early stages, ER-stress results in increased Ca^{2+} uptake by the mitochondria and a stimulation of OxPhos.[340] However, chronic ER-stress induces apoptosis.[341] In cancers, the PERK-ATF4 pathway is responsible for the orchestration of the prosurvival or proapoptotic pathways depending on whether the stress is acute (induction of prosurvival genes), or chronic (induction of proapoptotic genes such as *CHOP*).[342] In addition to stimulating OxPhos by Ca^{2+} uptake into the mitochondria, ER-stress also transcriptionally upregulates genes such as *Mitochondrial Nuclear Retrograde Regulator 1* (*MNRR1*) that, in turn, activates the mitochondrial Unfolded Protein Response (UPR^{mt}) to activate OxPhos.[343] MNRR1 is also upregulated in a variety of cancers[96] as well as cancer cell lines.[95]

Mitochondria and Nucleus: Other than the nucleus, mitochondria are the only organelle that possesses its own genome. The mitochondrial genome is a fraction of the size of the nuclear genome—16,569 bp in size—encoding 37 genes, including 13 proteins, 2 rRNAs, and 22 tRNAs[344] as well as some newly discovered active peptides.[345] The mitochondrial proteome comprises over 1000 proteins,[4] most of which are encoded by the nuclear genome and the translated proteins imported into the mitochondria. As a result, crosstalk between these two organelles is finely regulated. There are two distinct regulatory systems in play, anterograde and retrograde.[346] The former deals with the regulation of genes that encode mitochondrial proteins whereas the latter is associated with mitochondrial signaling that regulates nuclear function.

(a) *Anterograde Signaling:* Anterograde signaling can regulate four vital parameters of mitochondrial function: mitochondrial biogenesis, mitophagy, mitochondrial dynamics, and mitochondrial movement.[347] All these functions are known to promote cell growth and survival and, indeed, multiple studies have now shown that mitochondrial biogenesis promotes rather than inhibits cancer cell growth[348–350] and does so via Nrf-1/2-mediated induction of TFAM[351] and Myc, an oncogene.[54] Mitophagy is also essential for the proliferation of cancer cells. For example, oncogenic KRAS in pancreatic cancers induces mitophagy to support redox homeostasis and cancer cell growth.[352] Furthermore, FUNDC1-deficient tumor cells were shown to form smaller primary tumors[353] in xenograft models. FUNDC1 is a mediator of hypoxia-induced mitophagy.[354] With regard to mitochondrial dynamics, many cancer cells harbor a profission phenotype,[355–357] owing to the levels and functioning of the fission regulator DRP1.[358] The role of DRP1 in cancers has been reviewed extensively and includes regulatory roles in apoptosis as well as adaptation to metabolic phenotypes.[359] Mitochondrial trafficking and motility in cancer cells have thus been shown to be attractive therapeutic targets.[360] Intracellular mitochondrial movement is governed by the cytoskeleton and adaptor proteins.[361] Movement of mitochondria within cells occurs to support local energy demands and redox regulation.[362] Studies using PI3K inhibitors have shown that mitochondria move from the perinuclear region toward the cortical cytoskeleton to support the turnover of focal adhesion complexes resulting in cancer cell motility supporting invasion.[363]

(b) *Retrograde signaling:* Retrograde signaling is an adaptive mechanism wherein defects in the mitochondria are relayed to the nucleus in an attempt to achieve metabolic regulation. The first evidence of retrograde signaling was shown in *Saccharomyces cerevisiae.*[364] One critical retrograde signaling pathway is the increased levels of cytosolic Ca^{2+}. Studies in C2C12 myoblasts and human A549 cells have shown that mitochondrial inhibition via depletion of mtDNA or chemical inhibitors resulted in Ca^{2+} release into the cytosol, activating PKC, inducing Cathepsin L, and resulting in an invasive phenotype.[365] Although mitochondria are also generators of ROS, ROS generated as a result of electron flux is typically minimal.[366]

However, when the levels of ROS exceed the scavenging capacity of the cell, it can damage both mitochondrial and nuclear DNA.[367] The role of ROS in carcinogenesis has been extensively studied.[290, 368–370]

A relatively novel mitochondrial retrograde signaling mechanism is the mitochondrial unfolded protein response (UPR^{mt}). The initial evidence for the existence of the UPR^{mt} pathway was provided by studies using a deletion mutant of ornithine transcarbamylase that induces a transcriptional program for chaperones and proteases.[371] Studies in *C. elegans* identified ATFS-1 as the mediator of UPR^{mt}; upon stress, it translocates from the mitochondria to the nucleus to function as a transcription activator,[372, 373] similar to the ER-UPR program where proteins such as ATF6 translocate to the nucleus to activate transcription.[374] Also similar to the ER-UPR, UPR^{mt} is a homeostatic mechanism that restores the metabolic equilibrium required for growth and proliferation of cancer cells.[375] Cancer cells have been shown to have increased levels of chaperones and UPR^{mt} effectors induced in response to UPR^{mt}.[376–378] In addition to its damaging effect on nucleic acids, sustained high ROS levels also damage proteins and the accumulation of damaged proteins induces the UPR.[379] The UPR can be visualized in this context as a mechanism to keep ROS-induced damage to cancer cell growth to a minimum. MNRR1, which is upregulated in almost all cancers[95, 96, 380–383], is a bi-organellar protein transcriptionally induced in response to cellular stress. It has recently been shown to be upstream of ATF5 (mammalian ortholog of ATFS-1) and to be required for optimal induction of UPR^{mt}.[343] In contrast to, the canonical mechanism by which ATF5 functions via translocation into the nucleus, preliminary evidence shows that, upon stress, MNRR1 is proteolytically degraded in the mitochondrial intermembrane space (IMS) by the protease YME1L1, with the newly translated protein accumulating in the nucleus to function as a transcriptional activator for genes involved not only in the UPR^{mt} pathway but also in autophagy, ROS scavenging, and mitochondrial biogenesis.[343]

7 Mitochondrial metabolic pathways and potential therapeutic targets

One of the hallmarks of cancer is that tumor cells are resistant to apoptosis and that they have the flexibility to use various substrates for bioenergetic pathways to support their growth and proliferation.[384, 385] In this section, we will discuss the interplay of substrate metabolism between cancer cells and tumor stroma and how vulnerabilities have been exploited to evaluate novel therapies. Studies have shown that, within a tumor, metabolic pathways are distinct within cells (OxPhos vs glycolysis) depending on the distance of the cell from a blood vessel.[265] As shown in the Fig. 1, a gradient exists for both oxygen and nutrients within a tumor cell depending on its distance from the tumor vasculature. Cells in the immediate vicinity are exposed to an environment with relatively more oxygen and nutrients and therefore utilize OxPhos whereas those farther away rely on glycolysis. Glucose, glutamine, and fatty acids are all used as substrates for generation of ATP in addition to supporting other critical pathways required for cancer cell growth, proliferation, and metastasis.

Glucose: Glucose is the simplest and most abundant carbohydrate in the blood.[386] Cancer cells display enhanced uptake of glucose, a feature that underlies diagnostic imaging of cancers.[387] It is therefore not surprising that cancer cells also have enhanced expression of glucose transporters (GLUTs).[388] Glucose within the cancer cells generates pyruvate that is funneled into the mitochondria that utilize OxPhos for their bioenergetic requirements.[389] This is particularly true for cancer cells in the vicinity of blood vessels with a relatively higher oxygen tension. Tumor cells farther away from vasculature are exposed to lower oxygen tensions and stromal cells undergo glycolysis to form lactate[28, 390–393] that, via the monocarboxylate transporters MCT4 and MCT1,[394, 395] enters the cancer cell, where it gets converted to pyruvate that is utilized for OxPhos. This is called the Reverse Warburg Effect and is a potential therapeutic target.[396]

Additionally, glycolytic intermediates in cancer cells flux through the Pentose Phosphate Pathway to (a) generate NADPH for ROS scavenging, and (b) support nucleotide synthesis for anabolic demands of the proliferating cancer cell.[397] Glycolytic intermediate 3-phosphoglycerate, via action of phosphoglycerate dehydrogenase and other intermediates, can generate serine that, along with glycine, participates in one-carbon metabolism and protein synthesis.[398]

Glutamine: Glutamine is the most abundant amino acid in the blood.[399] ^{13}C labeling experiments have shown cancer cells to synthesize glutamine from glutamate via glucose oxidation involving the malate aspartate shuttle.[400] Glutamine from the blood is imported via its transporter SLC1A5, levels of which are higher in cancer cells.[401] Glutamine in cells is a substrate for enzymes involved in purine and pyrimidine biosynthesis.[402] Glutamine can also undergo glutaminolysis to generate glutamate.[403] Glutamate has

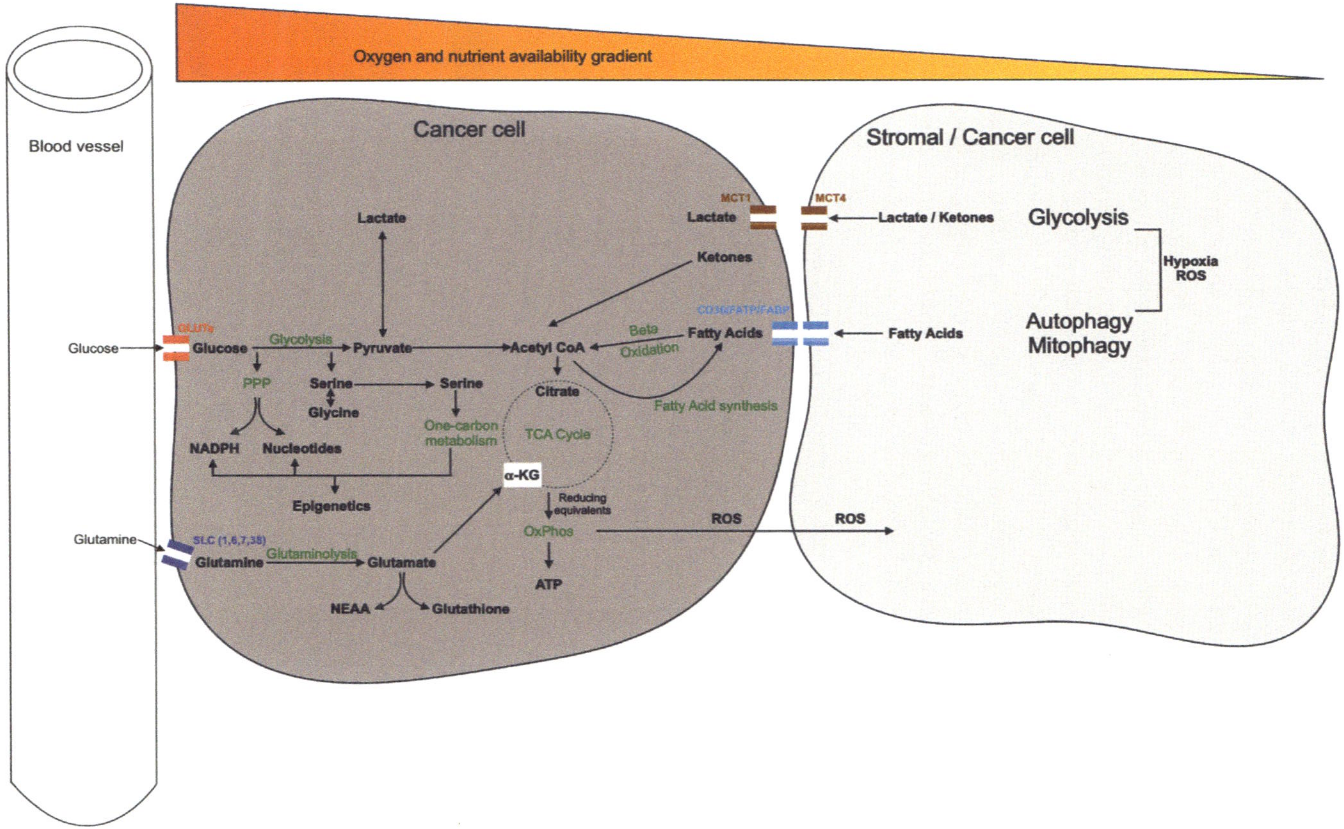

Fig. 1 Schematic representation of mitochondrial metabolism in cancer cells. The cancer cell (dark gray, left) lies in close proximity to a blood vessel and therefore is exposed to a relatively higher oxygen and nutrient concentration. In contrast, a stromal cell or another cancer cell (light gray, right), positioned farther away from the blood vessel, is relatively more hypoxic. These (light gray) cells are by and large catabolic in the presence of hypoxia and ROS and the major metabolic pathways are glycolysis, autophagy and mitophagy. Subsequently, the products of these are utilized by the cancer cell to support anabolic pathways for cell growth and metastasis. These are the cells in which mitochondria are functional. Green indicates pathways whereas transporters are highlighted in various colors on the cell surface.

been shown to maintain the TCA cycle in the absence of pyruvate,[284] suggesting a regulatory role in maintenance of mitochondrial function via flexibility of substrate utilization. Glutamate can also generate nonessential amino acids such as alanine and arginine. Alanine generated by tumor cells, along with lactate, is taken up by the liver (Cori cycle) to make more glucose that, in turn, can be imported into cancer cells.[404] Glutamine is also a critical precursor of glutathione,[405] which functions in maintenance of redox balance in cells[406] and thus plays a vital role in cancer progression.[407] Enhanced levels of reduced glutathione (GSH) are associated with an enhanced proliferative phenotype and cell cycle progression.[408] In one study, increased GSH levels were proportional to growth of hepatocarcinoma cells.[409] In B16 murine melanoma cells, exogenous GSH enhanced the metastatic potential of cells with a low GSH content.[410] GSH also promotes metastasis by resisting oxidative and nitrosative stress that in the absence of GSH would have a deleterious effect on the cells.[411]

Fatty Acids: Fatty acids are a vital source of energy especially under conditions of stress.[412–414] Cancer cells divert a majority of carbon from energy metabolism to fatty acids for the generation of membrane lipids for structure and to act as signaling molecules.[414] Thus, cancer cells have high levels of enzymes such as fatty acid synthase,[415, 416] again suggestive of an important role of fatty acid synthesis in cancer cell growth and proliferation.[417] Studies using labeled glucose in Erlich Ascites Tumor (EAT) have shown that cancer cells synthesize fatty acids de novo.[418] In addition, cancer cells also take up fatty acids from surrounding stromal cells where fatty acids are generated as a result of autophagic recycling.[419] For example, Fatty acid-binding protein 4 (FABP4) is upregulated in ovarian cancer cells that metastasize to the omentum but not in the primary tumor. This study suggested that fatty acids are provided for the growth of the tumor cells from adipocytes in the surrounding microenvironment.[420] Another such example is CD36, expressed on colorectal cancer cells, which assists in fatty acid transport from the surrounding cancer-associated fibroblasts.[421] Although fatty acids via beta-oxidation provide energy for growth and survival of cancer cells,[422] fatty acid oxidation also results in the generation of substrates for NADPH generating enzymes,[423, 424] suggesting the additional role of maintenance of cellular redox. Stromal cells in the vicinity of cancer cells have increased autophagy that generates substrates for metabolism of cancer cells.[425] Indeed, fatty acid pathways have been an attractive target for cancer therapy and have shown promising results in vitro.[426–428]

In summary, cellular and mitochondrial substrate flexibility and utilization have allowed cancer cells to survive harsh microenvironmental conditions, resulting in severe morbidity and mortality of the host. Targeting these pathways has been hypothesized to sensitize these cells to existing therapies.

8 Reactive oxygen species and mitochondria in cancer

One of the first in vitro studies correlating ROS and cancer was performed in the early 1990s when seven cancer cell lines originating from four different tissues generated ROS comparable to that produced by stimulated polymorphonuclear leucocytes.[429] Although ROS are well known to induce DNA damage, a potential oncogenic stimulus, a variety of defense systems exist to repair the damaged DNA. However, some of the mutations generated may accumulate with age and are risk factors for carcinogenesis.[430] Although ROS are procarcinogenic in the initial stages, increased oxidative stress may be detrimental for the growth and proliferation of cancer cells.[431, 432] These diametrically opposite roles of ROS in cancer have the potential to be targeted by either activating or inhibiting ROS, depending on the stage.

ROS generation occurs at two important sites in the cell: the mitochondrial ETC[433] and the cell surface NADPH oxidase (NOX) enzymes.[434] In the ETC, complexes I, II, and III are known to generate ROS[435] as the superoxide radical.[436] The superoxide radical is converted by the enzymes SOD1 (in the cytosol) and SOD2 (in the mitochondria) to hydrogen peroxide.[437] The peroxiredoxins, glutathione peroxidases,[438] and catalase[439] then detoxify the peroxide to water. Peroxide in the presence of cations generates hydroxyl radicals,[440, 441] resulting in damage to proteins, lipids, and DNA. However, peroxide via cysteine thiol modification is a vital cellular signaling mechanism.[442] Thus, an optimal level of peroxide is required for the maintenance of signaling homeostasis. In the presence of too low levels, ROS affect cellular signaling negatively[443] whereas at very high levels ROS may result in carcinogenesis[444] or even cell death.[445] ROS levels in cells are therefore a product of generation and scavenging and optimal ROS levels are required for cancer cell proliferation, survival, and metastasis. Although peroxide is generated by the mitochondria, it negatively affects mitochondrial function and also leads to apoptosis. Hydrogen peroxide has been shown in rat heart mitochondria to inhibit functioning via inactivation of α-ketoglutarate dehydrogenase and succinate dehydrogenase.[446] Increasing ROS levels in cells, therefore, has been an attractive target for cancer therapy. Indeed, most chemotherapeutic modalities induce cell death via overactivation of ROS.[447]

Acknowledgments

Supported by the Office of the Assistant Secretary of Defense for Health Affairs through the Peer-Reviewed Medical Research Program under Awards W81XWH-16-1-0516, the Perinatology Research Branch, Division of Obstetrics and Maternal-Fetal Medicine, Division of Intramural Research, Eunice Kennedy Shriver National Institute of Child Health and Human Development, National Institutes of Health, U.S. Department of Health and Human Services (NICHD/NIH/DHHS); and, in part, with Federal funds from NICHD/NIH/DHHS under Contract No. HHSN275201300006C (to S.A. and L.I.G.), and the Henry L. Brasza endowment at Wayne State University (to L.I.G.). The views expressed in this article are those of the authors and do not necessarily reflect the position or policy of the Department of Defense or the US Government.

References

1. Lane N, Martin W. The energetics of genome complexity. *Nature.* 2010;467(7318):929–934.
2. McBride HM, Neuspiel M, Wasiak S. Mitochondria: more than just a powerhouse. *Curr Biol.* 2006;16(14):R551–R560.
3. Taanman JW. The mitochondrial genome: structure, transcription, translation and replication. *Biochim Biophys Acta.* 1999;1410(2):103–123.
4. Rath S, et al. MitoCarta3.0: an updated mitochondrial proteome now with sub-organelle localization and pathway annotations. *Nucleic Acids Res.* 2021;49(D1):D1541–D1547.
5. Hara KY, Kondo A. ATP regulation in bioproduction. *Microb Cell Factories.* 2015;14:198.
6. Hatefi Y. The mitochondrial electron transport and oxidative phosphorylation system. *Annu Rev Biochem.* 1985;54:1015–1069.
7. Reid RA, Moyle J, Mitchell P. Synthesis of adenosine triphosphate by a protonmotive force in rat liver mitochondria. *Nature.* 1966;212(5059):257–258.
8. Finkel T. Signal transduction by reactive oxygen species. *J Cell Biol.* 2011;194(1):7–15.
9. Schieber M, Chandel NS. ROS function in redox signaling and oxidative stress. *Curr Biol.* 2014;24(10):R453–R462.
10. Cooke MS, et al. Oxidative DNA damage: mechanisms, mutation, and disease. *FASEB J.* 2003;17(10):1195–1214.
11. Xiao M, et al. Pathophysiology of mitochondrial lipid oxidation: role of 4-hydroxynonenal (4-HNE) and other bioactive lipids in mitochondria. *Free Radic Biol Med.* 2017;111:316–327.
12. Grimsrud PA, et al. Oxidative stress and covalent modification of protein with bioactive aldehydes. *J Biol Chem.* 2008;283(32):21837–21841.
13. Wang C, Youle RJ. The role of mitochondria in apoptosis*. *Annu Rev Genet.* 2009;43:95–118.
14. Fulda S. Shifting the balance of mitochondrial apoptosis: therapeutic perspectives. *Front Oncol.* 2012;2:121.
15. Bertram JS. The molecular biology of cancer. *Mol Asp Med.* 2000;21(6):167–223.
16. Bray F, et al. Global cancer statistics 2018: GLOBOCAN estimates of incidence and mortality worldwide for 36 cancers in 185 countries. *CA Cancer J Clin.* 2018;68(6):394–424.
17. Pitot HC, Dragan YP. Facts and theories concerning the mechanisms of carcinogenesis. *FASEB J.* 1991;5(9):2280–2286.
18. Zhang H, Rosdahl I. Ultraviolet A and B differently induce intracellular protein expression in human skin melanocytes - a speculation of separate pathways in initiation of melanoma. *Carcinogenesis.* 2003;24(12):1929–1934.

19. Gorodetska I, Kozeretska I, Dubrovska A. BRCA genes: the role in genome stability, cancer stemness and therapy resistance. *J Cancer.* 2019;10(9):2109–2127.
20. Trosko JE. Commentary: is the concept of "tumor promotion" a useful paradigm? *Mol Carcinog.* 2001;30(3):131–137.
21. Oliveira PA, et al. Chemical carcinogenesis. *An Acad Bras Cienc.* 2007;79(4):593–616.
22. Pitot HC. Progression: the terminal stage in carcinogenesis. *Jpn J Cancer Res.* 1989;80(7):599–607.
23. Warburg O. On the origin of cancer cells. *Science.* 1956;123(3191):309–314.
24. Cavalli LR, Varella-Garcia M, Liang BC. Diminished tumorigenic phenotype after depletion of mitochondrial DNA. *Cell Growth Differ.* 1997;8(11):1189–1198.
25. Weinberg F, et al. Mitochondrial metabolism and ROS generation are essential for Kras-mediated tumorigenicity. *Proc Natl Acad Sci U S A.* 2010;107(19):8788–8793.
26. Morais R, et al. Tumor-forming ability in athymic nude mice of human cell lines devoid of mitochondrial DNA. *Cancer Res.* 1994;54(14):3889–3896.
27. Zu XL, Guppy M. Cancer metabolism: facts, fantasy, and fiction. *Biochem Biophys Res Commun.* 2004;313(3):459–465.
28. Pavlides S, et al. The reverse Warburg effect: aerobic glycolysis in cancer associated fibroblasts and the tumor stroma. *Cell Cycle.* 2009;8(23):3984–4001.
29. Stratton MR, Campbell PJ, Futreal PA. The cancer genome. *Nature.* 2009;458(7239):719–724.
30. Krontiris TG, Cooper GM. Transforming activity of human tumor DNAs. *Proc Natl Acad Sci U S A.* 1981;78(2):1181–1184.
31. Shih C, et al. Transforming genes of carcinomas and neuroblastomas introduced into mouse fibroblasts. *Nature.* 1981;290(5803):261–264.
32. Reddy EP, et al. A point mutation is responsible for the acquisition of transforming properties by the T24 human bladder carcinoma oncogene. *Nature.* 1982;300(5888):149–152.
33. Tabin CJ, et al. Mechanism of activation of a human oncogene. *Nature.* 1982;300(5888):143–149.
34. Todd R, Wong DT. Oncogenes. *Anticancer Res.* 1999;19(6A):4729–4746.
35. Stanbridge EJ. Suppression of malignancy in human cells. *Nature.* 1976;260(5546):17–20.
36. Croce CM. Oncogenes and cancer. *N Engl J Med.* 2008;358(5):502–511.
37. Batlle E, Clevers H. Cancer stem cells revisited. *Nat Med.* 2017;23(10):1124–1134.
38. Chen X, et al. MLL-AF9 initiates transformation from fast-proliferating myeloid progenitors. *Nat Commun.* 2019;10(1):5767.
39. Brumatti G, et al. HoxA9 regulated Bcl-2 expression mediates survival of myeloid progenitors and the severity of HoxA9-dependent leukemia. *Oncotarget.* 2013;4(11):1933–1947.
40. Roychoudhury J, et al. MEIS1 regulates an HLF-oxidative stress axis in MLL-fusion gene leukemia. *Blood.* 2015;125(16):2544–2552.
41. Pulikkan JA, et al. CBFbeta-SMMHC inhibition triggers apoptosis by disrupting MYC chromatin dynamics in acute myeloid leukemia. *Cell.* 2018;174(1):172–186 e21.
42. Yuan X, Larsson C, Xu D. Mechanisms underlying the activation of TERT transcription and telomerase activity in human cancer: old actors and new players. *Oncogene.* 2019;38(34):6172–6183.
43. Bell RJ, et al. Cancer. The transcription factor GABP selectively binds and activates the mutant TERT promoter in cancer. *Science.* 2015;348(6238):1036–1039.
44. Yang ZF, Mott S, Rosmarin AG. The Ets transcription factor GABP is required for cell-cycle progression. *Nat Cell Biol.* 2007;9(3):339–346.
45. Virbasius JV, Scarpulla RC. Activation of the human mitochondrial transcription factor A gene by nuclear respiratory factors: a potential regulatory link between nuclear and mitochondrial gene expression in organelle biogenesis. *Proc Natl Acad Sci U S A.* 1994;91(4):1309–1313.

46. Carter RS, et al. The basal promoter elements of murine cytochrome c oxidase subunit IV gene consist of tandemly duplicated ets motifs that bind to GABP-related transcription factors. *J Biol Chem.* 1992;267(32):23418–23426.
47. Sucharov C, et al. A novel transcriptional initiator activity of the GABP factor binding ets sequence repeat from the murine cytochrome c oxidase Vb gene. *Gene Expr.* 1995;5(2):93–111.
48. Yang ZF, et al. GABP transcription factor (nuclear respiratory factor 2) is required for mitochondrial biogenesis. *Mol Cell Biol.* 2014;34(17):3194–3201.
49. Dalla-Favera R, et al. Human c-myc onc gene is located on the region of chromosome 8 that is translocated in Burkitt lymphoma cells. *Proc Natl Acad Sci U S A.* 1982;79(24):7824–7827.
50. Miller DM, et al. c-Myc and cancer metabolism. *Clin Cancer Res.* 2012;18(20):5546–5553.
51. Vafa O, et al. c-Myc can induce DNA damage, increase reactive oxygen species, and mitigate p53 function: a mechanism for oncogene-induced genetic instability. *Mol Cell.* 2002;9(5):1031–1044.
52. Gao P, et al. c-Myc suppression of miR-23a/b enhances mitochondrial glutaminase expression and glutamine metabolism. *Nature.* 2009;458(7239):762–765.
53. Wise DR, et al. Myc regulates a transcriptional program that stimulates mitochondrial glutaminolysis and leads to glutamine addiction. *Proc Natl Acad Sci U S A.* 2008;105(48):18782–18787.
54. Li F, et al. Myc stimulates nuclearly encoded mitochondrial genes and mitochondrial biogenesis. *Mol Cell Biol.* 2005;25(14):6225–6234.
55. Lee KM, et al. MYC and MCL1 cooperatively promote chemotherapy-resistant breast cancer stem cells via regulation of mitochondrial oxidative phosphorylation. *Cell Metab.* 2017;26(4):633–647 e7.
56. Gazit A, et al. Expression of the normal human sis/PDGF-2 coding sequence induces cellular transformation. *Cell.* 1984;39(1):89–97.
57. Andrae J, Gallini R, Betsholtz C. Role of platelet-derived growth factors in physiology and medicine. *Genes Dev.* 2008;22(10):1276–1312.
58. Waterfield MD, et al. Platelet-derived growth factor is structurally related to the putative transforming protein p28sis of simian sarcoma virus. *Nature.* 1983;304(5921):35–39.
59. Doolittle RF, et al. Simian sarcoma virus onc gene, v-sis, is derived from the gene (or genes) encoding a platelet-derived growth factor. *Science.* 1983;221(4607):275–277.
60. Koval OM, et al. Loss of MCU prevents mitochondrial fusion in G1-S phase and blocks cell cycle progression and proliferation. *Sci Signal.* 2019;12(579).
61. Salabei JK, Hill BG. Mitochondrial fission induced by platelet-derived growth factor regulates vascular smooth muscle cell bioenergetics and cell proliferation. *Redox Biol.* 2013;1:542–551.
62. Carracedo A, Cantley LC, Pandolfi PP. Cancer metabolism: fatty acid oxidation in the limelight. *Nat Rev Cancer.* 2013;13(4):227–232.
63. Xu L, et al. WISP-1 is a Wnt-1- and beta-catenin-responsive oncogene. *Genes Dev.* 2000;14(5):585–595.
64. Wang S, et al. Wnt1 inducible signaling pathway protein 1 (WISP1) blocks neurodegeneration through phosphoinositide 3 kinase/Akt1 and apoptotic mitochondrial signaling involving Bad, Bax, Bim, and Bcl-xL. *Curr Neurovasc Res.* 2012;9(1):20–31.
65. Venkatesan B, et al. WNT1-inducible signaling pathway protein-1 activates diverse cell survival pathways and blocks doxorubicin-induced cardiomyocyte death. *Cell Signal.* 2010;22(5):809–820.
66. Witsch E, Sela M, Yarden Y. Roles for growth factors in cancer progression. *Physiology (Bethesda).* 2010;25(2):85–101.

67. Olayioye MA, et al. The ErbB signaling network: receptor heterodimerization in development and cancer. *EMBO J.* 2000;19(13):3159–3167.
68. Gamett DC, Tracy SE, Robinson HL. Differences in sequences encoding the carboxyl-terminal domain of the epidermal growth factor receptor correlate with differences in the disease potential of viral erbB genes. *Proc Natl Acad Sci U S A.* 1986;83(16):6053–6057.
69. Boerner JL, Danielsen A, Maihle NJ. Ligand-independent oncogenic signaling by the epidermal growth factor receptor: v-ErbB as a paradigm. *Exp Cell Res.* 2003;284(1):111–121.
70. Velu TJ, et al. Epidermal-growth-factor-dependent transformation by a human EGF receptor proto-oncogene. *Science.* 1987;238(4832):1408–1410.
71. Danielsen AJ, Maihle NJ. Ligand-independent oncogenic transformation by the EGF receptor requires kinase domain catalytic activity. *Exp Cell Res.* 2002;275(1):9–16.
72. Che TF, et al. Mitochondrial translocation of EGFR regulates mitochondria dynamics and promotes metastasis in NSCLC. *Oncotarget.* 2015;6(35):37349–37366.
73. Lim SO, et al. EGFR signaling enhances aerobic glycolysis in triple-negative breast cancer cells to promote tumor growth and immune escape. *Cancer Res.* 2016;76(5):1284–1296.
74. Di Fiore PP, et al. erbB-2 is a potent oncogene when overexpressed in NIH/3T3 cells. *Science.* 1987;237(4811):178–182.
75. Chazin VR, et al. Transformation mediated by the human HER-2 gene independent of the epidermal growth factor receptor. *Oncogene.* 1992;7(9):1859–1866.
76. Ding Y, et al. Receptor tyrosine kinase ErbB2 translocates into mitochondria and regulates cellular metabolism. *Nat Commun.* 2012;3:1271.
77. Kurmi K, et al. Tyrosine phosphorylation of mitochondrial creatine kinase 1 enhances a druggable tumor energy shuttle pathway. *Cell Metab.* 2018;28(6):833–847 e8.
78. Chen FM, et al. Activation of mitochondrial unfolded protein response is associated with Her2-overexpression breast cancer. *Breast Cancer Res Treat.* 2020;183(1):61–70.
79. Munch C. The different axes of the mammalian mitochondrial unfolded protein response. *BMC Biol.* 2018;16(1):81.
80. McCormick F. GTP-binding proteins as oncogenes in human tumors. *Environ Health Perspect.* 1991;93:17–18.
81. Lahiry P, et al. Kinase mutations in human disease: interpreting genotype-phenotype relationships. *Nat Rev Genet.* 2010;11(1):60–74.
82. Prior IA, Lewis PD, Mattos C. A comprehensive survey of Ras mutations in cancer. *Cancer Res.* 2012;72(10):2457–2467.
83. Sahai E, Marshall CJ. RHO-GTPases and cancer. *Nat Rev Cancer.* 2002;2(2):133–142.
84. Siveen KS, et al. Role of non receptor tyrosine kinases in hematological malignances and its targeting by natural products. *Mol Cancer.* 2018;17(1):31.
85. Stacey DW, Kung HF. Transformation of NIH 3T3 cells by microinjection of Ha-ras p21 protein. *Nature.* 1984;310(5977):508–511.
86. Brown R, et al. Mechanism of activation of an N-ras gene in the human fibrosarcoma cell line HT1080. *EMBO J.* 1984;3(6):1321–1326.
87. Capon DJ, et al. Activation of Ki-ras2 gene in human colon and lung carcinomas by two different point mutations. *Nature.* 1983;304(5926):507–513.
88. Almoguera C, et al. Most human carcinomas of the exocrine pancreas contain mutant c-K-ras genes. *Cell.* 1988;53(4):549–554.
89. Rodenhuis S, et al. Incidence and possible clinical significance of K-ras oncogene activation in adenocarcinoma of the human lung. *Cancer Res.* 1988;48(20):5738–5741.
90. Bos JL, et al. Prevalence of ras gene mutations in human colorectal cancers. *Nature.* 1987;327(6120):293–297.

91. Yun J, et al. Glucose deprivation contributes to the development of KRAS pathway mutations in tumor cells. *Science*. 2009;325(5947):1555–1559.
92. Son J, et al. Glutamine supports pancreatic cancer growth through a KRAS-regulated metabolic pathway. *Nature*. 2013;496(7443):101–105.
93. Padanad MS, et al. Fatty acid oxidation mediated by Acyl-CoA synthetase long chain 3 is required for mutant KRAS lung tumorigenesis. *Cell Rep*. 2016;16(6):1614–1628.
94. Ozsvari B, Sotgia F, Lisanti MP. A new mutation-independent approach to cancer therapy: Inhibiting oncogenic RAS and MYC, by targeting mitochondrial biogenesis. *Aging (Albany NY)*. 2017;9(10):2098–2116.
95. Aras S, et al. Mitochondrial autoimmunity and MNRR1 in breast carcinogenesis. *BMC Cancer*. 2019;19(1):411.
96. Liu Y, et al. CHCHD2 inhibits apoptosis by interacting with Bcl-x L to regulate Bax activation. *Cell Death Differ*. 2015;22(6):1035–1046.
97. Aras S, et al. MNRR1 (formerly CHCHD2) is a bi-organellar regulator of mitochondrial metabolism. *Mitochondrion*. 2015;20:43–51.
98. Aras S, et al. Oxygen-dependent expression of cytochrome c oxidase subunit 4-2 gene expression is mediated by transcription factors RBPJ, CXXC5 and CHCHD2. *Nucleic Acids Res*. 2013;41(4):2255–2266.
99. Qiu RG, et al. A role for Rho in Ras transformation. *Proc Natl Acad Sci U S A*. 1995;92(25):11781–11785.
100. Qiu RG, et al. An essential role for Rac in Ras transformation. *Nature*. 1995;374(6521):457–459.
101. Fransson A, Ruusala A, Aspenstrom P. Atypical Rho GTPases have roles in mitochondrial homeostasis and apoptosis. *J Biol Chem*. 2003;278(8):6495–6502.
102. Guo X, et al. The GTPase dMiro is required for axonal transport of mitochondria to Drosophila synapses. *Neuron*. 2005;47(3):379–393.
103. Nemani N, et al. MIRO-1 determines mitochondrial shape transition upon GPCR activation and Ca(2+) stress. *Cell Rep*. 2018;23(4):1005–1019.
104. Kimmelman AC, White E. Autophagy and tumor metabolism. *Cell Metab*. 2017;25(5):1037–1043.
105. Chang JY, et al. Dysregulation of mitophagy in carcinogenesis and tumor progression. *Biochim Biophys Acta Bioenerg*. 2017;1858(8):633–640.
106. Fu H, et al. Wounding triggers MIRO-1 dependent mitochondrial fragmentation that accelerates epidermal wound closure through oxidative signaling. *Nat Commun*. 2020;11(1):1050.
107. Schuler MH, et al. Miro1-mediated mitochondrial positioning shapes intracellular energy gradients required for cell migration. *Mol Biol Cell*. 2017;28(16):2159–2169.
108. Hantschel O, Superti-Furga G. Regulation of the c-Abl and Bcr-Abl tyrosine kinases. *Nat Rev Mol Cell Biol*. 2004;5(1):33–44.
109. Colicelli J. ABL tyrosine kinases: evolution of function, regulation, and specificity. *Sci Signal*. 2010;3(139):re6.
110. Yokota A, et al. Constitutively active ABL family kinases, TEL/ABL and TEL/ARG, harbor distinct leukemogenic activities in vivo. *Leukemia*. 2017;31(12):2742–2751.
111. Raca G, et al. RCSD1-ABL2 fusion resulting from a complex chromosomal rearrangement in high-risk B-cell acute lymphoblastic leukemia. *Leuk Lymphoma*. 2015;56(4):1145–1147.
112. Keeshan K, Cotter TG, McKenna SL. High Bcr-Abl expression prevents the translocation of Bax and Bad to the mitochondrion. *Leukemia*. 2002;16(9):1725–1734.
113. Aras S, et al. Abl2 kinase phosphorylates Bi-organellar regulator MNRR1 in mitochondria, stimulating respiration. *Biochim Biophys Acta, Mol Cell Res*. 2017;1864(2):440–448.
114. Rikova K, et al. Global survey of phosphotyrosine signaling identifies oncogenic kinases in lung cancer. *Cell*. 2007;131(6):1190–1203.

115. Srinivasan D, Plattner R. Activation of Abl tyrosine kinases promotes invasion of aggressive breast cancer cells. *Cancer Res.* 2006;66(11):5648–5655.
116. Gil-Henn H, et al. Arg/Abl2 promotes invasion and attenuates proliferation of breast cancer in vivo. *Oncogene.* 2013;32(21):2622–2630.
117. Yeatman TJ. A renaissance for SRC. *Nat Rev Cancer.* 2004;4(6):470–480.
118. Djeungoue-Petga MA, et al. Intramitochondrial Src kinase links mitochondrial dysfunctions and aggressiveness of breast cancer cells. *Cell Death Dis.* 2019;10(12):940.
119. Hebert-Chatelain E, et al. Preservation of NADH ubiquinone-oxidoreductase activity by Src kinase-mediated phosphorylation of NDUFB10. *Biochim Biophys Acta.* 2012;1817(5):718–725.
120. Miyazaki T, et al. Regulation of cytochrome c oxidase activity by c-Src in osteoclasts. *J Cell Biol.* 2003;160(5):709–718.
121. Ogura M, et al. Mitochondrial c-Src regulates cell survival through phosphorylation of respiratory chain components. *Biochem J.* 2012;447(2):281–289.
122. Staal SP, Hartley JW, Rowe WP. Isolation of transforming murine leukemia viruses from mice with a high incidence of spontaneous lymphoma. *Proc Natl Acad Sci U S A.* 1977;74(7):3065–3067.
123. Staal SP. Molecular cloning of the akt oncogene and its human homologues AKT1 and AKT2: amplification of AKT1 in a primary human gastric adenocarcinoma. *Proc Natl Acad Sci U S A.* 1987;84(14):5034–5037.
124. Cheng JQ, et al. Transforming activity and mitosis-related expression of the AKT2 oncogene: evidence suggesting a link between cell cycle regulation and oncogenesis. *Oncogene.* 1997;14(23):2793–2801.
125. Cheng JQ, et al. Amplification of AKT2 in human pancreatic cells and inhibition of AKT2 expression and tumorigenicity by antisense RNA. *Proc Natl Acad Sci U S A.* 1996;93(8):3636–3641.
126. Aoki M, et al. The akt kinase: molecular determinants of oncogenicity. *Proc Natl Acad Sci U S A.* 1998;95(25):14950–14955.
127. Bijur GN, Jope RS. Rapid accumulation of Akt in mitochondria following phosphatidylinositol 3-kinase activation. *J Neurochem.* 2003;87(6):1427–1435.
128. Chae YC, et al. Mitochondrial Akt regulation of hypoxic tumor reprogramming. *Cancer Cell.* 2016;30(2):257–272.
129. Marchi S, et al. Akt-mediated phosphorylation of MICU1 regulates mitochondrial Ca(2+) levels and tumor growth. *EMBO J.* 2019;38(2).
130. Vidimar V, et al. Dysfunctional MnSOD leads to redox dysregulation and activation of prosurvival AKT signaling in uterine leiomyomas. *Sci Adv.* 2016;2(11), e1601132.
131. del Peso L, et al. Interleukin-3-induced phosphorylation of BAD through the protein kinase Akt. *Science.* 1997;278(5338):687–689.
132. Cuypers HT, et al. Murine leukemia virus-induced T-cell lymphomagenesis: integration of proviruses in a distinct chromosomal region. *Cell.* 1984;37(1):141–150.
133. Pasqualucci L, et al. Hypermutation of multiple proto-oncogenes in B-cell diffuse large-cell lymphomas. *Nature.* 2001;412(6844):341–346.
134. Chiang WF, et al. Up-regulation of a serine-threonine kinase proto-oncogene Pim-1 in oral squamous cell carcinoma. *Int J Oral Maxillofac Surg.* 2006;35(8):740–745.
135. Amson R, et al. The human protooncogene product p33pim is expressed during fetal hematopoiesis and in diverse leukemias. *Proc Natl Acad Sci U S A.* 1989;86(22):8857–8861.
136. Deutsch AJ, et al. MALT lymphoma and extranodal diffuse large B-cell lymphoma are targeted by aberrant somatic hypermutation. *Blood.* 2007;109(8):3500–3504.
137. Wang J, et al. Pim1 kinase synergizes with c-MYC to induce advanced prostate carcinoma. *Oncogene.* 2010;29(17):2477–2487.

138. Wang Z, et al. Phosphorylation of the cell cycle inhibitor p21Cip1/WAF1 by Pim-1 kinase. *Biochim Biophys Acta.* 2002;1593(1):45–55.
139. Aho TL, et al. Pim-1 kinase promotes inactivation of the pro-apoptotic Bad protein by phosphorylating it on the Ser112 gatekeeper site. *FEBS Lett.* 2004;571(1-3):43–49.
140. Lilly M, et al. The PIM-1 serine kinase prolongs survival and inhibits apoptosis-related mitochondrial dysfunction in part through a bcl-2-dependent pathway. *Oncogene.* 1999;18(27):4022–4031.
141. Wu Y, et al. Pim1 promotes cell proliferation and regulates glycolysis via interaction with MYC in ovarian cancer. *Onco Targets Ther.* 2018;11:6647–6656.
142. Peyssonnaux C, Eychene A. The Raf/MEK/ERK pathway: new concepts of activation. *Biol Cell.* 2001;93(1–2):53–62.
143. Rapp UR, et al. Structure and biological activity of v-raf, a unique oncogene transduced by a retrovirus. *Proc Natl Acad Sci U S A.* 1983;80(14):4218–4222.
144. Niault TS, Baccarini M. Targets of Raf in tumorigenesis. *Carcinogenesis.* 2010;31(7):1165–1174.
145. Emuss V, et al. Mutations of C-RAF are rare in human cancer because C-RAF has a low basal kinase activity compared with B-RAF. *Cancer Res.* 2005;65(21):9719–9726.
146. Pritchard CA, et al. Conditionally oncogenic forms of the A-Raf and B-Raf protein kinases display different biological and biochemical properties in NIH 3T3 cells. *Mol Cell Biol.* 1995;15(11):6430–6442.
147. Turski ML, et al. Genomically driven tumors and actionability across histologies: BRAF-mutant cancers as a paradigm. *Mol Cancer Ther.* 2016;15(4):533–547.
148. Lee MH, et al. Mitochondrial localization and regulation of BRAFV600E in thyroid cancer: a clinically used RAF inhibitor is unable to block the mitochondrial activities of BRAFV600E. *J Clin Endocrinol Metab.* 2011;96(1):E19–E30.
149. Yuryev A, et al. Isoform-specific localization of A-RAF in mitochondria. *Mol Cell Biol.* 2000;20(13):4870–4878.
150. Galmiche A, et al. Isoform-specific interaction of C-RAF with mitochondria. *J Biol Chem.* 2008;283(21):14857–14866.
151. Le Mellay V, et al. Regulation of glycolysis by Raf protein serine/threonine kinases. *Adv Enzym Regul.* 2002;42:317–332.
152. Hanahan D, Weinberg RA. The hallmarks of cancer. *Cell.* 2000;100(1):57–70.
153. Cory S, Adams JM. The Bcl2 family: regulators of the cellular life-or-death switch. *Nat Rev Cancer.* 2002;2(9):647–656.
154. Frenzel A, et al. Bcl2 family proteins in carcinogenesis and the treatment of cancer. *Apoptosis.* 2009;14(4):584–596.
155. Tsujimoto Y, et al. Involvement of the bcl-2 gene in human follicular lymphoma. *Science.* 1985;228(4706):1440–1443.
156. Vaux DL, Cory S, Adams JM. Bcl-2 gene promotes haemopoietic cell survival and cooperates with c-myc to immortalize pre-B cells. *Nature.* 1988;335(6189):440–442.
157. Sabin AB. Suppression of malignancy in human cancer cells: issues and challenges. *Proc Natl Acad Sci U S A.* 1981;78(11):7129–7133.
158. Lee EY, Muller WJ. Oncogenes and tumor suppressor genes. *Cold Spring Harb Perspect Biol.* 2010;2(10):a003236.
159. Zilfou JT, Lowe SW. Tumor suppressive functions of p53. *Cold Spring Harb Perspect Biol.* 2009;1(5):a001883.
160. Dyson NJ. RB1: a prototype tumor suppressor and an enigma. *Genes Dev.* 2016;30(13):1492–1502.
161. Sherr CJ. Principles of tumor suppression. *Cell.* 2004;116(2):235–246.
162. Rivlin N, et al. Mutations in the p53 tumor suppressor gene: important milestones at the various steps of tumorigenesis. *Genes Cancer.* 2011;2(4):466–474.

163. Mello SS, Attardi LD. Deciphering p53 signaling in tumor suppression. *Curr Opin Cell Biol.* 2018;51:65–72.
164. Humpton TJ, Vousden KH, et al. Regulation of cellular metabolism and hypoxia by p53. *Cold Spring Harb Perspect Med.* 2016;6(7).
165. Matoba S, et al. p53 regulates mitochondrial respiration. *Science.* 2006;312(5780):1650–1653.
166. Hu W, et al. Glutaminase 2, a novel p53 target gene regulating energy metabolism and antioxidant function. *Proc Natl Acad Sci U S A.* 2010;107(16):7455–7460.
167. Schwartzenberg-Bar-Yoseph F, Armoni M, Karnieli E. The tumor suppressor p53 down-regulates glucose transporters GLUT1 and GLUT4 gene expression. *Cancer Res.* 2004;64(7):2627–2633.
168. Bensaad K, et al. TIGAR, a p53-inducible regulator of glycolysis and apoptosis. *Cell.* 2006;126(1):107–120.
169. Zhang C, et al. Tumour-associated mutant p53 drives the Warburg effect. *Nat Commun.* 2013;4:2935.
170. Berkers CR, et al. Metabolic regulation by p53 family members. *Cell Metab.* 2013;18(5):617–633.
171. Freed-Pastor WA, et al. Mutant p53 disrupts mammary tissue architecture via the mevalonate pathway. *Cell.* 2012;148(1–2):244–258.
172. Williams AB, Schumacher B. p53 in the DNA-damage-repair process. *Cold Spring Harb Perspect Med.* 2016;6(5).
173. Wong TS, et al. Physical and functional interactions between human mitochondrial single-stranded DNA-binding protein and tumour suppressor p53. *Nucleic Acids Res.* 2009;37(2):568–581.
174. Kenzelmann Broz D, et al. Global genomic profiling reveals an extensive p53-regulated autophagy program contributing to key p53 responses. *Genes Dev.* 2013;27(9):1016–1031.
175. Wai T, Langer T. Mitochondrial dynamics and metabolic regulation. *Trends Endocrinol Metab.* 2016;27(2):105–117.
176. Youle RJ, van der Bliek AM. Mitochondrial fission, fusion, and stress. *Science.* 2012;337(6098):1062–1065.
177. Wang W, et al. Mitofusin-2 is a novel direct target of p53. *Biochem Biophys Res Commun.* 2010;400(4):587–592.
178. Zhou S, Kachhap S, Singh KK. Mitochondrial impairment in p53-deficient human cancer cells. *Mutagenesis.* 2003;18(3):287–292.
179. Chellappan SP, et al. The E2F transcription factor is a cellular target for the RB protein. *Cell.* 1991;65(6):1053–1061.
180. Bagchi S, Weinmann R, Raychaudhuri P. The retinoblastoma protein copurifies with E2F-I, an E1A-regulated inhibitor of the transcription factor E2F. *Cell.* 1991;65(6):1063–1072.
181. Chittenden T, Livingston DM, Kaelin Jr WG. The T/E1A-binding domain of the retinoblastoma product can interact selectively with a sequence-specific DNA-binding protein. *Cell.* 1991;65(6):1073–1082.
182. Giacinti C, Giordano A. RB and cell cycle progression. *Oncogene.* 2006;25(38):5220–5227.
183. Hilgendorf KI, et al. The retinoblastoma protein induces apoptosis directly at the mitochondria. *Genes Dev.* 2013;27(9):1003–1015.
184. Collard TJ, et al. The retinoblastoma protein (Rb) as an anti-apoptotic factor: expression of Rb is required for the anti-apoptotic function of BAG-1 protein in colorectal tumour cells. *Cell Death Dis.* 2012;3, e408.
185. Nicolay BN, et al. Proteomic analysis of pRb loss highlights a signature of decreased mitochondrial oxidative phosphorylation. *Genes Dev.* 2015;29(17):1875–1889.

186. Zacksenhaus E, et al. Mitochondrial OXPHOS induced by RB1 deficiency in breast cancer: implications for anabolic metabolism, stemness, and metastasis. *Trends Cancer.* 2017;3(11):768–779.
187. Reynolds MR, et al. Control of glutamine metabolism by the tumor suppressor Rb. *Oncogene.* 2014;33(5):556–566.
188. Nicolay BN, et al. Loss of RBF1 changes glutamine catabolism. *Genes Dev.* 2013;27(2):182–196.
189. Muranaka H, et al. A distinct function of the retinoblastoma protein in the control of lipid composition identified by lipidomic profiling. *Oncogenesis.* 2017;6(6), e350.
190. Macleod KF. The role of the RB tumour suppressor pathway in oxidative stress responses in the haematopoietic system. *Nat Rev Cancer.* 2008;8(10):769–781.
191. Pollard PJ, et al. Accumulation of Krebs cycle intermediates and over-expression of HIF1alpha in tumours which result from germline FH and SDH mutations. *Hum Mol Genet.* 2005;14(15):2231–2239.
192. Bardella C, et al. Cells lacking the fumarase tumor suppressor are protected from apoptosis through a hypoxia-inducible factor-independent, AMPK-dependent mechanism. *Mol Cell Biol.* 2012;32(15):3081–3094.
193. Tong WH, et al. The glycolytic shift in fumarate-hydratase-deficient kidney cancer lowers AMPK levels, increases anabolic propensities and lowers cellular iron levels. *Cancer Cell.* 2011;20(3):315–327.
194. Scott RE, Wille Jr JJ, Wier ML. Mechanisms for the initiation and promotion of carcinogenesis: a review and a new concept. *Mayo Clin Proc.* 1984;59(2):107–117.
195. Dvorak HF. Tumors: wounds that do not heal. Similarities between tumor stroma generation and wound healing. *N Engl J Med.* 1986;315(26):1650–1659.
196. Colotta F, et al. Cancer-related inflammation, the seventh hallmark of cancer: links to genetic instability. *Carcinogenesis.* 2009;30(7):1073–1081.
197. Coussens LM, et al. Inflammatory mast cells up-regulate angiogenesis during squamous epithelial carcinogenesis. *Genes Dev.* 1999;13(11):1382–1397.
198. Hudson JD, et al. A proinflammatory cytokine inhibits p53 tumor suppressor activity. *J Exp Med.* 1999;190(10):1375–1382.
199. Dougan M, Dranoff G. Inciting inflammation: the RAGE about tumor promotion. *J Exp Med.* 2008;205(2):267–270.
200. Grivennikov SI, Greten FR, Karin M. Immunity, inflammation, and cancer. *Cell.* 2010;140(6):883–899.
201. Troll W, Wiesner R. The role of oxygen radicals as a possible mechanism of tumor promotion. *Annu Rev Pharmacol Toxicol.* 1985;25:509–528.
202. Guerra C, et al. Chronic pancreatitis is essential for induction of pancreatic ductal adenocarcinoma by K-Ras oncogenes in adult mice. *Cancer Cell.* 2007;11(3):291–302.
203. Sieweke MH, et al. Mediation of wound-related Rous sarcoma virus tumorigenesis by TGF-beta. *Science.* 1990;248(4963):1656–1660.
204. Yu H, Pardoll D, Jove R. STATs in cancer inflammation and immunity: a leading role for STAT3. *Nat Rev Cancer.* 2009;9(11):798–809.
205. Lewis CE, Pollard JW. Distinct role of macrophages in different tumor microenvironments. *Cancer Res.* 2006;66(2):605–612.
206. Program NT. *14th report on carcinogens. Research Triangle Park: National Toxicology Program, US DUSEPArtment of Health and Human Services*; 2019. June; 2016. Available at URL http://ntp.niehs.nih.gov/ntp/roc/twelfth/roc12.pdf.
207. Takahashi H, et al. Tobacco smoke promotes lung tumorigenesis by triggering IKKbeta- and JNK1-dependent inflammation. *Cancer Cell.* 2010;17(1):89–97.
208. Kamp DW. Asbestos-induced lung diseases: an update. *Transl Res.* 2009;153(4):143–152.
209. Park EJ, et al. Dietary and genetic obesity promote liver inflammation and tumorigenesis by enhancing IL-6 and TNF expression. *Cell.* 2010;140(2):197–208.

210. Khasawneh J, et al. Inflammation and mitochondrial fatty acid beta-oxidation link obesity to early tumor promotion. *Proc Natl Acad Sci U S A*. 2009;106(9):3354–3359.
211. Scaffidi P, Misteli T, Bianchi ME. Release of chromatin protein HMGB1 by necrotic cells triggers inflammation. *Nature*. 2002;418(6894):191–195.
212. Brunetto E, et al. The IL-1/IL-1 receptor axis and tumor cell released inflammasome adaptor ASC are key regulators of TSLP secretion by cancer associated fibroblasts in pancreatic cancer. *J Immunother Cancer*. 2019;7(1):45.
213. Vakkila J, Lotze MT. Inflammation and necrosis promote tumour growth. *Nat Rev Immunol*. 2004;4(8):641–648.
214. He S, et al. HMGB1 released by irradiated tumor cells promotes living tumor cell proliferation via paracrine effect. *Cell Death Dis*. 2018;9(6):648.
215. Sakurai T, et al. Hepatocyte necrosis induced by oxidative stress and IL-1 alpha release mediate carcinogen-induced compensatory proliferation and liver tumorigenesis. *Cancer Cell*. 2008;14(2):156–165.
216. Kim S, et al. Carcinoma-produced factors activate myeloid cells through TLR2 to stimulate metastasis. *Nature*. 2009;457(7225):102–106.
217. Huang B, et al. Listeria monocytogenes promotes tumor growth via tumor cell toll-like receptor 2 signaling. *Cancer Res*. 2007;67(9):4346–4352.
218. Zhang XY, Zhang PY, Aboul-Soud MA. From inflammation to gastric cancer: role of Helicobacter pylori. *Oncol Lett*. 2017;13(2):543–548.
219. Echizen K, et al. Inflammation in gastric cancer: Interplay of the COX-2/prostaglandin E2 and Toll-like receptor/MyD88 pathways. *Cancer Sci*. 2016;107(4):391–397.
220. Wu S, et al. A human colonic commensal promotes colon tumorigenesis via activation of T helper type 17 T cell responses. *Nat Med*. 2009;15(9):1016–1022.
221. de Oliveria Andrade LJ, et al. Association between hepatitis C and hepatocellular carcinoma. *J Global Infect Dis*. 2009;1(1):33–37.
222. Toomer KH, Chen Z. Autoimmunity as a double agent in tumor killing and cancer promotion. *Front Immunol*. 2014;5:116.
223. Franks AL, Slansky JE. Multiple associations between a broad spectrum of autoimmune diseases, chronic inflammatory diseases and cancer. *Anticancer Res*. 2012;32(4):1119–1136.
224. Yu X, et al. FGF19 sustains the high proliferative ability of keratinocytes in psoriasis through the regulation of Wnt/GSK-3beta/beta-catenin signalling via FGFR4. *Clin Exp Pharmacol Physiol*. 2019;46(8):761–769.
225. Gartung A, et al. Suppression of chemotherapy-induced cytokine/lipid mediator surge and ovarian cancer by a dual COX-2/sEH inhibitor. *Proc Natl Acad Sci U S A*. 2019;116(5):1698–1703.
226. Kozin SV, et al. Recruitment of myeloid but not endothelial precursor cells facilitates tumor regrowth after local irradiation. *Cancer Res*. 2010;70(14):5679–5685.
227. Chen FH, et al. Radiotherapy decreases vascular density and causes hypoxia with macrophage aggregation in TRAMP-C1 prostate tumors. *Clin Cancer Res*. 2009;15(5):1721–1729.
228. de Visser KE, Eichten A, Coussens LM. Paradoxical roles of the immune system during cancer development. *Nat Rev Cancer*. 2006;6(1):24–37.
229. Swann JB, et al. Demonstration of inflammation-induced cancer and cancer immunoediting during primary tumorigenesis. *Proc Natl Acad Sci U S A*. 2008;105(2):652–656.
230. Balkwill F. Tumour necrosis factor and cancer. *Nat Rev Cancer*. 2009;9(5):361–371.
231. Leibovich SJ, et al. Macrophage-induced angiogenesis is mediated by tumour necrosis factor-alpha. *Nature*. 1987;329(6140):630–632.
232. Kastl L, et al. TNF-alpha mediates mitochondrial uncoupling and enhances ROS-dependent cell migration via NF-kappaB activation in liver cells. *FEBS Lett*. 2014;588(1):175–183.

233. Suematsu N, et al. Oxidative stress mediates tumor necrosis factor-alpha-induced mitochondrial DNA damage and dysfunction in cardiac myocytes. *Circulation*. 2003;107(10):1418–1423.
234. Doll DN, et al. Rapid mitochondrial dysfunction mediates TNF-alpha-induced neurotoxicity. *J Neurochem*. 2015;132(4):443–451.
235. Bell C, et al. Quantitative proteomics reveals the induction of mitophagy in tumor necrosis factor-alpha-activated (TNFalpha) macrophages. *Mol Cell Proteomics*. 2013;12(9):2394–2407.
236. Jung J, et al. Mitochondrial NIX promotes tumor survival in the hypoxic niche of glioblastoma. *Cancer Res*. 2019;79(20):5218–5232.
237. Bernardini JP, Lazarou M, Dewson G. Parkin and mitophagy in cancer. *Oncogene*. 2017;36(10):1315–1327.
238. Mantovani A, Barajon I, Garlanda C. IL-1 and IL-1 regulatory pathways in cancer progression and therapy. *Immunol Rev*. 2018;281(1):57–61.
239. Litmanovich A, Khazim K, Cohen I. The role of interleukin-1 in the pathogenesis of cancer and its potential as a therapeutic target in clinical practice. *Oncol Ther*. 2018;6(2):109–127.
240. Kaplanov I, et al. Blocking IL-1beta reverses the immunosuppression in mouse breast cancer and synergizes with anti-PD-1 for tumor abrogation. *Proc Natl Acad Sci U S A*. 2019;116(4):1361–1369.
241. Zhou H, et al. IL-1 induces mitochondrial translocation of IRAK2 to suppress oxidative metabolism in adipocytes. *Nat Immunol*. 2020;21(10):1219–1231.
242. Hollinshead KER, et al. Respiratory supercomplexes promote mitochondrial efficiency and growth in severely hypoxic pancreatic cancer. *Cell Rep*. 2020;33(1):108231.
243. Wang Q, et al. Mitochondrial function is a critical determinant of IL-1-induced ERK activation. *FASEB J*. 2005;19(7):837–839.
244. Zell R, et al. TNF-alpha and IL-1 alpha inhibit both pyruvate dehydrogenase activity and mitochondrial function in cardiomyocytes: evidence for primary impairment of mitochondrial function. *Mol Cell Biochem*. 1997;177(1–2):61–67.
245. Heinrich PC, et al. Interleukin-6-type cytokine signalling through the gp130/Jak/STAT pathway. *Biochem J*. 1998;334(Pt 2):297–314.
246. Silva EM, et al. High systemic IL-6 is associated with worse prognosis in patients with non-small cell lung cancer. *PLoS One*. 2017;12(7), e0181125.
247. Tian G, et al. Circulating interleukin-6 and cancer: a meta-analysis using Mendelian randomization. *Sci Rep*. 2015;5:11394.
248. Gao SP, et al. Mutations in the EGFR kinase domain mediate STAT3 activation via IL-6 production in human lung adenocarcinomas. *J Clin Invest*. 2007;117(12):3846–3856.
249. Sansone P, et al. IL-6 triggers malignant features in mammospheres from human ductal breast carcinoma and normal mammary gland. *J Clin Invest*. 2007;117(12):3988–4002.
250. Kumari N, et al. Role of interleukin-6 in cancer progression and therapeutic resistance. *Tumour Biol*. 2016;37(9):11553–11572.
251. Ji C, et al. IL-6 induces lipolysis and mitochondrial dysfunction, but does not affect insulin-mediated glucose transport in 3T3-L1 adipocytes. *J Bioenerg Biomembr*. 2011;43(4):367–375.
252. Buoncervello M, et al. Inflammatory cytokines associated with cancer growth induce mitochondria and cytoskeleton alterations in cardiomyocytes. *J Cell Physiol*. 2019;234(11):20453–20468.
253. Forcina L, et al. Increased circulating levels of interleukin-6 affect the redox balance in skeletal muscle. *Oxidative Med Cell Longev*. 2019;2019:3018584.
254. Marasco MR, et al. Interleukin-6 reduces beta-cell oxidative stress by linking autophagy with the antioxidant response. *Diabetes*. 2018;67(8):1576–1588.

255. Nowell PC. Mechanisms of tumor progression. *Cancer Res.* 1986;46(5):2203–2207.
256. Williams GM. Mechanisms of chemical carcinogenesis and application to human cancer risk assessment. *Toxicology.* 2001;166(1–2):3–10.
257. Kerr EM, et al. Mutant Kras copy number defines metabolic reprogramming and therapeutic susceptibilities. *Nature.* 2016;531(7592):110–113.
258. Ames BN, Gold LS. Chemical carcinogenesis: too many rodent carcinogens. *Proc Natl Acad Sci U S A.* 1990;87(19):7772–7776.
259. Frisch SM, Francis H. Disruption of epithelial cell-matrix interactions induces apoptosis. *J Cell Biol.* 1994;124(4):619–626.
260. Massague J, Obenauf AC. Metastatic colonization by circulating tumour cells. *Nature.* 2016;529(7586):298–306.
261. Chen J, et al. Gain of glucose-independent growth upon metastasis of breast cancer cells to the brain. *Cancer Res.* 2015;75(3):554–565.
262. LeBleu VS, et al. PGC-1alpha mediates mitochondrial biogenesis and oxidative phosphorylation in cancer cells to promote metastasis. *Nat Cell Biol.* 2014;16(10):992–1003. 1–15.
263. Porporato PE, et al. A mitochondrial switch promotes tumor metastasis. *Cell Rep.* 2014;8(3):754–766.
264. Jiang L, et al. Metabolic reprogramming during TGFbeta1-induced epithelial-to-mesenchymal transition. *Oncogene.* 2015;34(30):3908–3916.
265. Kumar S, et al. Intra-tumoral metabolic zonation and resultant phenotypic diversification are dictated by blood vessel proximity. *Cell Metab.* 2019;30(1):201–211 e6.
266. Lamb R, et al. Mitochondria as new therapeutic targets for eradicating cancer stem cells: quantitative proteomics and functional validation via MCT1/2 inhibition. *Oncotarget.* 2014;5(22):11029–11037.
267. Dong LF, et al. Horizontal transfer of whole mitochondria restores tumorigenic potential in mitochondrial DNA-deficient cancer cells. *eLife.* 2017;6.
268. Piskounova E, et al. Oxidative stress inhibits distant metastasis by human melanoma cells. *Nature.* 2015;527(7577):186–191.
269. Le Gal K, et al. Antioxidants can increase melanoma metastasis in mice. *Sci Transl Med.* 2015;7(308):308re8.
270. Kamarajugadda S, et al. Manganese superoxide dismutase promotes anoikis resistance and tumor metastasis. *Cell Death Dis.* 2013;4, e504.
271. Qu Y, et al. Thioredoxin-like 2 regulates human cancer cell growth and metastasis via redox homeostasis and NF-kappaB signaling. *J Clin Invest.* 2011;121(1):212–225.
272. Lucas S, et al. Serine catabolism is essential to maintain mitochondrial respiration in mammalian cells. *Life Sci Alliance.* 2018;1(2), e201800036.
273. Ye J, et al. Serine catabolism regulates mitochondrial redox control during hypoxia. *Cancer Discov.* 2014;4(12):1406–1417.
274. Locasale JW. Serine, glycine and one-carbon units: cancer metabolism in full circle. *Nat Rev Cancer.* 2013;13(8):572–583.
275. Thiery JP, et al. Epithelial-mesenchymal transitions in development and disease. *Cell.* 2009;139(5):871–890.
276. Chen S, et al. Conversion of epithelial-to-mesenchymal transition to mesenchymal-to-epithelial transition is mediated by oxygen concentration in pancreatic cancer cells. *Oncol Lett.* 2018;15(5):7144–7152.
277. Fischer KR, et al. Epithelial-to-mesenchymal transition is not required for lung metastasis but contributes to chemoresistance. *Nature.* 2015;527(7579):472–476.
278. Ye X, Weinberg RA. Epithelial-mesenchymal plasticity: a central regulator of cancer progression. *Trends Cell Biol.* 2015;25(11):675–686.
279. Schild T, et al. Unique metabolic adaptations dictate distal organ-specific metastatic colonization. *Cancer Cell.* 2018;33(3):347–354.

280. McKeown SR. Defining normoxia, physoxia and hypoxia in tumours-implications for treatment response. *Br J Radiol.* 2014;87(1035):20130676.
281. Mergenthaler P, et al. Sugar for the brain: the role of glucose in physiological and pathological brain function. *Trends Neurosci.* 2013;36(10):587–597.
282. Hamberger A, et al. Amino acids in the neuronal microenvironment of focal human epileptic lesions. *Epilepsy Res.* 1991;9(1):32–43.
283. Shelton LM, Huysentruyt LC, Seyfried TN. Glutamine targeting inhibits systemic metastasis in the VM-M3 murine tumor model. *Int J Cancer.* 2010;127(10):2478–2485.
284. Yang C, et al. Glutamine oxidation maintains the TCA cycle and cell survival during impaired mitochondrial pyruvate transport. *Mol Cell.* 2014;56(3):414–424.
285. Dupuy F, et al. PDK1-dependent metabolic reprogramming dictates metastatic potential in breast cancer. *Cell Metab.* 2015;22(4):577–589.
286. Pond CM, Mattacks CA. The source of fatty acids incorporated into proliferating lymphoid cells in immune-stimulated lymph nodes. *Br J Nutr.* 2003;89(3):375–383.
287. Lee CK, et al. Tumor metastasis to lymph nodes requires YAP-dependent metabolic adaptation. *Science.* 2019;363(6427):644–649.
288. Shay JW, Werbin H. Are mitochondrial DNA mutations involved in the carcinogenic process? *Mutat Res.* 1987;186(2):149–160.
289. Petros JA, et al. mtDNA mutations increase tumorigenicity in prostate cancer. *Proc Natl Acad Sci U S A.* 2005;102(3):719–724.
290. Ishikawa K, et al. ROS-generating mitochondrial DNA mutations can regulate tumor cell metastasis. *Science.* 2008;320(5876):661–664.
291. Kantorovich V, King KS, Pacak K. SDH-related pheochromocytoma and paraganglioma. *Best Pract Res Clin Endocrinol Metab.* 2010;24(3):415–424.
292. Briere JJ, et al. Mitochondrial succinate is instrumental for HIF1alpha nuclear translocation in SDHA-mutant fibroblasts under normoxic conditions. *Hum Mol Genet.* 2005;14(21):3263–3269.
293. Guzy RD, et al. Loss of the SdhB, but Not the SdhA, subunit of complex II triggers reactive oxygen species-dependent hypoxia-inducible factor activation and tumorigenesis. *Mol Cell Biol.* 2008;28(2):718–731.
294. Yan H, et al. IDH1 and IDH2 mutations in gliomas. *N Engl J Med.* 2009;360(8):765–773.
295. Amary MF, et al. IDH1 and IDH2 mutations are frequent events in central chondrosarcoma and central and periosteal chondromas but not in other mesenchymal tumours. *J Pathol.* 2011;224(3):334–343.
296. Borger DR, et al. Frequent mutation of isocitrate dehydrogenase (IDH)1 and IDH2 in cholangiocarcinoma identified through broad-based tumor genotyping. *Oncologist.* 2012;17(1):72–79.
297. Stein EM. Molecular pathways: IDH2 mutations-co-opting cellular metabolism for malignant transformation. *Clin Cancer Res.* 2016;22(1):16–19.
298. Xu W, et al. Oncometabolite 2-hydroxyglutarate is a competitive inhibitor of alpha-ketoglutarate-dependent dioxygenases. *Cancer Cell.* 2011;19(1):17–30.
299. Agledal L, Niere M, Ziegler M. The phosphate makes a difference: cellular functions of NADP. *Redox Rep.* 2010;15(1):2–10.
300. Gelman SJ, et al. Consumption of NADPH for 2-HG synthesis increases pentose phosphate pathway flux and sensitizes cells to oxidative stress. *Cell Rep.* 2018;22(2):512–522.
301. Tomlinson IP, et al. Germline mutations in FH predispose to dominantly inherited uterine fibroids, skin leiomyomata and papillary renal cell cancer. *Nat Genet.* 2002;30(4):406–410.
302. Yang M, et al. The succinated proteome of FH-mutant tumours. *Metabolites.* 2014;4(3):640–654.

303. Xiao M, et al. Corrigendum: inhibition of alpha-KG-dependent histone and DNA demethylases by fumarate and succinate that are accumulated in mutations of FH and SDH tumor suppressors. *Genes Dev.* 2015;29(8):887.
304. Isaacs JS, et al. HIF overexpression correlates with biallelic loss of fumarate hydratase in renal cancer: novel role of fumarate in regulation of HIF stability. *Cancer Cell.* 2005;8(2):143–153.
305. Nunnari J, Suomalainen A. Mitochondria: in sickness and in health. *Cell.* 2012;148(6):1145–1159.
306. Islinger M, Schrader M. Peroxisomes. *Curr Biol.* 2011;21(19):R800–R801.
307. Islinger M, et al. The peroxisome: an update on mysteries 2.0. *Histochem Cell Biol.* 2018;150(5):443–471.
308. Novikoff AB, Goldfischer S. Visualization of peroxisomes (microbodies) and mitochondria with diaminobenzidine. *J Histochem Cytochem.* 1969;17(10):675–680.
309. Shai N, et al. Systematic mapping of contact sites reveals tethers and a function for the peroxisome-mitochondria contact. *Nat Commun.* 2018;9(1):1761.
310. Sugiura A, et al. Newly born peroxisomes are a hybrid of mitochondrial and ER-derived pre-peroxisomes. *Nature.* 2017;542(7640):251–254.
311. Wanders RJ, Poll-The BT. Role of peroxisomes in human lipid metabolism and its importance for neurological development. *Neurosci Lett.* 2017;637:11–17.
312. Chen C, et al. Peroxisomal membrane contact sites in mammalian cells. *Front Cell Dev Biol.* 2020;8:512.
313. Walton PA, Pizzitelli M. Effects of peroxisomal catalase inhibition on mitochondrial function. *Front Physiol.* 2012;3:108.
314. Fransen M, et al. Role of peroxisomes in ROS/RNS-metabolism: implications for human disease. *Biochim Biophys Acta.* 2012;1822(9):1363–1373.
315. Walbrecq G, et al. Antioxidant cytoprotection by peroxisomal peroxiredoxin-5. *Free Radic Biol Med.* 2015;84:215–226.
316. Boveris A, Oshino N, Chance B. The cellular production of hydrogen peroxide. *Biochem J.* 1972;128(3):617–630.
317. Okumoto K, et al. The peroxisome counteracts oxidative stresses by suppressing catalase import via Pex14 phosphorylation. *elife.* 2020;9.
318. Sheikh FG, et al. Abnormality in catalase import into peroxisomes leads to severe neurological disorder. *Proc Natl Acad Sci U S A.* 1998;95(6):2961–2966.
319. Zhou M, et al. Alpha-methylacyl-CoA racemase: a novel tumor marker over-expressed in several human cancers and their precursor lesions. *Am J Surg Pathol.* 2002;26(7):926–931.
320. Benjamin DI, et al. Ether lipid generating enzyme AGPS alters the balance of structural and signaling lipids to fuel cancer pathogenicity. *Proc Natl Acad Sci U S A.* 2013;110(37):14912–14917.
321. Sugiura A, et al. A new pathway for mitochondrial quality control: mitochondrial-derived vesicles. *EMBO J.* 2014;33(19):2142–2156.
322. Chiarugi A, et al. The NAD metabolome—a key determinant of cancer cell biology. *Nat Rev Cancer.* 2012;12(11):741–752.
323. Liou GY, Storz P. Reactive oxygen species in cancer. *Free Radic Res.* 2010;44(5):479–496.
324. Kim JA. Peroxisome metabolism in cancer. *Cells.* 2020;9(7):1692.
325. Vance JE. Phospholipid synthesis in a membrane fraction associated with mitochondria. *J Biol Chem.* 1990;265(13):7248–7256.
326. Giorgi C, et al. Mitochondria-associated membranes: composition, molecular mechanisms, and physiopathological implications. *Antioxid Redox Signal.* 2015;22(12):995–1019.
327. Csordas G, et al. Structural and functional features and significance of the physical linkage between ER and mitochondria. *J Cell Biol.* 2006;174(7):915–921.

328. Annunziata I, Sano R, d'Azzo A. Mitochondria-associated ER membranes (MAMs) and lysosomal storage diseases. *Cell Death Dis.* 2018;9(3):328.
329. Csordas G, et al. Imaging interorganelle contacts and local calcium dynamics at the ER-mitochondrial interface. *Mol Cell.* 2010;39(1):121–132.
330. de Brito OM, Scorrano L. Mitofusin 2 tethers endoplasmic reticulum to mitochondria. *Nature.* 2008;456(7222):605–610.
331. Wu W, et al. FUNDC1 regulates mitochondrial dynamics at the ER-mitochondrial contact site under hypoxic conditions. *EMBO J.* 2016;35(13):1368–1384.
332. Gelmetti V, et al. PINK1 and BECN1 relocalize at mitochondria-associated membranes during mitophagy and promote ER-mitochondria tethering and autophagosome formation. *Autophagy.* 2017;13(4):654–669.
333. Ahn CS, Metallo CM. Mitochondria as biosynthetic factories for cancer proliferation. *Cancer Metab.* 2015;3(1):1.
334. Cardenas C, et al. Cancer cells with defective oxidative phosphorylation require endoplasmic reticulum-to-mitochondria Ca^{2+} transfer for survival. *Sci Signal.* 2020;13(640).
335. Kameoka S, et al. Phosphatidic acid and cardiolipin coordinate mitochondrial dynamics. *Trends Cell Biol.* 2018;28(1):67–76.
336. Vance JE. MAM (mitochondria-associated membranes) in mammalian cells: lipids and beyond. *Biochim Biophys Acta.* 2014;1841(4):595–609.
337. Hetz C. The unfolded protein response: controlling cell fate decisions under ER stress and beyond. *Nat Rev Mol Cell Biol.* 2012;13(2):89–102.
338. Li X, Zhang K, Li Z. Unfolded protein response in cancer: the physician's perspective. *J Hematol Oncol.* 2011;4:8.
339. Romero-Ramirez L, et al. XBP1 is essential for survival under hypoxic conditions and is required for tumor growth. *Cancer Res.* 2004;64(17):5943–5947.
340. Bravo R, et al. Increased ER-mitochondrial coupling promotes mitochondrial respiration and bioenergetics during early phases of ER stress. *J Cell Sci.* 2011;124(Pt 13):2143–2152.
341. Kaufman RJ. Orchestrating the unfolded protein response in health and disease. *J Clin Invest.* 2002;110(10):1389–1398.
342. Siwecka N, et al. Dual role of endoplasmic reticulum stress-mediated unfolded protein response signaling pathway in carcinogenesis. *Int J Mol Sci.* 2019;20(18):4354.
343. Aras S, et al. Mitochondrial Nuclear Retrograde Regulator 1 (MNRR1) rescues the cellular phenotype of MELAS by inducing homeostatic mechanisms. *Proc Natl Acad Sci U S A.* 2020;117(50):32056–32065.
344. Anderson S, et al. Sequence and organization of the human mitochondrial genome. *Nature.* 1981;290(5806):457–465.
345. Merry TL, et al. Mitochondrial-derived peptides in energy metabolism. *Am J Physiol Endocrinol Metab.* 2020;319(4):E659–E666.
346. Ng S, et al. Anterograde and retrograde regulation of nuclear genes encoding mitochondrial proteins during growth, development, and stress. *Mol Plant.* 2014;7(7):1075–1093.
347. Soledad RB, Charles S, Samarjit D. The secret messages between mitochondria and nucleus in muscle cell biology. *Arch Biochem Biophys.* 2019;666:52–62.
348. Salem AF, et al. Mitochondrial biogenesis in epithelial cancer cells promotes breast cancer tumor growth and confers autophagy resistance. *Cell Cycle.* 2012;11(22):4174–4180.
349. Torrens-Mas M, et al. Sirtuin 3 silencing impairs mitochondrial biogenesis and metabolism in colon cancer cells. *Am J Phys Cell Phys.* 2019;317(2):C398–C404.
350. Sotgia F, et al. Mitochondria "fuel" breast cancer metabolism: fifteen markers of mitochondrial biogenesis label epithelial cancer cells, but are excluded from adjacent stromal cells. *Cell Cycle.* 2012;11(23):4390–4401.

351. Lee CH, et al. Aberrant cell proliferation by enhanced mitochondrial biogenesis via mtTFA in arsenical skin cancers. *Am J Pathol.* 2011;178(5):2066–2076.
352. Humpton TJ, et al. Oncogenic KRAS induces NIX-mediated mitophagy to promote pancreatic cancer. *Cancer Discov.* 2019;9(9):1268–1287.
353. Li J, et al. The mitophagy effector FUNDC1 controls mitochondrial reprogramming and cellular plasticity in cancer cells. *Sci Signal.* 2020;13(642).
354. Liu L, et al. Mitochondrial outer-membrane protein FUNDC1 mediates hypoxia-induced mitophagy in mammalian cells. *Nat Cell Biol.* 2012;14(2):177–185.
355. Rehman J, et al. Inhibition of mitochondrial fission prevents cell cycle progression in lung cancer. *FASEB J.* 2012;26(5):2175–2186.
356. Kashatus JA, et al. Erk2 phosphorylation of Drp1 promotes mitochondrial fission and MAPK-driven tumor growth. *Mol Cell.* 2015;57(3):537–551.
357. Serasinghe MN, et al. Mitochondrial division is requisite to RAS-induced transformation and targeted by oncogenic MAPK pathway inhibitors. *Mol Cell.* 2015;57(3):521–536.
358. Qian W, Wang J, Van Houten B. The role of dynamin-related protein 1 in cancer growth: a promising therapeutic target? *Expert Opin Ther Targets.* 2013;17(9):997–1001.
359. Lima AR, et al. Dynamin-related protein 1 at the crossroads of cancer. *Genes (Basel).* 2018;9(2):115.
360. Altieri DC. Mitochondria on the move: emerging paradigms of organelle trafficking in tumour plasticity and metastasis. *Br J Cancer.* 2017;117(3):301–305.
361. Lin MY, Sheng ZH. Regulation of mitochondrial transport in neurons. *Exp Cell Res.* 2015;334(1):35–44.
362. Sheng ZH. The interplay of axonal energy homeostasis and mitochondrial trafficking and anchoring. *Trends Cell Biol.* 2017;27(6):403–416.
363. Caino MC, et al. PI3K therapy reprograms mitochondrial trafficking to fuel tumor cell invasion. *Proc Natl Acad Sci U S A.* 2015;112(28):8638–8643.
364. Parikh VS, et al. Interaction between the yeast mitochondrial and nuclear genomes influences the abundance of novel transcripts derived from the spacer region of the nuclear ribosomal DNA repeat. *Mol Cell Biol.* 1989;9(5):1897–1907.
365. Amuthan G, et al. Mitochondria-to-nucleus stress signaling induces phenotypic changes, tumor progression and cell invasion. *EMBO J.* 2001;20(8):1910–1920.
366. Balaban RS, Nemoto S, Finkel T. Mitochondria, oxidants, and aging. *Cell.* 2005;120(4):483–495.
367. Yakes FM, Van Houten B. Mitochondrial DNA damage is more extensive and persists longer than nuclear DNA damage in human cells following oxidative stress. *Proc Natl Acad Sci U S A.* 1997;94(2):514–519.
368. Formentini L, et al. The mitochondrial ATPase inhibitory factor 1 triggers a ROS-mediated retrograde prosurvival and proliferative response. *Mol Cell.* 2012;45(6):731–742.
369. Woo DK, et al. Mitochondrial genome instability and ROS enhance intestinal tumorigenesis in APC(Min/+) mice. *Am J Pathol.* 2012;180(1):24–31.
370. Hanna SC, et al. HIF1alpha and HIF2alpha independently activate SRC to promote melanoma metastases. *J Clin Invest.* 2013;123(5):2078–2093.
371. Zhao Q, et al. A mitochondrial specific stress response in mammalian cells. *EMBO J.* 2002;21(17):4411–4419.
372. Nargund AM, et al. Mitochondrial import efficiency of ATFS-1 regulates mitochondrial UPR activation. *Science.* 2012;337(6094):587–590.
373. Nargund AM, et al. Mitochondrial and nuclear accumulation of the transcription factor ATFS-1 promotes OXPHOS recovery during the UPR(mt). *Mol Cell.* 2015;58(1):123–133.

374. Li M, et al. ATF6 as a transcription activator of the endoplasmic reticulum stress element: thapsigargin stress-induced changes and synergistic interactions with NF-Y and YY1. *Mol Cell Biol*. 2000;20(14):5096–5106.
375. Haynes CM, Ron D. The mitochondrial UPR - protecting organelle protein homeostasis. *J Cell Sci*. 2010;123(Pt 22):3849–3855.
376. Ghosh JC, et al. Hsp60 regulation of tumor cell apoptosis. *J Biol Chem*. 2008;283(8):5188–5194.
377. Kang BH, et al. Regulation of tumor cell mitochondrial homeostasis by an organelle-specific Hsp90 chaperone network. *Cell*. 2007;131(2):257–270.
378. Kenny TC, et al. Selected mitochondrial DNA landscapes activate the SIRT3 axis of the UPR(mt) to promote metastasis. *Oncogene*. 2017;36(31):4393–4404.
379. Cao SS, Kaufman RJ. Endoplasmic reticulum stress and oxidative stress in cell fate decision and human disease. *Antioxid Redox Signal*. 2014;21(3):396–413.
380. Yao Y, et al. CHCHD2 promotes hepatocellular carcinoma and indicates poor prognosis of hepatocellular carcinoma patients. *J Cancer*. 2019;10(27):6822–6828.
381. Ma L, et al. CHCHD2 decreases docetaxel sensitivity in breast cancer via activating MMP2. *Eur Rev Med Pharmacol Sci*. 2020;24(11):6426–6433.
382. Lai X, et al. Identification of novel biomarker and therapeutic target candidates for diagnosis and treatment of follicular carcinoma. *J Proteome*. 2017;166:59–67.
383. Wei Y, et al. CHCHD2 is coamplified with EGFR in NSCLC and regulates mitochondrial function and cell migration. *Mol Cancer Res*. 2015;13(7):1119–1129.
384. Strickaert A, et al. Cancer heterogeneity is not compatible with one unique cancer cell metabolic map. *Oncogene*. 2017;36(19):2637–2642.
385. Hanahan D, Weinberg RA. Hallmarks of cancer: the next generation. *Cell*. 2011;144(5):646–674.
386. McMillin JM. Blood glucose. In: Walker H Kenneth, Hall W Dallas, Hurst J Willis, eds. *Clinical Methods: The History, Physical, and Laboratory Examinations*. Boston: Buttersworth; 1990.
387. Adekola K, Rosen ST, Shanmugam M. Glucose transporters in cancer metabolism. *Curr Opin Oncol*. 2012;24(6):650–654.
388. Macheda ML, Rogers S, Best JD. Molecular and cellular regulation of glucose transporter (GLUT) proteins in cancer. *J Cell Physiol*. 2005;202(3):654–662.
389. Zheng J. Energy metabolism of cancer: glycolysis versus oxidative phosphorylation (Review). *Oncol Lett*. 2012;4(6):1151–1157.
390. Zhang D, et al. Metabolic reprogramming of cancer-associated fibroblasts by IDH3alpha downregulation. *Cell Rep*. 2015;10(8):1335–1348.
391. Roy A, Bera S. CAF cellular glycolysis: linking cancer cells with the microenvironment. *Tumour Biol*. 2016;37(7):8503–8514.
392. Chiarugi P, Cirri P. Metabolic exchanges within tumor microenvironment. *Cancer Lett*. 2016;380(1):272–280.
393. Fiaschi T, et al. Reciprocal metabolic reprogramming through lactate shuttle coordinately influences tumor-stroma interplay. *Cancer Res*. 2012;72(19):5130–5140.
394. Whitaker-Menezes D, et al. Evidence for a stromal-epithelial "lactate shuttle" in human tumors: MCT4 is a marker of oxidative stress in cancer-associated fibroblasts. *Cell Cycle*. 2011;10(11):1772–1783.
395. Baltazar F, et al. Monocarboxylate transporters as targets and mediators in cancer therapy response. *Histol Histopathol*. 2014;29(12):1511–1524.
396. Fu Y, et al. The reverse Warburg effect is likely to be an Achilles' heel of cancer that can be exploited for cancer therapy. *Oncotarget*. 2017;8(34):57813–57825.
397. Patra KC, Hay N. The pentose phosphate pathway and cancer. *Trends Biochem Sci*. 2014;39(8):347–354.

398. Amelio I, et al. Serine and glycine metabolism in cancer. *Trends Biochem Sci.* 2014;39(4):191–198.
399. Roth E. Nonnutritive effects of glutamine. *J Nutr.* 2008;138(10):2025S–2031S.
400. Marin-Valencia I, et al. Analysis of tumor metabolism reveals mitochondrial glucose oxidation in genetically diverse human glioblastomas in the mouse brain in vivo. *Cell Metab.* 2012;15(6):827–837.
401. Bhutia YD, Ganapathy V. Glutamine transporters in mammalian cells and their functions in physiology and cancer. *Biochim Biophys Acta.* 2016;1863(10):2531–2539.
402. Cory JG, Cory AH. Critical roles of glutamine as nitrogen donors in purine and pyrimidine nucleotide synthesis: asparaginase treatment in childhood acute lymphoblastic leukemia. *In Vivo.* 2006;20(5):587–589.
403. Yang L, Venneti S, Nagrath D. Glutaminolysis: a hallmark of cancer metabolism. *Annu Rev Biomed Eng.* 2017;19:163–194.
404. DeBerardinis RJ, Cheng T. Q's next: the diverse functions of glutamine in metabolism, cell biology and cancer. *Oncogene.* 2010;29(3):313–324.
405. Sedlak TW, et al. The glutathione cycle shapes synaptic glutamate activity. *Proc Natl Acad Sci U S A.* 2019;116(7):2701–2706.
406. Bansal A, Simon MC. Glutathione metabolism in cancer progression and treatment resistance. *J Cell Biol.* 2018;217(7):2291–2298.
407. Traverso N, et al. Role of glutathione in cancer progression and chemoresistance. *Oxidative Med Cell Longev.* 2013;2013:972913.
408. Messina JP, Lawrence DA. Cell cycle progression of glutathione-depleted human peripheral blood mononuclear cells is inhibited at S phase. *J Immunol.* 1989;143(6):1974–1981.
409. Huang ZZ, et al. Mechanism and significance of increased glutathione level in human hepatocellular carcinoma and liver regeneration. *FASEB J.* 2001;15(1):19–21.
410. Carretero J, et al. Growth-associated changes in glutathione content correlate with liver metastatic activity of B16 melanoma cells. *Clin Exp Metastasis.* 1999;17(7):567–574.
411. Ortega A, et al. Tumoricidal activity of endothelium-derived NO and the survival of metastatic cells with high GSH and Bcl-2 levels. *Nitric Oxide.* 2008;19(2):107–114.
412. Kamphorst JJ, et al. Hypoxic and Ras-transformed cells support growth by scavenging unsaturated fatty acids from lysophospholipids. *Proc Natl Acad Sci U S A.* 2013;110(22):8882–8887.
413. Young RM, et al. Dysregulated mTORC1 renders cells critically dependent on desaturated lipids for survival under tumor-like stress. *Genes Dev.* 2013;27(10):1115–1131.
414. Currie E, et al. Cellular fatty acid metabolism and cancer. *Cell Metab.* 2013;18(2):153–161.
415. Swinnen JV, et al. Selective activation of the fatty acid synthesis pathway in human prostate cancer. *Int J Cancer.* 2000;88(2):176–179.
416. Yoon S, et al. Up-regulation of acetyl-CoA carboxylase alpha and fatty acid synthase by human epidermal growth factor receptor 2 at the translational level in breast cancer cells. *J Biol Chem.* 2007;282(36):26122–26131.
417. Menendez JA, Lupu R. Fatty acid synthase and the lipogenic phenotype in cancer pathogenesis. *Nat Rev Cancer.* 2007;7(10):763–777.
418. Ookhtens M, et al. Liver and adipose tissue contributions to newly formed fatty acids in an ascites tumor. *Am J Phys.* 1984;247(1 Pt 2):R146–R153.
419. Mowers EE, Sharifi MN, Macleod KF. Functions of autophagy in the tumor microenvironment and cancer metastasis. *FEBS J.* 2018;285(10):1751–1766.
420. Nieman KM, et al. Adipocytes promote ovarian cancer metastasis and provide energy for rapid tumor growth. *Nat Med.* 2011;17(11):1498–1503.
421. Gong J, et al. Reprogramming of lipid metabolism in cancer-associated fibroblasts potentiates migration of colorectal cancer cells. *Cell Death Dis.* 2020;11(4):267.

422. Viale A, et al. Oncogene ablation-resistant pancreatic cancer cells depend on mitochondrial function. *Nature.* 2014;514(7524):628–632.
423. Qu Q, et al. Fatty acid oxidation and carnitine palmitoyltransferase I: emerging therapeutic targets in cancer. *Cell Death Dis.* 2016;7, e2226.
424. Pike LS, et al. Inhibition of fatty acid oxidation by etomoxir impairs NADPH production and increases reactive oxygen species resulting in ATP depletion and cell death in human glioblastoma cells. *Biochim Biophys Acta.* 2011;1807(6):726–734.
425. Rabinowitz JD, White E. Autophagy and metabolism. *Science.* 2010;330(6009):1344–1348.
426. Zaugg K, et al. Carnitine palmitoyltransferase 1C promotes cell survival and tumor growth under conditions of metabolic stress. *Genes Dev.* 2011;25(10):1041–1051.
427. Flavin R, et al. Fatty acid synthase as a potential therapeutic target in cancer. *Future Oncol.* 2010;6(4):551–562.
428. Chajes V, et al. Acetyl-CoA carboxylase alpha is essential to breast cancer cell survival. *Cancer Res.* 2006;66(10):5287–5294.
429. Szatrowski TP, Nathan CF. Production of large amounts of hydrogen peroxide by human tumor cells. *Cancer Res.* 1991;51(3):794–798.
430. Ames BN, Shigenaga MK, Hagen TM. Oxidants, antioxidants, and the degenerative diseases of aging. *Proc Natl Acad Sci U S A.* 1993;90(17):7915–7922.
431. Trachootham D, et al. Selective killing of oncogenically transformed cells through a ROS-mediated mechanism by beta-phenylethyl isothiocyanate. *Cancer Cell.* 2006;10(3):241–252.
432. Sabharwal SS, Schumacker PT. Mitochondrial ROS in cancer: initiators, amplifiers or an Achilles' heel? *Nat Rev Cancer.* 2014;14(11):709–721.
433. Goncalves RL, et al. Sites of superoxide and hydrogen peroxide production by muscle mitochondria assessed ex vivo under conditions mimicking rest and exercise. *J Biol Chem.* 2015;290(1):209–227.
434. Franchini AM, et al. FcgammaR-driven release of IL-6 by macrophages requires NOX2-dependent production of reactive oxygen species. *J Biol Chem.* 2013;288(35):25098–25108.
435. Murphy MP. How mitochondria produce reactive oxygen species. *Biochem J.* 2009;417(1):1–13.
436. Muller F. The nature and mechanism of superoxide production by the electron transport chain: its relevance to aging. *J Am Aging Assoc.* 2000;23(4):227–253.
437. Buettner GR. Superoxide dismutase in redox biology: the roles of superoxide and hydrogen peroxide. *Anti Cancer Agents Med Chem.* 2011;11(4):341–346.
438. Mailloux RJ. Mitochondrial antioxidants and the maintenance of cellular hydrogen peroxide levels. *Oxidative Med Cell Longev.* 2018;2018:7857251.
439. Bai J, et al. Overexpression of catalase in cytosolic or mitochondrial compartment protects HepG2 cells against oxidative injury. *J Biol Chem.* 1999;274(37):26217–26224.
440. Puppo A, Halliwell B. Formation of hydroxyl radicals from hydrogen peroxide in the presence of iron. Is haemoglobin a biological Fenton reagent? *Biochem J.* 1988;249(1):185–190.
441. Florence TM. The production of hydroxyl radical from the reaction between hydrogen peroxide and NADH. *J Inorg Biochem.* 1986;28(1):33–37.
442. Ying J, et al. Thiol oxidation in signaling and response to stress: detection and quantification of physiological and pathophysiological thiol modifications. *Free Radic Biol Med.* 2007;43(8):1099–1108.
443. Mittler R. ROS Are Good. *Trends Plant Sci.* 2017;22(1):11–19.
444. Ziech D, et al. Reactive oxygen species (ROS) - induced genetic and epigenetic alterations in human carcinogenesis. *Mutat Res.* 2011;711(1–2):167–173.

445. Simon HU, Haj-Yehia A, Levi-Schaffer F. Role of reactive oxygen species (ROS) in apoptosis induction. *Apoptosis*. 2000;5(5):415–418.
446. Nulton-Persson AC, Szweda LI. Modulation of mitochondrial function by hydrogen peroxide. *J Biol Chem*. 2001;276(26):23357–23361.
447. Perillo B, et al. ROS in cancer therapy: the bright side of the moon. *Exp Mol Med*. 2020;52(2):192–203.

CHAPTER 7

Mitochondria and cancer therapy

Somayeh Handali[a], Mohsen Rezaei[b], and Jalal Pourahmad[c]
[a]Medical Biomaterial Research Center (MBRC), Tehran University of Medical Sciences, Tehran, Iran
[b]Department of Toxicology, Faculty of Medical Sciences, Tarbiat Modares University, Tehran, Iran
[c]Department of Pharmacology & Toxicology, School of Pharmacy, Shahid Beheshti University of Medical Sciences, Tehran, Iran

Contents

1 Introduction

Normal cells exploit more efficient oxidative phosphorylation (OXPHOS) pathway and generate 36 ATPs per glucose for their metabolic process, while cancer cells mostly metabolize glucose into lactate (Warburg effect) and produce two ATPs per glucose.[1] This pattern for production of ATP referred as aerobic glycolysis was initially highlighted by Otto Warburg in 1920s. He described that tumor slices even at normoxic condition consume glucose in such a way that overproduction of lactate is observed. Intensive experiments following the Warburg description indicated that some tumors are entirely dependent on aerobic glycolysis for their metabolic requirements, but they may also include reprogrammed mitochondrial metabolic pathways for providing TCA (tricarboxylic acid) cycle intermediates that are needed during their increased proliferation.

Normal cells undergo apoptosis induced by different stimuli that eventually lead to BAK/BAX activation and outer membrane (MOM) permeabilization (MOMP) with consequent cytochrome *c* release into the cytosol, apoptosome formation, and induction of apoptotic pathway. Changes in metabolism of cancer cells can be attained by gain or loss of function of oncogenes, including TP53, MYC, and HIF1α. Glucose is converted via glycolysis to lactate for generation of ATP.[2] In cancer cells, glycolysis is a suitable way for production of ATP, particularly in hypoxic conditions since it is more appropriate for growth and results in higher production of energy.

Mitochondrial Metabolism
https://doi.org/10.1016/B978-0-12-822416-8.00010-5

While less ATP is generated in glycolysis than OXPHOS, this metabolic pathway is faster and more appropriate for rapid proliferating of cancer cells.[1]

In this chapter, we will discuss the differences between cancer cells and healthy cells from mitochondrial perspective and evaluate the studies in which mitochondria have been targeted. For this purpose, mitochondria-related biochemical signaling pathways can be targeted in cancer cells using small molecules or based on the design of new mitochondriotropic formulations. Based on these approaches, cancer cells are more vulnerable to these new formulations. However, using either of these methods requires recognizing the important differences between mitochondria in normal and cancer cells. In the following section, we discuss the use of small molecules to target the mitochondria of cancer cells.

2 Small molecules interfering with mitochondrial signaling

In recent decade, many characteristics of reprogrammed mitochondrial metabolism have been revealed and consequently targeted by synthetic or natural small molecules. For this propose, reprogrammed aspects of mitochondrial metabolism in cancer cells, including electron transport chain (ETC), TCA, glutamine, and nucleotide metabolism, attracted more attention. In some studies, these mitochondrial metabolism inhibitors combined with other antitumor drugs resulted in improved efficacy and toxicity.

Iridium (III)-based complexes show high uptake and anticancer activity in tumor cells. Three iridium (III) complexes (Ir-1, Ir-2, and Ir-3) were synthesized and encapsulated in liposomes (Ir-1-Lipo, Ir-2-Lipo, and Ir-3-Lipo). These metal complexes loaded in liposomes exhibited more significant anticancer activity than free metal on A549 cells. Ir-1-Lipo, Ir-2-Lipo, and Ir-3-Lipo interacted with lysosomes, enhanced intracellular ROS level, reduced mitochondrial membrane potential, released cytochrome *c*, and arrested cell growth at G0/G1 phase in cancer cells. According to the obtained results, iridium (III) complexes loaded liposome-triggered apoptosis in cancer cells via ROS-mediated lysosomal and mitochondrial dysfunction pathway and microtubules disruption.[3]

Observed higher glycolytic rate in cancer cells provides higher energy and metabolic requirements demanded for rapid proliferation and migration. The activation of AKT (protein kinase B) leads to binding to mitochondria, enhancing transcription level of glucose transporter 1 (GLUT1) on the membrane cells, which increased the glucose entrances and aerobic glycolysis in cancer cells. Consequently, rapid glucose oxidation enhances the required ATP level and finally increases tumorigenesis and cell growth. It has been reported that ginsenoside Rh4 (from *Panax notoginseng*) inhibited esophageal cancer growth through targeting AKT, which leads to decreasing lactate dehydrogenase A (LDHA) and GLUT1 expression and inhibition of aerobic glycolysis.[4,5] Moreover, it was reported that FV-429 (a flavonoid extracted from Wogonin) inhibited glycolysis and induced apoptosis in prostate cancer cells via inhibition of the AR (an androgen-activated transcription factor)-AKT-HK2 (Hexokinase 2) signaling pathway. In glycolysis, HK2 is the rate-limiting enzyme that is involved in the high rate of glycolysis and is reportedly overexpressed in prostate cancer.[5]

Sodium selenite inhibited bladder cancer cell (RT-112/D21) growth through inducing superoxide production and oxidative stress, reducing mitochondrial membrane potential and ATP synthesis, and triggering necrotic, apoptotic, and autophagic cell death. These events were not observed in normal urothelial cells (UROtsa).[6]

It was reported that [1,2,4]-triazolo[1,5-a]pyrimidine derivatives as small-molecule anticancer agents displayed its anticancer activity on gastric cancer cells through damaging mitochondria and increasing ROS production, which led to autophagy and apoptosis.[7]

Aspirin shows anticancer activity by inhibition of cyclooxygenase. However, the clinical application of aspirin is restricted due to its high dosage requirement and side effects such as renal and gastrointestinal toxicity. NCX4040 (a nitric oxide-releasing aspirin) shows a protective effect on gastric mucosa and has potential anticancer activity. It was found that NCX4040 induced oxidative stress in prostate cancer cells with inhibition of mitochondrial ETC via the release of nitric oxide group and H_2O_2 and peroxynitrite radical generation. Increased H_2O_2 levels led to depolarization of mitochondrial membrane and cytochrome *c* release, resulting in apoptosis.[8]

Aspirin also interferes in the biosynthesis and transferring of acetyl-coenzyme A (acetyl-CoA) into the mitochondria. Acetyl-CoA is metabolically essential in eukaryotic organisms, since it is the major component of the TCA cycle activity needed for energy generation. As a model of cancer cells, manganese superoxide dismutase (MnSOD)-deficient *S. cerevisiae* EG110 cells were used. This mitochondrial manganese MnSOD deficiency reprogrammed the cells to be metabolically susceptible for uncontrolled proliferation. According to the results, aspirin inhibited the genes involved in the synthesis and transport of acetyl-CoA to the mitochondria.

These findings illustrated that mitochondria can be a promising therapeutic target for aspirin in cancer cells.[9]

B lymphocyte-induced maturation protein-1 (Blimp1) regulates the terminal differentiation of B lymphocytes into plasma cells and is found to be critical for development of multiple myeloma. Blimp1 also inhibits β-interferon secretion. It was observed that aspirin exhibits cytotoxic activity against multiple myeloma through downregulation of Blimp1 expression and triggering of mitochondrial and ERS pathway of apoptosis.[10]

Celecoxib (as another NSAID) also displayed cytotoxicity upon murine and human metastatic cancer cells by inhibiting of mitochondrial O_2 consumption, increasing ROS production, reducing $\Delta\Psi_m$, and inducing apoptosis.[11]

Oleanolic acid methyl ester (OAME) as mitocan induced cell death in concentration- and time-dependent manner in prostate cancer PC-3 cells. OAME triggered apoptotic pathway through collapse of $\Delta\Psi_m$, increased release of cytochrome *c*, as well as caspases 3/7 activity. Moreover, OAME induced apoptosis in vivo and blocked tumor angiogenesis.[12]

Irinotecan (CPT-11) is a derivative of camptothecin that is extensively used in the treatment of solid tumors. CPT-11 is a water-soluble prodrug and hydrolyzed by liver enzymes to the hydrophobic SN38. Tocopheryl oxyacetate (TOA) is a hydrophobic mitocan that targets the mitochondria by the production of ROS, which results in apoptosis. For increasing intratumoral delivery of SN38, it was encapsulated in NPs, and TOA was linked to SN38 to considerably enhance hydrophobicity, reduce its side effects, and increase NPs retention. It was observed that SN38-TOA-containing NPs increased antitumor efficacy in neuroblastoma xenograft model.[13]

Uncoupling of oxidation/phosphorylation contributed to enhanced cytotoxic activity of specific agents. It was observed that mitochondria-targeted 1,4-naphthoquinone (SkQN) is a potent prooxidant and cytotoxic agent. In addition to increasing the ROS production, SkQN facilitates the transfer of protons across the inner membrane via cyclic transport of endogenous free fatty acids and therefore uncoupling of oxidative phosphorylation can be occurred.[14]

Canagliflozin, as a member of gliflozins (sodium/glucose cotransport inhibitors), is used for the treatment of type 2 diabetes. Canagliflozin was observed to inhibit the mitochondrial ETC complex I and glutamate dehydrogenase (GDH). This combined inhibition of GDH and ETC prevented glutamine entry into the TCA cycle. These results indicated that canagliflozin by interfering with the above-mentioned critical metabolic pathways in glutamine-dependent human cells and altering the intracellular amino acid pool conferred appropriate tumor inhibiting inactivity. Moreover, it was found that canagliflozin treatment for 24 h did not reduce ATP levels in differentiated cells, probably due to insignificant effect on complex I and adequate compensatory glycolysis for ATP production. In cancer cells, canagliflozin inhibited the availability of complex I substrates pyruvate, fatty acids, and glutamine in cancer cells and induced necrotic pathway during proliferation; however, complex II and IV were not affected.[15]

Abnormal vasculature and mitochondrial respiration alteration in tumors contribute to lower oxygenation and hypoxia, which decrease the efficiency of radiation therapy. Increased tumoral oxygenation prior to radiation therapy considerably improves the eradication of tumor cells after radiation. Therefore mitochondrial respiration inhibition leads to considerable decrease for oxygen demand. It was observed that papaverine inhibited the mitochondrial complex I, enhanced oxygenation, and increased radiation response.[16]

Dihydroorotate dehydrogenase (DHODH) is located in the outer surface of inner mitochondrial membrane[17] and is involved in the de novo pyrimidine synthesis by catalyzing of ubiquinone-mediated conversion of dihydroorotate to orotate. DHODH activity is dependent on ubiquinone availability and the proper function of complex III and its inhibition have reportedly been effective in some different mouse models of cancer.[18] BAY 2402234 as inhibitor of DHODH attaches to the ubiquinone binding site of DHODH between the N-terminal helices, helix A and B and triggered differentiation in acute myeloid leukemia cells (THP-1 cells) and inhibited the proliferation of the cells with IC_{50} of 2.6 nM.[17]

Glutamine is involved in nucleotide, amino acid, and glutathione production for maintaining antioxidant capacity and is also used as fuel for TCA cycle. It was observed that the capability of cells to overwhelm glutamine deprivation is compensated by aspartate metabolism, which is provided by the aspartate/glutamate transporter (SLC1A3). In this condition, p53 as tumor suppressor increases the expression of SLC1A3. Upregulated expression of SLC1A3 is evident in tumor cells which make them more resistant to glutamine starvation. Downregulation of the SLC1A3 inhibits the cell growth. Therefore SLC1A3 inhibition can be a potential therapeutic target for cancer treatment.[19]

Under aerobic glycolysis, mitochondrial uncoupling protein 2 (UCP2) is highly expressed in cancer cells. It was shown that UCP2 is rapidly and reversibly downregulated in neuroblastoma cells under the glutamine deficiency, so that the metabolic activity and proliferation of these cells are reduced. These results showed a quick shift (within 1 h) in metabolism that permits the cells to recruit alternative pathway instead of glutamine consuming or entering a reversible more quiescent state. These findings indicate that UCP2 facilitates using of glutamine as an energetic fuel supply, consequently, providing metabolic flexibility during glucose deficiency. Therefore UCP2 targeting can be a novel approach for cancer therapy.[20]

3 Mitochondrially targeted formulations

MITO Porter (a liposomal delivery system to mitochondria) was developed for delivery of doxorubicin (DOX) to mitochondria in OS-RC-2 cells. Stearyl octaarginine (STR-R8), a cationic peptide was employed to modify the DOX-MITO-Porter for increasing intracellular accumulation and mitochondrial targeting. Moreover, DSPE-PEG 2000 was used for achieving stability of DOX-MITO-Porter in the plasma. DOX-MITO-Porter accumulated in tumor cells through enhanced permeability and retention (EPR) effect and introduced DOX into mitochondria via fusion of membrane, which results in cell death. In addition, tumor size in treated mice with DOX-MITO-Porter was significantly reduced compared with free DOX-received groups.[21]

A dual-targeting proapoptotic peptide (DTP) contained folic acid (FA) and triphenyl phosphonium (TPP) cation was applied for targeting of cancer cells mitochondria. FA was used for enhancing of proapoptotic peptide cellular uptake through receptor-mediated endocytosis and TPP cation as mitochondrial-targeting agent that delivers the proapoptotic peptide to mitochondria. It was observed that DTP decorated with folic acid and TPP cation led to loss of $\Delta\Psi_m$, release of cytochrome *c*, and enhanced expression of caspase-3 and caspase-9. These findings indicated that DTP induced cell death mainly via mitochondria-originated apoptosis pathway.[22]

It has been found that targeting mitochondrial DNA using DNA-damaging anticancer drugs is more suitable than nuclear DNA since the repair pathways of mitochondrial DNA are not as efficient as those of nuclear DNA. However, targeting mitochondrial DNA is difficult owing to thickness of outer and inner mitochondrial membranes, which prevent the entry of drugs. For overcoming these drawbacks, dual targeting and dual responsive polymer micelles were designed. In a study, DOX containing micelles papered by the self-assembly of TPP-grafted poly(ethylene glycol) (PEG)-poly (D,L-lactide) (PLA) copolymers (TPP-PEG-ss-PLA) using disulfide bonds and coated with chondroitin sulfate (CS) layer. CS layer prevents the removal of NPs by the reticuloendothelial system (RES) in the blood circulation, and as a result increases blood circulation time of NPs. The CS layer with its negative charge prevents the exposing of TPP in response to the lysosomes/endosomes acidic microenvironment so the NPs target and anchor into mitochondrial outer membrane. Consequently, mitochondrial membrane depolarizes, $\Delta \Psi_m$ collapses, and the permeability of mitochondria membrane enhances. NPs break down occurs through the initial removing of CS layer in acidic condition that permits for further cleavage of the disulfide bonds by glutathione (GSH) response. The concentration of GSH in the cytoplasm of tumor cells is higher than normal cells. Mitochondrial targeting through TPP resulted in reduced $\Delta \Psi_m$, overproduction of the ROS, and drug release (by GSH response) into the mitochondria, and consequently apoptosis.[23] In a similar study conducted by Tan et al., a mitochondria-targeted stimulus-responsive micelle was developed for codelivery of DOX and evodiamine (EVO). EVO (a plant alkaloid) can disrupt the mitochondrial membrane, enhance the permeability of mitochondrial membrane, resulting in drug uptake into the mitochondria. These micelles were composed of TPP, Dex (dextran), and SPCL (starch-polycaprolactone) (TPP-Dex-ss-SPCL). Dex, as a hydrophilic agent, gives longer blood circulation and durability for micelles. According to the results, GSH cleaved the disulfide bonds in the micelles which led to release of DOX and EVO. Lowered $\Delta \Psi_m$ by EVO led to high DOX amount entering into the mitochondria and subsequently damaging mitochondrial DNA.[24]

Dequalinium (DQA), a lipophilic cation that is frequently used for mitochondria-targeted formulation, highly accumulates inside the mitochondria. Previous studies have reported that lipophilic cations decreased the mitochondrial membrane potential (MMP) by accumulating inside the mitochondria. In a study, mitochondria-targeted glycol chitosan (GC)-DQA NPs containing curcumin were developed as anticancer formulation. Curcumin can disturb the MMP, decrease energy production, and trigger apoptosis in cancer cells. Hydrophobic property of curcumin limits its therapeutic applications. Several approaches have been employed for overcoming this problem including encapsulating in NPs. The GC structure increases cell internalization and consequently endosomal escape avoiding any considerable toxicity. Results showed that curcumin containing GC-DQA NPs exhibited higher cytotoxicity than free curcumin on HeLa cells. It was observed that GC-DQA-curcumin led to depolarization of the MMP. These findings indicated that there is a synergistic effect between the mitochondria-targeting NPs in combination with curcumin.[25]

Guanidinium derivative is also used as mitochondria-targeting ligand. The polycationic guanidinium-rich NPs form hydrogen bonding with phosphate groups on the surface of the cell to make a complex. This complex formation neutralizes the charge and consequently decreases hydrophilicity of the guanidinium groups so that the nonpolar complex can easily cross the biological membrane. In a research, a guanidinium derivative (GA) conjugated to Fe_3O_4 core-mesoporous silica NP shell (Fe_3O_4@ MSNs) was employed for mitochondria targeting and delivery of DOX. MSNs have been widely employed as drug delivery due to biocompatibility, simple synthesis, and controllable size and pore volume. Magnetic properties of Fe_3O_4 facilitated the separation and elimination of particles. It was found that only during the first 5 min of exposure, uptake of DOX/GA-Fe_3O_4@MSNs into the mitochondria was high. Moreover, the mitochondria targeting of DOX/GA-Fe_3O_4@MSNs has been higher than DOX/TPP-Fe_3O_4@MSNs in HeLa cells. Moreover, GA-Fe_3O_4@MSNs without DOX did not display significant cytotoxicity, indicating proper biocompatibility for these NPs.[26]

Mitochondrial-targeting sequences (MTS) are a group of endogenous peptides consisting of 15–55 amino acids that can selectively transfer proteins from cytoplasm into mitochondria. MTS usually enter into mitochondria through attachment with outer membrane (OMM). Then, MTS by interaction with inner import machinery or remaining in intermembrane space (IMS) specifically transport proteins to the matrix. Based on this, a mitochondrial-targeted hybrid peptide R8MTS was designed, which was composed of a cell-penetrating peptide octaarginine (R8) and a mitochondrial-targeting sequence ALD5 ^{MTS} conjugated with N-(2-hydroxypropyl) methacrylamide (HPMA) copolymers for inhibition of breast cancer. HPMA copolymer, a water-soluble nanocarrier, is used for drug (DOX) delivery due to its lack of toxicity and biocompatibility. R8 is a cell-penetrating peptide that is widely employed for enhanced cellular drug delivery. In the low pH of lysosomes environment of cancer cells, DOX-R8MTS is released from HPMA copolymer and consequently enters the mitochondria. Then, DOX-R8MTS produced considerable ROS in mitochondria and induced apoptosis in cancer cells. Moreover, low expression of matrix metalloproteinases-2 (MMP-2), transforming growth factor-β (TGF-β), and VEGF (vascular endothelial growth factor) (typical proteins associated with tumor metastasis and invasion) was observed in the treated mice.[27,28]

Using antibiotic is another approach for targeting mitochondria in cancer cells. Conjugation of ciprofloxacin to a TPP group (Mt-CFX) was employed to increase the antibiotic concentration in the cancer cell mitochondria. Results showed that Mt-CFX induced ROS, which resulted in damaging of DNA, proteins, and lipids. Furthermore, it was found that Mt-CFX induced mtDNA damage over nuclear DNA as compared with 5-fluorouracil (5-FU), cisplatin, DOX, temozolomide, camptothecin, and chlorambucil. Mt-CFX also reduced growth of cancer cells in a xenograft mouse model.[29]

Glycyrrhetinic acid and its derivatives exhibit potential anticancer activity. However, their high hydrophobicity restricts their further development as anticancer drugs. It has been suggested that mitochondria-targeted glycyrrhetinic acid derivatives to cancer cells considerably improved their anticancer activity. Glycyrrhetinic acid conjugated with TTP^+ specifically targeted the mitochondria of tumor cells and induced apoptosis through the loss of $\Delta\Psi_m$, production of ROS, and activation of caspase-3 and caspase-9.[30]

Polycationic glycolipid polymer (CSOSA) (as a skeletal nanomaterial) was conjugated with 4-carboxybutyl triphenylphosphonium bromide (CTPP, a lipophilic cation) to form the mitochondria-targeted polymer C-P-CSOSA. It was observed that C-P-CSOSA polymer micelles were further uptaken in tumor cells than normal cells and showed quick lysosomal escape ability owing to their high proton-buffering capability. Moreover, encapsulation of DOX in the C-P-CSOSA/DOX micelles accumulated drug in the mitochondria of cancer cells and induced apoptosis cell death. The C-P-CSOSA polymer displayed more mitochondrial colocalization in vivo.[31]

Selenium NPs (SeNPs) can target cancer cells through EPR effect that increased production of ROS and mediated cell death. However, their application is restricted due to low toxicity on cancer cells. For better efficacy, human serum albumin (HSA)-coated SeNPs conjugated with TPP were designed to target cancer cell mitochondria. HSA, an abundant transport protein in blood, is biocompatible and is extensively employed for improving the stability of formulations. Moreover, functionalization of HAS is easy due to high number of reactive groups like the carboxyl and amino groups, which provide appropriate binding potential for TPP with high density. It was found that HSA-TPP@SeNPs exhibited higher cytotoxicity than HSA@SeNPs on HeLa cells (IC_{50} values of 10.82 μg/mL vs 90.78 μg/mL), which confirmed higher cytotoxic efficiency of TPP-conjugated SeNPs. Furthermore, SeNPs targeted to mitochondria increased ROS generation in cancer cells.[32]

Enzyme-responsive mesoporous silica NPs were developed for targeting tumor cells and mitochondria. Mesoporous silica NPs were bound to TPP for mitochondrial targeting and conjugated with hyaluronic acid (HA) for improving the CD44 receptor-mediated cancer cells. DOX was loaded into the pore of NPs. In tumor tissues with high acidic microenvironment, more HA is degraded by hyaluronidase (HAase), while it is stable in neutral pH of normal. Results showed that conjugation of HA enhanced cellular uptake of NPs in cancer cells through CD44 receptor-mediated endocytosis. In turn, TPP-grafted NPs accumulated in mitochondria. Cleavage of HA by overexpressed HAase led to elevated release of DOX in MGC-803 (as cancer cells), while it displayed much lower cytotoxicity to Cos-7 (as normal cells).[33]

Liposomes composed of cholesterol and dioleoyl-sn-glycero-3-phosphoethanolamine (DOPE) were prepared for delivery of antimycin A to cancer cells. Liposomes were conjugated with mitochondria-penetrating peptide (phenylalanine-arginine-phenylalanine-lysine; FRFK peptide sequence). Cholesterol-conjugated NPs are interned in the cells using low-density lipoprotein cholesterol receptors. According to the obtained results, liposomal formulation exhibited higher cellular accumulation and mitochondrial damage than control groups.[34]

Some tumor tissues such as breast cancer have higher levels of MUC1 (a glycoprotein with extensive O-linked glycosylation of its extracellular domain) than normal cells. In a study, PTX-TPP combination was loaded in albumin NPs and conjugated with MUC1 binding aptamer for selectively targeting of cancer cells. TPP was conjugated to PTX through pH-sensitive ester bond. Mitochondrial aldehyde dehydrogenase degrades this ester bond that is relatively resistive to cellular esterases. Results showed that the targeted NPs significantly enhanced the cellular uptake of PTX in cancer cells due to EPR effect, active targeting through aptamer modification, and targeting of mitochondria by TPP conjugation. Moreover, targeted NPs triggered further apoptosis in breast cancer cells.[35]

It has been reported that mitochondrial citrate transporter (CTP) and plasma membrane citrate transporter (PMCT) are overexpressed in cancer cells. These transporters make available acetyl-CoA precursors for the de novo lipogenesis (DNL) to provide required metabolic intermediates for cancer cells with high rate of proliferation. Apoptosis of cancer cells can be promoted by inhibition of DNL pathway without any adverse effect to normal cells. This treatment has been shown to be selective and strong strategy for targeting of cancer cells. It was observed that combination treatments with CTP inhibitor (4-chloro-3-[(3-nitrophenyl) amino] sulfonyl benzoic acid; CTPi) and PMCT inhibitor (2-(benzylsulfanyl)-N-[(pyridin-2-yl) methyl] propanamide; PMCTi) inhibited DNL pathway and induced apoptosis through lowering of citrate transportation and as a result inhibiting the fatty acid availability in cancer cells. Furthermore, it was found that combination of these inhibitors triggering apoptosis is associated with enhanced ROS production which inhibited the Bcl-2 expression.[36]

Mitochondria-targeted oxidants such as TPP-metformin and TPP-paraquat mediated ROS generation through disturbing of membrane potential and activation of Ca^{2+}/Cn (Calcium/Calcineuri) pathway that was found to be independent of ROS-induced HIF1α or AMPK pathways.[37]

Metformin is an inhibitor for complex I that reportedly prevented the growth of cancer cells and triggered mitochondrial apoptosis. Moreover, metformin can enhance the activation of sirtuin 3 (SIRT3, a mitochondrial deacetylase). It was observed that overexpressed SIRT3 exacerbated metformin-induced apoptosis, ATP decline, and mitochondrial damage.[38]

The functionalization of NPs with PEG is a classic approach to improve their stability in blood. However, it has been revealed that systemically administered PEGylated NPs can initiate immunogenic reactions. In a study, IR780 encapsulated sulfobetaine methacrylate (SBMA) functionalized bovine serum albumin (BSA) NPs (IR/SBMA-BSA NPs) were prepared for improving the efficacy of NPs. SBMA was conjugated to NPs since it can decrease protein adsorption. IR780 was employed for cancer phototherapy. It was observed that IR/SBMA-BSA NPs exhibited higher uptake and cytotoxicity in breast cancer cells, which may be related to the higher accumulation of IR780 in cancer cells mitochondria.[39]

In a study, L-arginine/5-FU combination was employed for the treatment of breast cancer cells. According to the results, L-arginine inhibited glycolysis pathway and ATP production through OXPHOS pathway under hypoxic condition. Consequently, increased ROS generation mediated by NO in mitochondria triggered death in cancer cells and also reduced the side effects of 5-FU on normal cells.[40]

Glutamine is an essential amino acid for regulating energy production, redox balance, and signaling in cancer cells. As a result, mitochondrial glutamine transporter can be considered as a target for cancer therapy. It has been shown that SLC1A5 (a mitochondrial glutamine transporter) variant with an N-terminal targeting signal localized in mitochondria and through the transportation of glutamine into mitochondria plays a significant role in cancer metabolic requirement for reprogramming. SLC1A5 variant overexpression under hypoxia is mediated by hypoxia-inducible factor-2α (HIF-2α). This overexpression manifested by an increase in ATP production and glutathione synthesis and also enhancement of gemcitabine resistance in pancreatic cancer cells.[41] Given the critical role of this transporter for cancer cell metabolism, its suppression can be used as a tool against the proliferation of cancer cells.

References

1. Cazzaniga M, Bonanni B. Breast cancer metabolism and mitochondrial activity: the possibility of chemoprevention with metformin. *Biomed Res Int.* 2015;2015:1–9. https://doi.org/10.1155/2015/972193.
2. Ribas V, García-Ruiz C, Fernández-Checa JC. Mitochondria, cholesterol and cancer cell metabolism. *Clin Transl Med.* 2016;5:1–24.
3. Bai L, Fei W-D, Gu Y-Y, et al. Liposomes encapsulated iridium (III) polypyridyl complexes enhance anticancer activity in vitro and in vivo. *J Inorg Biochem.* 2020;205:111014.
4. Deng X, Zhao J, Qu L, et al. Ginsenoside Rh4 suppresses aerobic glycolysis and the expression of PD-L1 via targeting AKT in esophageal cancer. *Biochem Pharmacol.* 2020;114038.
5. Chen X, Wei L, Yang L, Guo W, Guo Q, Zhou Y. Glycolysis inhibition and apoptosis induction in human prostate cancer cells by FV-429-mediated regulation of AR-AKT-HK2 signaling network. *Food Chem Toxicol.* 2020;111517.
6. Soukupova K, Rudolf E. Suppression of proliferation and activation of cell death by sodium selenite involves mitochondria and lysosomes in chemoresistant bladder cancer cells. *J Trace Elem Med Biol.* 2019;52:58–67.
7. Wang S, Ma X-B, Yuan X-H, Yu B, Xu Y-C, Liu H-M. Discovery of new [1, 2, 4] triazolo [1, 5-a] pyrimidine derivatives that kill gastric cancer cells via the mitochondria pathway. *Eur J Med Chem.* 2020;112630.
8. Chinnapaka S, Zheng G, Chen A, Munirathinam G. Nitro aspirin (NCX4040) induces apoptosis in PC3 metastatic prostate cancer cells via hydrogen peroxide (H2O2)-mediated oxidative stress. *Free Radic Biol Med.* 2019;143:494–509.
9. Gianluca F, Azzopardi M, Saliba C, et al. Aspirin impairs acetyl-coenzyme A metabolism in redox-compromised yeast cells. *Sci Rep.* 2019;9:1–13. https://doi.org/10.1038/s41598-019-39489-4.
10. Liu H, Xiong C, Liu J, et al. Aspirin exerts anti-tumor effect through inhibiting Blimp1 and activating ATF4/CHOP pathway in multiple myeloma. *Biomed Pharmacother.* 2020;125:110005.
11. Pritchard R, Rodríguez-Enríquez S, Pacheco-Velázquez SC, Bortnik V, Moreno-Sánchez R, Ralph S. Celecoxib inhibits mitochondrial O2 consumption, promoting ROS dependent death of murine and human metastatic cancer cells via the apoptotic signalling pathway. *Biochem Pharmacol.* 2018;154:318–334.
12. Abdelmageed N, Morad SA, Elghoneimy AA, et al. Oleanolic acid methyl ester, a novel cytotoxic mitocan, induces cell cycle arrest and ROS-mediated cell death in castration-resistant prostate cancer PC-3 cells. *Biomed Pharmacother.* 2017;96:417–425.
13. Nguyen F, Alferiev I, Guan P, et al. Enhanced intratumoral delivery of SN38 as a tocopherol oxyacetate prodrug using nanoparticles in a neuroblastoma xenograft model. *Clin Cancer Res.* 2018;24:2585–2593.
14. Goleva TN, Lyamzaev KG, Rogov AG, et al. Mitochondria-targeted 1,4-naphthoquinone (SkQN) is a powerful prooxidant and cytotoxic agent. *Biochim Biophys Acta Bioenerg.* 2020;1861(8):148210. https://doi.org/10.1016/j.bbabio.2020.148210.

15. Secker PF, Beneke S, Schlichenmaier N, et al. Canagliflozin mediated dual inhibition of mitochondrial glutamate dehydrogenase and complex I: an off-target adverse effect. *Cell Death Dis.* 2018;9:1–13.
16. Benej M, Hong X, Vibhute S, et al. Papaverine and its derivatives radiosensitize solid tumors by inhibiting mitochondrial metabolism. *Proc Natl Acad Sci USA.* 2018;115:10756–10761.
17. Christian S, Merz C, Evans L, et al.The novel dihydroorotate dehydrogenase (DHODH) inhibitor BAY 2402234 triggers differentiation and is effective in the treatment of myeloid malignancies. *Leukemia.* 2019;33:2403–2415.
18. Letts JA, Fiedorczuk K, Degliesposti G, Skehel M, Sazanov LA. Structures of respiratory supercomplex I+ III2 reveal functional and conformational crosstalk. *Mol Cell.* 2019;75. 1131–1146.e6.
19. Tajan M, Hock AK, Blagih J, et al.A role for p53 in the adaptation to glutamine starvation through the expression of SLC1A3. *Cell Metab.* 2018;28. 721–736.e6.
20. Rupprecht A, Moldzio R, Mödl B, Pohl EE. Glutamine regulates mitochondrial uncoupling protein 2 to promote glutaminolysis in neuroblastoma cells. *Biochim Biophys Acta Bioenerg.* 2019;1860:391–401.
21. Yamada Y, Munechika R, Kubota F, Sato Y, Sakurai Y, Harashima H. Mitochondrial delivery of an anticancer drug via systemic administration using a mitochondrial delivery system that inhibits the growth of drug-resistant cancer engrafted on mice. *J Pharm Sci.* 2020;109(8):2493–2500. https://doi.org/10.1016/j.xphs.2020.04.020.
22. Chen W-H, Xu X-D, Luo G-F, et al. Dual-targeting pro-apoptotic peptide for programmed cancer cell death via specific mitochondria damage. *Sci Rep.* 2013;3:3468.
23. Zhang X, Wang Y, Wei G, Zhao J, Yang G, Zhou S. Stepwise dual targeting and dual responsive polymer micelles for mitochondrion therapy. *J Control Release.* 2020.
24. Tan X, Zhou Y, Shen L, Jia H, Tan X. A mitochondria-targeted delivery system of doxorubicin and evodiamine for the treatment of metastatic breast cancer. *RSC Adv.* 2019;9:37067–37078.
25. Mallick S, Song SJ, Bae Y, Choi JS. Self-assembled nanoparticles composed of glycol chitosan-dequalinium for mitochondria-targeted drug delivery. *Int J Biol Macromol.* 2019;132:451–460.
26. Ahn J, Lee B, Choi Y, et al. Non-peptidic guanidinium-functionalized silica nanoparticles as selective mitochondria-targeting drug nanocarriers. *J Mater Chem B.* 2018;6:5698–5707.
27. Li Q, Huang Y. Mitochondrial targeted strategies and their application for cancer and other diseases treatment. *J Pharm Investig.* 2020;1–23.
28. Li Q, Yang J, Chen C, et al. A novel mitochondrial targeted hybrid peptide modified HPMA copolymers for breast cancer metastasis suppression. *J Control Release.* 2020;325:38–51. https://doi.org/10.1016/j.jconrel.2020.06.010.
29. Sunwoo K, Won M, Ko K-P, et al. Mitochondrial relocation of a common synthetic antibiotic: a non-genotoxic approach to cancer therapy. *Chem.* 2020;6(6):1408–1419. https://doi.org/10.1016/j.chempr.2020.03.004.
30. Jin L, Dai L, Ji M, Wang H. Mitochondria-targeted triphenylphosphonium conjugated glycyrrhetinic acid derivatives as potent anticancer drugs. *Bioorg Chem.* 2019;85:179–190.
31. Tan Y, Yang X, Dai S, et al. In vivo programming of tumor mitochondria-specific doxorubicin delivery by a cationic glycolipid polymer for enhanced antitumor activity. *Polym Chem.* 2019;10:512–525.
32. Zhuang Y, Li L, Feng L, et al. Mitochondrion-targeted selenium nanoparticles enhance reactive oxygen species-mediated cell death. *Nanoscale.* 2020;12:1389–1396.
33. Naz S, Wang M, Han Y, et al. Enzyme-responsive mesoporous silica nanoparticles for tumor cells and mitochondria multistage-targeted drug delivery. *Int J Nanomedicine.* 2019;14:2533.

34. Mallick S, Lee S, Park J-I, Choi JS. Liposomes containing cholesterol and mitochondria-penetrating peptide (MPP) for targeted delivery of antimycin A to A549 cells. *Colloids Surf B Biointerfaces.* 2018;161:356–364.
35. Esfandyari-Manesh M, Mohammadi A, Atyabi F, et al. Enhancement mitochondrial apoptosis in breast cancer cells by paclitaxel-triphenylphosphonium conjugate in DNA aptamer modified nanoparticles. *J Drug Deliv Sci Technol.* 2019;54:101228.
36. Poolsri W-a, Phokrai P, Suwankulanan S, et al. Combination of mitochondrial and plasma membrane citrate transporter inhibitors inhibits de novo lipogenesis pathway and triggers apoptosis in hepatocellular carcinoma cells. *Biomed Res Int.* 2018;2018:1–15. https://doi.org/10.1155/2018/3683026.
37. Chowdhury AR, Zielonka J, Kalyanaraman B, Hartley RC, Murphy MP, Avadhani NG. Mitochondria-targeted paraquat and metformin mediate ROS production to induce multiple pathways of retrograde signaling: a dose-dependent phenomenon. *Redox Biol.* 2020;101606.
38. Wu Y, Gao W-N, Xue Y-N, et al. SIRT3 aggravates metformin-induced energy stress and apoptosis in ovarian cancer cells. *Exp Cell Res.* 2018;367:137–149.
39. Alves CG, de Melo-Diogo D, Lima-Sousa R, Correia IJ. IR780 loaded sulfobetaine methacrylate-functionalized albumin nanoparticles aimed for enhanced breast cancer phototherapy. *Int J Pharm.* 2020;119346:1–9. https://doi.org/10.1016/j.ijpharm.2020.119346.
40. Jahani M, Azadbakht M, Rasouli H, et al. L-arginine/5-fluorouracil combination treatment approaches cells selectively: rescuing endothelial cells while killing MDA-MB-468 breast cancer cells. *Food Chem Toxicol.* 2019;123:399–411.
41. Yoo HC, Park SJ, Nam M, et al. A variant of SLC1A5 is a mitochondrial glutamine transporter for metabolic reprogramming in cancer cells. *Cell Metab.* 2020;31. 267–283.e12.

CHAPTER 8

Mitochondrial metabolism in diabetes

Alfredo Saavedra-Molina, Christian Cortés-Rojo, Rocío Montoya-Pérez, Rafael Salgado-Garciglia, Donovan J. Peña-Montes, and Maribel Huerta-Cervantes
Instituto de Investigaciones Químico Biológicas, Universidad Michoacana de San Nicolás de Hidalgo, Morelia, Michoacan, Mexico

Contents

1 Introduction

Diabetes mellitus (DM) is a chronic degenerative disease characterized by chronic high blood glucose levels (hyperglycemia). The International Diabetes Federation[1] classifies DM into type 1 diabetes mellitus, type 2 diabetes mellitus, gestational diabetes, and other specific types. Chronic hyperglycemia in diabetes is associated with long-term complications that affect various organs and tissues, including the eyes, nerves, liver, muscle, heart, and kidneys.[2] The incidence and prevalence of type 2 diabetes mellitus (DM2) has increased in recent years. DM2 is the most common form of diabetes (90%–95%) and is characterized by glucose intolerance or insulin resistance and abnormalities in insulin production in pancreatic β cells or insulin-dependent tissues. Abnormalities in the metabolism of lipids are one of the most frequent complications in diabetes.[3] Dyslipidemia (disruption in the normal levels of lipids, mainly cholesterol, and triglycerides) is considered a major cardiovascular risk factor in diabetes.[3] Therefore the detection of dyslipidemia and its treatment to reduce the cardiovascular risk and its consequences are required in diabetic patients.[3] Hyperglycemia can also stimulate the generation of reactive oxygen species (ROS) from different sources, such as oxidative phosphorylation, NADPH oxidase, xanthine oxidase, among others.[4] The development of diabetic encephalopathy is a multifactorial process in which both vascular and metabolic

Mitochondrial Metabolism
https://doi.org/10.1016/B978-0-12-822416-8.00007-5

factors are involved.[5] The direct toxicity of glucose in neurons is due to increased production of ROS by NADPH oxidases and the electron transport chain (ETC).[6] Diabetes can lead to renal dysfunction.[7] Diabetic kidney disease (DKD) or diabetic nephropathy is a complication that occurs in both type 1 diabetes and type 2 diabetes and a common cause of death in diabetic patients is also one of the leading causes of end-stage kidney disease today (≈80%). DKD is characterized by a loss in glomerular filtration rate and structural abnormalities in the kidneys.[2] However, the overall development of DKD pathogenesis is the result of metabolic and hemodynamic abnormalities during diabetes, closely related to oxidative stress and humorous response that interact in the diabetic environment, triggering functional and structural alterations.[8] DKD characterized by proteinuria and loss of renal function is the most common global cause of chronic kidney disease.[9] Additionally, mitochondrial dysfunctionality is a common component in disease physiopathology, as mitochondria are particularly important for kidney and other organs functions.[8] Given the high energy and oxygen demands, mitochondria and renal function are closely related. In addition, the mitochondrial electron transport system (ETS) is one of the main sources of ROS during diabetes. Several studies have reported that modulation in mitochondrial functionality as well as decreased production of ROS and oxidative stress, greatly improve complications of DM. There are several reports where kidneys of diabetic rats' exhibit modifications in the redox state and in oxidative and nitrosative stress, accompanied by changes in mitochondrial function.[10, 11] Likewise, dysfunction of mitochondrial complex activities in the kidney is associated with mitochondrial iron accumulation in diabetes.[12] Diabetic rats also reportedly show dysregulation of mitochondrial complex activities, for example, increased complex I activity, which may occur due to an NADH/NAD+ redox imbalance caused by NADH overproduction in complex I subunits.[13, 14] Other studies demonstrated increased complex I and II activities in diabetic rats; increases that were prevented by an iron-restricted diet.[12] Likewise, in diabetes, decreased activity of mitochondrial complex III in the kidneys is reportedly caused by posttranslational modifications related to oxidative stress, which is closely related to increased mitochondrial ROS production.[15, 16] Therapies aimed at regulating mitochondrial functionality have shown promising results in the treatment of chronic degenerative diseases. The growing need to find alternatives for the treatment of diabetes justify the study of medicinal plants used in traditional medicine. In traditional medicine, for example, there are a great variety of plants with importance for diabetes.[10]

2 Medicinal plants in diabetes

Ethnobotanical findings are important area of research for new pharmaceuticals produced from natural sources, mainly plants. Plants are one of the most promising sources for discovering new antioxidant agents.[4] Since antiquity, humans have used the seeds, roots, stems, flowers, and fruits of plants to eat and alleviate disorders due to the effectiveness of their healing, availability, and low cost. Secondary metabolites present in plants with hypoglycemic properties belong to these groups: terpenes, carotenoids, alkaloids, sterols, tannins, polysaccharides, coumarins, steroids, phenolic acids, lignans, and flavonoids.[17, 18] A large number of active chemical compounds (from a therapeutic point of view) have been extracted from plants and more than 23 uses for different conditions have been identified, including as a blood tonic, a stimulant, as an antidysentery, antiinflammatory, and antispasmodic agent, among others.[18] Among the plant secondary metabolites prescribed to patients who present painful diseases and cancer include taxol, an active principle from *Taxus cuspidata*[19]; or reserpine, synthesized by the species of *Rauwolfia serpentina* and *Rauwolfia vomitoria*.[20] *Turnera diffusa* is recognized for its medicinal properties including kidney disorders and diabetes.[21] Another study claims that *Eryngium carlinae* hexanic extract possesses hypoglycemic, hypolipidemic effects in serum, and antioxidant effects in heart mitochondria.[22] As well, other protective antioxidant effects in brain, kidneys, heart, and liver in streptozotocin-induced diabetic male rats were reported.[23] Ethanol extract from *E. carlinae* possesses hypolipidemic and hypoglycemic effects in Wistar rats with experimental diabetes.[24] Another study related to metabolic syndrome demonstrated that Nopal (*Opuntia ficus indica*) a domestic species of the Cactaceae family had lower levels of triglycerides and diastolic arterial pressure.[25] Phytochemical analysis of ethanolic extract of *E. carlinae* in rats with experimental diabetes showed sesquiterpenes[24] as major components and the most abundant was β-Selinene, reported to have antiinflammatory,[26] antiproliferative,[27] and hypolipidemic effects.[28] Another study using *n*-hexane as a solvent extracted from *E. carlinae* mainly terpenes and sesquiterpenes that showed hypoglycemic, diminished peroxidation of lipids, and antioxidant effects in diabetic rats.[23] Importantly, major chemical components to determine the pharmacological effects, the stability of its products, and its clinical behavior, have to be evaluated to find new compounds with medicinal properties or validate its folk medicinal use.

2.1 Avocado oil in diabetes

2.1.1 Avocado oil improves the function of the electron transport chain in target organs of diabetes

2.1.1.1 Overview of the electron transport chain

The ETC is at the heart of the conversions of the chemical energy provided by metabolic fuels like carbohydrates, lipids, and proteins, into usable energy in the form of ATP to carry out a myriad of processes that decrease entropy in living organisms. The metabolism captures the energy from fuels by successive redox reactions producing NADH and $FADH_2$, which are burned with oxygen inside mitochondria to produce ATP, while also a part of the energy can be used by mitochondria to maintain the transport of metabolites, ions, and proteins across the inner mitochondrial membrane. Moreover, energy can be wasted in a controlled manner to dissipate excessive ROS production in the ETC or to produce heat for adaptive processes like e.g., adaptation of newborns to its new environment.[29] The ETC consists of an arrangement of proteins that oxidize NADH and $FADH_2$, and that transfer electrons from these species to diatomic oxygen, using intermediate electron carriers like ubiquinone and cytochrome *c*. The free energy obtained from these reactions is used by three of the four ETC complexes to pump protons across the inner mitochondrial membrane to generate a membrane potential ($\Delta\Psi$) whose energy drives ATP synthesis by the F_1F_0-ATP synthase.[30]

The multimeric proteins that constitute the ETC are the complex I, which oxidizes NADH and transfer electrons to ubiquinone, with concomitant proton pump from the matrix to the intermembrane space. The complex II transfer electrons from $FADH_2$ to ubiquinone and this is the only complex that does not pump protons. The complex III oxidizes the ubiquinol produced in both the complexes I and II, and this complex also pumps protons, as electrons are transferred to cytochrome *c*. Reduced cytochrome *c* in turn is oxidized by complex IV and reduces diatomic oxygen to water, with concomitant proton pump to the matrix. The buildup of protons stores the energy from substrates oxidation in the $\Delta\Psi$. The energy from the $\Delta\Psi$ is used for ATP synthesis when protons enter back to the matrix by the F_1F_0-ATP synthase, causing conformational changes catalyzing the phosphorylation of ADP to ATP.[31]

2.1.2 Diabetes impairs the function of the ETC and increases ROS levels

Oxidative stress can impair the activity of the ETC by peroxidizing the phospholipid cardiolipin, which is essential for the structure and function of all the complexes of the ETC,[32] oxidizing reactive thiols in some complexes,[33] destroying prosthetic groups like iron-sulfur (Fe-S) clusters[34] and heme groups,[35] and inducing protein carbonylation.[36] Oxidative damage in the ETC is caused not only by the ROS produced in the ETC itself, but also by cytosolic sources of ROS-like activated NADPH oxidases,[37] which even can translocate into kidney mitochondria.[38]

The damage to ETC complexes blocks electron transfer and leads to accumulation of electrons in electron carriers, stimulating in this way the partial reduction of oxygen into superoxide radical. Examples of this is the blocking of electron transfer by antimycin A in complex III, which inhibits electron transfer in cytochrome *b* from the heme b_H to ubiquinone, leading to buildup of ubisemiquinone in the Rieske protein and its reaction with dioxygen to yield superoxide.[39] Another example is the inhibition of complex I by rotenone, where this inhibitor blocks the electron transfer from the Fe-S cluster N2 to ubiquinone, allowing monoelectronic oxygen reduction by N2.[40]

Diabetes has been shown to inhibit complex I activity and increase ROS production in mitochondria from heart[41,42] and liver,[43–45] and it inhibits complex III activity and increases ROS production in kidney mitochondria.[15,16,46–48] ROS production due to the inhibition of these ETC complexes in kidney and liver is critical, as mitochondrial ROS generation has been established as a main cause of inflammation and injury in these organs.[49,50]

Diverse plants are a source of compounds that exert effects by regulating the mitochondrial function under different mechanisms. Mu et al.[51] have proposed the development of plant extracts and their compounds to remedy mitochondrial dysfunction and found that many natural products diminish loss of mitochondrial function in diabetes mellitus type 2 by protecting the structure and morphology of the mitochondria, regulating the mitochondria apoptotic pathway, increasing mitochondrial energy metabolism, and inhibiting oxidative stress.

Studies have shown that there is a link between mitochondrial permeability transition (MPT) pore opening and diabetes[52] and different plant extracts have inhibitory effect on opening MPT pore as methanol leave extract of Blighia sapida,[53] methanol leave extract of Clerodendrum volubile[54] and methanol fraction of Ficus mucoso.[55]

Natural antioxidant compounds such as polyphenols have shown a protective effect of mitochondrial dysfunction through inhibition of MPT pore opening in diseases such as diabetes. Quercetin and galic acid are some of these compounds that act as MPT pore opening inhibitors during diabetes mellitus.[56,57]

2.1.3 Avocado oil protects the function of the ETC and decreases ROS production

Avocado oil, a dietary source of bioactive molecules, including oleic acid, antioxidants, and phytosterols, counteracts impaired ETC function in mitochondria from liver, kidney, and brain of diabetic rats.[16, 43, 58] In kidney mitochondria from diabetic rats, 90-days oral supplementation of avocado oil prevented a twofold decrease in complex III activity and increased ROS generation. It was identified that impaired complex III activity was due to decreased levels of cytochromes $c + c_1$, which was also prevented by avocado oil.[16] In liver mitochondria of diabetic rats, 90-day intake of avocado oil by diabetic rats counteracted a 50% inhibition of complex III activity and a 1.7-fold-higher ROS production. The protection of complex I activity might be related to the positive effects of avocado oil on GSH levels.[43] The redox state of glutathione modulates complex I activity: at reduced glutathione (GSH)/oxidized glutathione (GSSG) ratios of 2.0, complex I activity decreases ~50% due to glutathionylation of reactive cysteine thiols.[33] These numbers fit well with the GSH/GSSG ratio of 1.5 and the 50% inhibition of complex I in diabetic mitochondria, and with the full recovery of complex I activity and the increase in the GSH/GSSH ratio to ~4.0.[43]

Oxidative phosphorylation can be impaired also in the absence of a specific defect in the ETC complexes, as substrate oxidation is not under the control of a single ETC complex, but is controlled mainly by the $\Delta\Psi$, which, in turn, varies in response to ATP demand, H^+ leak, and other energy-depending processes.[30] Avocado oil also improves oxidative phosphorylation even in the absence of a specific inhibition in the ETC, as demonstrated in brain mitochondria of diabetic rats.[58] In this case, it was not detected any impaired activity in the ETC complexes. Nevertheless, respiration in both nonphosphorylating state 4 and in phosphorylating state 3 was decreased by 50%, with a concomitant decrease in $\Delta\Psi$ and exacerbated ROS production. All these alterations were fully prevented by avocado oil.[58]

2.1.4 Avocado oil increases antioxidant capacity in mitochondria

The beneficial effects of avocado oil on the ETC activity and, overall, on oxidative phosphorylation, can be attributed to the enhanced resistance to oxidative stress elicited in mitochondria. The main finding supporting this notion is the improvement of the redox state of glutathione (i.e., the GSH/GSSG ratio) by avocado oil. As mentioned earlier, the decrease in GSH/GSSG ratio in mitochondria from liver, brain, and kidney was prevented by oral avocado oil intake.[15, 43, 59] Of note, avocado oil increased the GSH/

GSSG ratio in brain mitochondria in diabetic rats several-fold above the levels of normoglycemic rats.[43] Moreover, avocado oil has also long-lasting effects not only on the redox state of glutathione but also on the pool size of this antioxidant, as avocado oil consumption by up to 12-months augmented the levels of GSH and total glutathione (GSH + GSSG) in kidney mitochondria of type 2 diabetic rats.[59] The higher antioxidant capacity in mitochondria elicited by avocado oil consumption was confirmed by a higher resistance of liver mitochondria from both normoglycemic and diabetic rats to in vitro stimulated lipid peroxidation by Fe^{2+}, which catalyzes the production of hydroxyl radical.[43] Moreover, impaired activities of the complexes II and III by in vitro treatment of kidney mitochondria with Fe^{2+} were improved by the treatment with a lipophilic antioxidant only in diabetic rats consuming avocado oil, but not in diabetic rats.[16] This suggest that avocado oil might reinforce the effects of therapeutic approaches aimed to decrease oxidative stress at the mitochondrial milieu in diabetes.

Avocado oil might be a safe dietary approach to cope with diabetes and its complications along with pharmacological therapeutics currently in use, since avocado oil has hypoglycemic effect on type 2 diabetic rats that last up to 12-months,[59] and also has hypolipidemic effects by normalizing hypertriglyceridemia and hypercholesterolemia in type 1 diabetic rats, without inducing liver steatosis.[43] Indeed, preliminary results show that the effects of avocado oil on mitochondria are linked to amelioration of hepatic and renal complications of diabetes as avocado oil consumption delayed the beginning of diabetic nephropathy in type 2 diabetic rats,[60] and reverses the development of NAFLD in rats feed with a high-fat, high fructose diet.[61]

2.1.5 Involvement of mitochondria in skeletal muscle during diabetes

It has been considered that the function of the mitochondria is an important factor in the toxicity of the accumulation of lipid metabolites and, consequently, the triggering of insulin resistance. Some reports assure that mitochondrial dysfunction does not play an essential role at the beginning of the process, but rather a complication associated with the production of ROS caused by dyslipidemia in skeletal muscle, which promotes mitochondrial alterations, accumulation of lipids, and inhibition of the effect of insulin.[62]

Although the topic continues with several questions, the general line indicates that insulin resistance is due to the inability of adipose tissue to store fat, which causes it to be circulating in the bloodstream and eventually is deposited in other tissues, as in skeletal muscle.[63] It has been proposed that the mitochondria participate in the process by reducing fatty acid oxidation and the intracellular accumulation of acetyl-CoA and diacylglycerol.[64] However, it has also been proposed that it is due to increased oxidation as it is forced to produce energy by the presence of energy components, which would activate different signaling pathways to reduce carbohydrate transport.[65]

Insulin resistance in skeletal muscle and impaired function in the pancreas beta cells are the first detectable symptoms before hyperglycemia; they can even be identified up to 10 years before the diagnosis of diabetes.[62] Skeletal muscle, by its mass and insulin-stimulated glucose transport, represents one of the most important organs for the development of insulin resistance and a key element in maintaining normal blood glucose levels,[66, 67] which will eventually develop into type 2 diabetes mellitus.

Since the mitochondria are the primary site of lipid oxidation and utilization, reduced activity by the mitochondria has been considered an essential factor in the toxicity of lipid accumulation and its consequent toxicity, eventually, insulin resistance. It has been documented that during DM2, there is a lower activity of the ETC[68] probably due to the reduction of the expression of genes that code for the subunits of mitochondrial enzymes.[69]

A considerable transductional and translational decrease in carnitine acylcarnitine translocase has been identified in patients with diabetes, leading to increased triglycerides, diacylglycerol, and acyl-coenzyme A, characteristic of insulin resistance.[70] Similarly, a lower amount and activity of subsarcolemmal mitochondria has been seen, suggesting that this deficit may contribute to insulin resistance pathogenesis during diabetes.[68] This issue has been a wide discrepancy since, in animal models, the same has not been observed as in human patients. These data report that at least in rodents, the consumption of a high-fat diet is not always accompanied by mitochondrial dysfunction. In fact, oxidative phosphorylation and/or biogenesis has been found to be increased[71, 72] at the beginning of the experiments (2–3 weeks of treatment) and a subsequent decrease in the long term, even in the presence of RI.

Much has been said about the deleterious effects of oxidative stress on various tissues and organs. However, it is also known that moderate amounts of ROS tend to have favorable effects that lead to adaptive cellular responses. ROS regulates growth processes, apoptosis, and signaling pathways, while at the systemic level, they contribute to complex functions such as the regulation of blood pressure, the immune system, and cognition.

It is worth mentioning that antioxidant systems play an important role in the restoration of functions depleted by free radicals, in such a way, the overexpression of enzymes such as MnSOD has been documented, which improves sensitivity to dependent glucose uptake insulin in muscle tissue and eventually insulin sensitivity in rodents on high-fat diets.[73]

It has been widely documented that excess oxidation by mitochondria under conditions of abundant fuel, such as hyperglycemia, contributes to the development of insulin resistance in skeletal muscle,[65] a process which will eventually cause mitochondrial dysfunction, which is linked to the opening of the MTP.[74] In turn, a clear link between mitochondrial fusion and beta-oxidation with insulin resistance and impaired glucose metabolism has been documented.[62]

Currently, there is still much debate about mitochondrial dysfunction in skeletal muscle during diabetes. Studies by Boushel et al.[75] state that during diabetes, there is a reduced activity and number of mitochondria, as well as damage to lipid oxidation[69, 76] and excessive production of ROS.[77, 78] According to Pettersen et al.[79] during diabetes, the mitochondria exhibit a lower oxidative capacity in individuals with insulin resistance resulting from excessive fat accumulation in the muscle tissue. On the other hand, Pereira et al.[80] propose that one of the causes of the increase in metabolism is due to the blockage in mitochondrial dynamics (fusion and fission), which at a systemic level causes a deterioration in insulin sensitivity.

In this context, differences between the different types of mitochondria have been documented at the muscle level. The study carried out by Lai et al.[81] observed that the subsarcolemmal mitochondria show minor damage, while the intermyofibrillar ones preserve their function. Changes in subsarcolemmal mitochondria are characterized by a decline in oxidative phosphorylation efficiency related to ATP synthase activity and alterations in the inner mitochondrial membrane. However, it is generally considered that mitochondrial function is not affected by insulin resistance in skeletal muscle in type 2 diabetes not associated with obesity.

In conclusion, more studies are needed to ameliorate diabetes disease with the knowledge that bioactive compounds from medicinal plants exert beneficial effects as antioxidants, hypoglycemic, and antilipidemic due to the fact that mitochondria from several organs are the main producers of ROS.

References

1. International Diabetes Federation. *IDF Diabetes Atlas.* 9th ed; 2019. http://www.idf.org/diabetesatlas. Accessed on 24 December 2019.
2. Cao Z, Cooper ME. Pathogenesis of diabetic nephropathy. *J Diabetes Invest.* 2011;2(4):243–247.
3. Punthakee Z, Goldenberg R, Katz P. Definition, classification and diagnosis of diabetes, prediabetes and metabolic syndrome. *Can J Diabetes.* 2018;42:S10–S15.
4. Saavedra-Molina A, Salgado-Garciglia R, Noriega-Cisneros R, et al. Oxidative stress in diabetes and hypertension treated with alternative therapy of medicinal plants. In: Catalá A, ed. *Reactive Oxygen Species, Lipid Peroxidation and Protein Oxidation.* 1st ed. Nova Publishers; 2015:205–217.
5. Mastrocola R, Restivo F, Vercellinato I, et al. Oxidative and nitrosative stress in brain mitochondria of diabetic rats. *J Endocrinol.* 2005;187:37–44.
6. Rice-Evans C, Burdon R. Free radical-lipid interactions and their pathological consequences. *Prog Lipid Res.* 1993;32:71–110.
7. Yokozawa T, Yamabe N, Kim HY, et al. Protective effects of morroniside isolated from Corni fructus against renal damage in streptozotocin-induced diabetic rats. *Biol Pharm Bull.* 2008;31(7):1422–1428.
8. Schiffer TA, Friederich-Persson M. Mitochondrial reactive oxygen species and kidney hypoxia in the development of diabetic nephropathy. *Front Physiol.* 2017;8:211. https://doi.org/10.3389/fphys.2017.00211.
9. Park CW. Diabetic kidney disease from epidemiology to clinical perspectives. *Diabetes Metab.* 2014;38(4):252–260.
10. Noriega-Cisneros R, Cortés-Rojo C, Manzo-Avalos S, et al. Mitochondrial response to oxidative and nitrosative stress in early stages of diabetes. *Mitochondrion.* 2013;13(6):835–840.
11. Pérez-Gallardo RV, Noriega-Cisneros R, Esquivel-Gutiérrez E, et al. Effects of diabetes on oxidative and nitrosative stress in kidney mitochondria from aged rats. *J Bioenerg Biomembr.* 2014;46(6):511–518.
12. Peña-Montes DJ, Huerta-Cervantes M, Ríos-Silva M, et al. Effects of dietary iron restriction on kidney mitochondria function and oxidative stress in streptozotocin-diabetic rats. *Mitochondrion.* 2020;54:41–48.
13. Wu J, Luo X, Yan LJ. Two dimensional blue native/SDS-PAGE to identify mitochondrial complex I subunits modified by 4-hydroxynonenal (HNE). *Front Physiol.* 2015;6:98.
14. Wu J, Luo X, Thangthaeng N, et al. Pancreatic mitochondrial complex I exhibits aberrant hyperactivity in diabetes. *Biochem Biophys Rep.* 2017;11:119–129.
15. Munusamy S, Saba H, Mitchell T, Megyesi JK, Brock RW, MacMillan-Crow LA. Alteration of renal respiratory complex-III during experimental type-1 diabetes. *BMC Endocr Disord.* 2009;9(1):2.
16. Ortiz-Avila O, Sámano-García CA, Calderón-Cortés E, et al. Dietary avocado oil supplementation attenuates the alterations induced by type I diabetes and oxidative stress in electron transfer at the complex II-complex III segment of the electron transport chain in rat kidney mitochondria. *J Bioenerg Biomembr.* 2013;45(3):271–287.
17. Andrade-Cetto A, Heinrich M. Mexican plants with hypoglycaemic effect used in the treatment of diabetes. *J Ethnopharmacol.* 2005;99:325–348.
18. Esquivel-Gutiérrez ER, Noriega-Cisneros R, Saavedra-Molina A, Salgado-Garciglia R. Plants used in Mexican folk medicine with antidiabetic and antihypertensive properties. *Pharmacol Online.* 2013;2:15–23.
19. Wani MC, Taylor HL, Wall ME, Coggon P, McPhail AT. Plant antitumor agents. VI. The isolation and structure of taxol, a novel antileukemic and antitumor agent from Taxus brevifolia. *J Am Chem Soc.* 1971;93:2325–2327.

20. Vakil RJ. Rauwolfia serpentina in the treatment of high blood pressure: a review of the literature. *Circulation.* 1955;12:220–229.
21. Esquivel-Gutiérrez ER, Alcaráz-Meléndez L, Salgado-Garciglia R, Saavedra-Molina A. Antioxidant effects of damiana (*Turnera diffusa* Willd. ex Schult.) in kidney mitochondria from streptozotocin-diabetic rats. *Nat Prod Res.* 2017. https://doi.org/10.1080/14786419.2017.1380009.
22. García-Cerrillo D, Noriega-Cisneros R, Peña-Montes D, et al. Antioxidant effects of Eryngium carlinae in diabetic rats. *Asian J Appl Sci.* 2018;6(5):308–314.
23. Peña-Montes DJ, Huerta-Cervantes M, Ríos-Silva M, et al. Protective effect of the hexanic extract of Eryngium carlinae inflorescences in vitro, in yeast, and in streptozotocin-induced diabetic male rats. *Antioxidants.* 2019;8:73. https://doi.org/10.3390/antiox8030073.
24. Noriega-Cisneros R, Peña-Montes DJ, Huerta-Cervantes M, et al. Eryngium carlinae ethanol extract corrects lipid abnormalities in Wistar rats with experimental diabetes. *J Med Food.* 2019;1–9. https://doi.org/10.1089/jmf.2019.0189.
25. Cárdenas Y, Ríos-Silva M, Huerta M, et al. The comparative effect of nopal and mucilage in metabolic parameters in rats with high-fructose diet. *J Med Food.* 2019;1–4. https://doi.org/10.1089/jmf.2018.0124.
26. Fernandes ES, Passos GF, Medeiros R, et al. Anti-inflammatory effects of compounds alpha-humulene and (−)-trans-caryophyllene isolated from the essential oil of Cordia verbenacea. *Eur J Pharmacol.* 2007;569:228–236.
27. Zhan YH, Liu J, Qu XJ, et al. β-Elemene induces apoptosis in human renal-cell carcinoma 786-0 cells through inhibition of MAPK/ERK and PI3K/Akt/mTOR signalling pathways. *Asian Pac J Cancer Prev.* 2012;13:2739–2744.
28. Chung IM, Kim MY, Park WH, et al. Antiatherogenic activity of Dendropanax morbifera essential oil in rats. *Pharmazie.* 2009;64:547–549.
29. Ravussin E, Galgani JE. The implication of brown adipose tissue for humans. *Annu Rev Nutr.* 2011;31:33–47.
30. Brand MD, Nicholls DG. Assessing mitochondrial dysfunction in cells. *Biochem J.* 2011;435(2):297–312.
31. Saraste M. Oxidative phosphorylation at the fin de siècle. *Science.* 1999;283(5407):1488–1493.
32. Panov AV, Dikalov SI. Cardiolipin, perhydroxyl radicals, and lipid peroxidation in mitochondrial dysfunctions and aging. *Oxid Med Cell Longev.* 2020;2020:1323028.
33. Beer SM, Taylor ER, Brown SE, et al. Glutaredoxin 2 catalyzes the reversible oxidation and glutathionylation of mitochondrial membrane thiol proteins: implications for mitochondrial redox regulation and antioxidant defense. *J Biol Chem.* 2004;279(46):47939–47951.
34. Gomez M, Pérez-Gallardo RV, Sánchez LA, et al. Malfunctioning of the iron-sulfur cluster assembly machinery in Saccharomyces cerevisiae produces oxidative stress via an iron-dependent mechanism, causing dysfunction in respiratory complexes. *PLoS One.* 2014;9(10), e111585.
35. Harel S, Salan MA, Kanner J. Iron release from metmyoglobin, methaemoglobin and cytochrome c by a system generating hydrogen peroxide. *Free Radic Res Commun.* 1988;5(1):11–19.
36. Choksi KB, Boylston WH, Rabek JP, Widger WR, Papaconstantinou J. Oxidatively damaged proteins of heart mitochondrial electron transport complexes. *Biochim Biophys Acta.* 2004;1688(2):95–101.
37. Dikalov SI, Nazarewicz RR, Bikineyeva A, et al. Nox2-induced production of mitochondrial superoxide in angiotensin II-mediated endothelial oxidative stress and hypertension. *Antioxid Redox Signal.* 2014;20(2):281–294.
38. Kim SM, Kim YG, Jeong KH, et al. Angiotensin II-induced mitochondrial Nox4 is a major endogenous source of oxidative stress in kidney tubular cells. *PLoS One.* 2012;7(7), e39739.

39. Sun J, Trumpower BL. Superoxide anion generation by the cytochrome bc1 complex. *Arch Biochem Biophys.* 2003;419(2):198–206.
40. Fato R, Bergamini C, Bortolus M, et al. Differential effects of mitochondrial complex I inhibitors on production of reactive oxygen species. *Biochim Biophys Acta.* 2009;1787(5):384–392.
41. Vazquez EJ, Berthiaume JM, Kamath V, et al. Mitochondrial complex I defect and increased fatty acid oxidation enhance protein lysine acetylation in the diabetic heart. *Cardiovasc Res.* 2015;107(4):453–465.
42. Niemann B, Chen Y, Teschner M, Li L, Silber RE, Rohrbach S. Obesity induces signs of premature cardiac aging in younger patients: the role of mitochondria. *J Am Coll Cardiol.* 2011;57(5):577–585.
43. Ortiz-Avila O, Gallegos-Corona MA, Sánchez-Briones LA, et al. Protective effects of dietary avocado oil on impaired electron transport chain function and exacerbated oxidative stress in liver mitochondria from diabetic rats. *J Bioenerg Biomembr.* 2015;47(4):337–353.
44. Lukivskaya O, Patsenker E, Buko VU. Protective effect of ursodeoxycholic acid on liver mitochondrial function in rats with alloxan-induced diabetes: link with oxidative stress. *Life Sci.* 2007;80(26):2397–2402.
45. Bouderba S, Sanz MN, Sánchez-Martín C, et al. Hepatic mitochondrial alterations and increased oxidative stress in nutritional diabetes-prone *Psammomys obesus* model. *Exp Diabetes Res.* 2012;2012:430176.
46. Rosca MG, Mustata TG, Kinter MT, et al. Glycation of mitochondrial proteins from diabetic rat kidney is associated with excess superoxide formation. *Am J Physiol Renal Physiol.* 2005;289(2):F420–F430.
47. Zhang H, Zhang HM, Wu LP, et al. Impaired mitochondrial complex III and melatonin responsive reactive oxygen species generation in kidney mitochondria of db/db mice. *J Pineal Res.* 2011;51(3):338–344.
48. Yu HT, Fu XY, Liang B, et al. Oxidative damage of mitochondrial respiratory chain in different organs of a rat model of diet-induced obesity. *Eur J Nutr.* 2018;57(5):1957–1967.
49. Forbes JM, Thorburn DR. Mitochondrial dysfunction in diabetic kidney disease. *Nat Rev Nephrol.* 2018;14(5):291–312.
50. Simões ICM, Fontes A, Pinton P, Zischka H, Wieckowski MR. Mitochondria in non-alcoholic fatty liver disease. *Int J Biochem Cell Biol.* 2018;95:93–99.
51. Mu JK, Li YQ, Shi TT, et al. Remedying the mitochondria to cure human diseases by natural products. *Oxid Med Cell Longev.* **2020**;2020:5232614.
52. Lowell BB, Shulman GI. Mitochondrial dysfunction and type 2 diabetes. *Science.* 2005;307(5708):384–387.
53. Olowofolahan AO, Nwaokolo EO, Olorunsogo OO. Effect of methanol leaf extract of Blighia Sapida on mitochondrial membrane permeability transition pore opening and blood glucose levels in normal and streptozotocin-induced diabetic rats. *Biomed J Sci Tech Res.* 2019;19(1):14047–14057.
54. Ajayi EIO, Molehin OR, Oloyede OI, et al. Liver mitochondrial membrane permeability modulation in insulin-resistant, uninephrectomised male rats by Clerodendrum volubile P. Beauv and *Manihot esculenta* Crantz. *Clin Phytosci.* 2019;5(31).
55. Oyebode OT, Ogunbiyi FO, Olorunsogo OO. Opening of liver mitochondrial permeability transition pore in streptozotocin-induced diabetic rats and its inhibition by methanol fraction of *Ficus mucoso* (Welw) root bark. *J Integr Med.* 2019;17(6):446–454.
56. Sun J, Ren DD, Wan JY, et al. Desensitizing mitochondrial permeability transition by ERK-Cyclophilin D axis contributes to the neuroprotective effect of gallic acid against cerebral ischemia/reperfusion injury. *Front Pharmacol.* 2017.

57. Gorbenko NI, Borikov OY, Ivanova OV, et al. The effect of quercetin on oxidative stress markers and mitochondrial permeability transition in the heart of rats with type 2 diabetes. *Ukr Biochem J.* 2019;91(5):46–54.
58. Ortiz-Avila O, Esquivel-Martínez M, Olmos-Orizaba BE, Saavedra-Molina A, Rodriguez-Orozco AR, Cortés-Rojo C. Avocado oil improves mitochondrial function and decreases oxidative stress in brain of diabetic rats. *J Diabetes Res.* 2015;2015:485759.
59. Ortiz-Avila O, Figueroa-García MC, García-Berumen CI, et al. Avocado oil induces long-term alleviation of oxidative damage in kidney mitochondria from type 2 diabetic rats by improving glutathione status. *J Bioenerg Biomembr.* 2017;49:204–214.
60. Ortiz-Avila O, Saavedra-Molina A, Cortés-Rojo C. Effect of avocado oil on metabolic profile and development of diabetic nephropathy in Goto-Kakizaki rats. *FASEB J.* 2019;33(S1):487–515.
61. Garcia-Berumen CI, Olmos-Orizaba BE, Márquez-Ramírez CA, et al. Avocado oil ameliorates non-alcoholic fatty liver disease by down-regulating inflammatory cytokines and improving mitochondrial dynamics. *FASEB J.* 2019;33(S1):660–666.
62. Di Meo S, Iossa S, Venditti P. Skeletal muscle insulin resistance: role of mitochondria and other ROS sources. *J Endocrinol.* 2017;233(1):R15–R42. https://doi.org/10.1530/JOE-16-0598. 28232636. Review.
63. Sethi JK, Vidal-Puig AJ. Thematic review series: adipocyte biology. Adipose tissue function and plasticity orchestrate nutritional adaptation. *J Lipid Res.* 2007;48:1253–1262. https://doi.org/10.1194/jlr.R700005-JLR200.
64. Lowell BB, Shulman GI. Mitochondrial dysfunction and type 2 diabetes. *Science.* 2005;307(5708):384–387.
65. Rindler PM, Crewe CL, Fernandes J, Kinter M, Szweda LI. Redox regulation of insulin sensitivity due to enhanced fatty acid utilization in the mitochondria. *Am J Physiol.* 2013;305:H634–H643. https://doi.org/10.1152/ajpheart.00799.2012.
66. Caro JF, Dohm LG, Pories WJ, Sinha MK. Cellular alterations in liver, skeletal muscle, and adipose tissue responsible for insulin resistance in obesity and type II diabetes. *Diabetes Metab Rev.* 1989;5:665–689. https://doi.org/10.1002/dmr.5610050804.
67. Zierath JR, Krook A, Wallberg-Henriksson H. Insulin action and insulin resistance in human skeletal muscle. *Diabetologia.* 2000;43:821–835. https://doi.org/10.1007/s001250051457.
68. Ritov VB, Menshikova EV, He J, Ferrell RE, Goodpaster BH, Kelley DE. Deficiency of subsarcolemmal mitochondria in obesity and type 2 diabetes. *Diabetes.* 2005;54:8–14. https://doi.org/10.2337/diabetes.54.1.8.
69. Patti ME, Butte AJ, Crunkhorn S, et al. Coordinated reduction of genes of oxidative metabolism in humans with insulin resistance and diabetes: potential role of PGC-1 and NRF1. *Proc Natl Acad Sci USA.* 2003;100:8466–8471. https://doi.org/10.1073/pnas.1032913100.
70. Peluso G, Petillo O, Margarucci S, et al. Decreased mitochondrial carnitine translocase in skeletal muscles impairs utilization of fatty acids in insulin-resistant patients. *Front Biosci.* 2002;7:a109–a116. https://doi.org/10.2741/A745.
71. Gómez-Pérez Y, Capllonch-Amer G, Gianotti M, Lladó I, Proenza AM. Long-term high-fat-diet feeding induces skeletal muscle mitochondrial biogenesis in rats in a sex-dependent and muscle-type specific manner. *Nutr Metab (London).* 2012;9:15. https://doi.org/10.1186/1743-7075-9-15.
72. Wessels B, van den Broek NM, Ciapaite J, et al. Carnitine supplementation in high fat diet-fed rats does not ameliorate lipid-induced skeletal muscle mitochondrial dysfunction in vivo. *Am J Physiol.* 2015;309:E670–E678. https://doi.org/10.1152/ajpendo.00144.2015.
73. Boden MJ, Brandon AE, Tid-Ang JD, et al. Overexpression of manganese superoxide dismutase ameliorates high-fat diet-induced insulin resistance in rat skeletal muscle. *Am J Physiol.* 2012;303:E798–E805. https://doi.org/10.1152/ajpendo.00577.2011.

74. Taddeo EP, Laker RC, Breen DS, et al. Opening of the mitochondrial permeability transition pore links mitochondrial dysfunction to insulin resistance in skeletal muscle. *Mol Metab.* 2014;3:124–134. https://doi.org/10.1016/j.molmet.2013.11.003.
75. Boushel R, Gnaiger E, Schjerling P, Skovbro M, Kraunsoe R, Dela F. Patients with type 2 diabetes have normal mitochondrial function in skeletal muscle. *Diabetologia.* 2007;50:790–796. https://doi.org/10.1007/s00125-007-0594-3.
76. Davidsen PK, Gallagher IJ, Hartman JW, et al. High responders to resistance exercise training demonstrate differential regulation of skeletal muscle microRNA expression. *J Appl Physiol.* 2011;110:309–317.
77. Baur JA, Pearson KJ, Price NL, et al. Resveratrol improves health and survival of mice on a high-calorie diet. *Nature.* 2006;444:337–342.
78. Nishikawa T, Kukidome D, Sonoda K, et al. Impact of mitochondrial ROS production in the pathogenesis of insulin resistance. *Diabetes Res Clin Pract.* 2007;77:S161–S164.
79. Pettersen KF, Befroy D, Dufour S, et al. Mitochondrial dysfunction in the elderly: possible role in insulin resistance. *Science.* 2003;300(5622):1140–1142.
80. Pereira RO, Tadinada SM, Zasadny FM, et al. OPA1 deficiency promotes secretion of FGF21 from muscle that prevents obesity and insulin resistance. *EMBO J.* 2017;36:2126–2145.
81. Lai N, Kummitha C, Hoppel C. Defects in skeletal muscle subsarcolemmal mitochondria in a non-obese model of type 2 diabetes mellitus. *PLoS One.* 2017;12(8). https://doi.org/10.1371/journal.pone.0183978, e0183978.

CHAPTER 9

Mitochondrial abnormalities in neurological disorders

Ana Flávia Fernandes Ferreira, Monique Patricio Singulani, and Luiz Roberto G. Britto
Laboratory of Cellular Neurobiology, Department of Physiology and Biophysics, Biomedical Sciences Institute, University of São Paulo, São Paulo, SP, Brazil

Contents

1 Mitochondrial function

1.1 Normal mitochondrial metabolism and function

Mitochondria are key organelles to energy metabolism. They provide adenosine triphosphate (ATP) for eukaryotic cells, which is extremally necessary to diverse cellular functions. In addition, mitochondria participate in calcium (Ca^{2+}) homeostasis, in reactive oxygen species (ROS) regulation, and in apoptotic processes.[1] In the first part of this chapter, the main characteristic of this organelle will be revised. Fig. 1 illustrates the mitochondrial general aspects that will be discussed in the following sections.

1.1.1 Structure

Mitochondria structure is characterized by a double-membrane system that delimits the organelle in three compartments. From the outside in, there is the outer mitochondrial membrane that faces the cytosol, and the inner

Mitochondrial Metabolism
https://doi.org/10.1016/B978-0-12-822416-8.00012-9

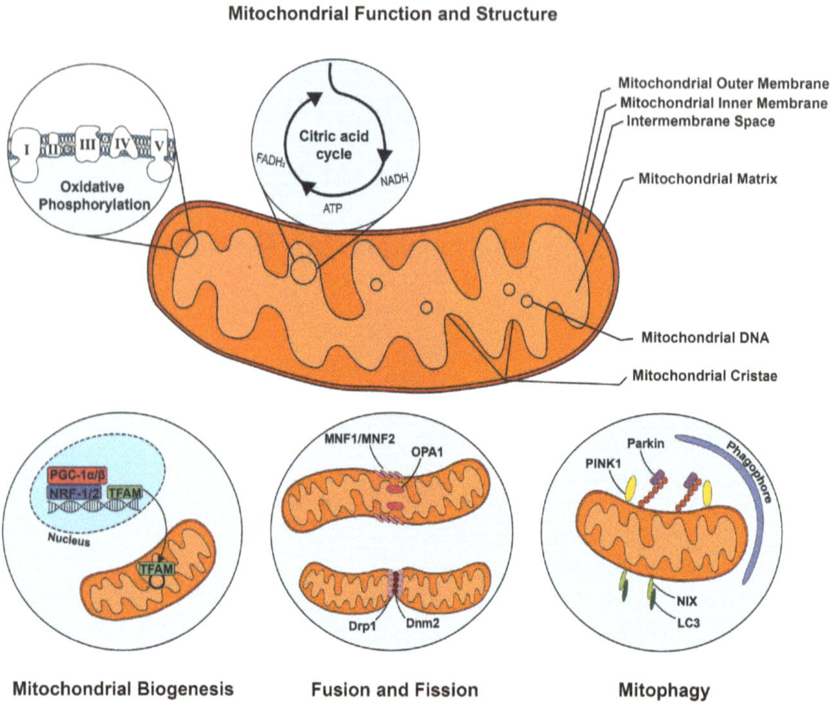

Fig. 1 Basic aspects of mitochondria. Mitochondria are organelles that have an outer and inner membrane. The space between the mitochondrial inner and outer membrane is called the intermembrane space. The inner membrane surrounds the mitochondrial matrix, where mtDNA is found. Folds in inner membrane form the mitochondrial cristae. Five protein complexes (complexes I–V) are located in the mitochondrial inner membrane and are responsible for ATP production through oxidative phosphorylation. Citric acid cycle takes place in the mitochondrial matrix, where NADH, ATP, and $FADH_2$ are formed. Mitochondrial integrity is maintained through mitochondrial biogenesis, fusion and fission, and mitophagy. In mitochondrial biogenesis, important transcription factors, such as PGC-1α/β, NRF-1/2, and TFAM, act in nuclear and mitochondrial DNA and participate in the transcription of mitochondrial proteins. In fusion, the proteins MNF1/MNF2 and OPA1 are involved in the fusion of outer and inner mitochondrial membranes, respectively. In fission, Drp1 is responsible for membrane constriction, while Dnm2 scissions the membrane. In mitophagy, PINK1 accumulates in damaged mitochondria and recruits Parkin that ubiquitinates key proteins in the outer membrane. There is also the involvement of NIX, which interacts with LC3. Either way, phagophores recognize these proteins and lead mitochondria to degradation.

mitochondrial membrane. The outer mitochondrial membrane has unspecific pores that allow free crossing of ions and molecules weighing less than 10 kDa. The inner mitochondrial membrane has a distinct composition that acts as a tight diffusion barrier to all ions and molecules, making it a selective barrier. This membrane delimits the inner compartment, the mitochondrial matrix.

Mitochondrial DNA replication, transcription, protein biosynthesis, and citric acid cycle biochemical reactions take place in the mitochondrial matrix. A second compartment is formed by invaginations of inner membrane that invade the matrix; they are the mitochondrial cristae. Proteins involved in ATP synthesis are mainly located in the cristae. Lastly, between the inner and the outer membrane there is the intermembrane space.[2,3]

Mitochondria also have a small amount of their own DNA, which is called the mitochondrial DNA (mtDNA). Mitochondrial genome is circular and is located in the matrix. It contains 37 genes that encode 13 proteins, 22 tRNAs, and 2 rRNAs. The encoded proteins are enzymes involved in ATP synthesis pathway. Due to its small size, mtDNA cannot produce all mitochondrial proteins, so that the other proteins are coded by nuclear DNA (nDNA) and need to be imported to mitochondria.[4]

1.1.2 Function

Mitochondria are also known as the powerhouse of the cell because of their essential energy production function. Despite being important in energy, mitochondria have other functions that are not always explored. They play a role in ROS production, carbon metabolism, sugars, fats, and protein oxidation, nitrogen metabolism, heat production, and storage of Ca^{2+}.[5] Here, mitochondrial involvement in ATP synthesis will be discussed (for more information about the other mitochondrial functions, see review article[6]).

Two main biochemical pathways that occur in mitochondrial compartments are necessary for ATP synthesis: the citric acid cycle (also known as the tricarboxylic acid cycle or Krebs cycle) and the oxidative phosphorylation. The first one takes place in the mitochondrial matrix. The citric acid cycle links the metabolism of carbohydrates, fats, and proteins with the production of molecules that will be used in oxidative phosphorylation to synthetize ATP. A molecule of acetyl-CoA, derived from carbohydrate, fat, and protein metabolism, is combined with oxaloacetate to form citrate. The cycle continues with a series of other reactions that ends with the formation of oxaloacetate that combines to another acetyl-CoA molecule. In this cycle, ATP and the byproducts NADH and $FADH_2$ are produced and used in oxidative phosphorylation.[7]

In the oxidative phosphorylation process, electrons flow through four protein complexes that are in the cristae of the mitochondrial inner membrane. They are also known as the electron transport chain (ETC), as these complexes transport electrons from electron donors to electron acceptors via reduction and oxidation reactions. Complex I (NADH/ubiquinone oxidoreductase) represents the entry point of electrons from NADH.

The electron is transferred to ubiquinol, also present in the inner membrane, and the energy released in this process is used for pumping protons from the matrix into the intermembrane space. Complex III (cytochrome *c* reductase) transfers the electrons from ubiquinol to cytochrome *c*. Then, complex IV (cytochrome *c* oxidase) transfers the electron to a molecule of oxygen, making two molecules of water. Complexes III and IV also contribute to pumping protons into the intermembrane space, creating an electrochemical gradient. Complex II (succinate dehydrogenase) transfers electrons from a molecule of succinate to ubiquinol. A fifth complex, the complex V or ATP synthase, uses this electrochemical gradient to produce ATP by moving protons from intermembrane space back to the mitochondrial matrix. The respiratory chain comprises the electron transport chain and the ATP synthase.[2]

During this process, electrons can be released before being transported to oxygen by complex IV. These free electrons can react with oxygen and nitrogen and generate ROS and reactive nitrogen species (RNS), respectively. Complexes I and III represent two of the most relevant locations of ROS production. ROS and RNS have important roles in signaling pathways. However, they can cause damage to cells, as they react and alter protein, lipid, and DNA structure and function.[8]

1.1.3 Biogenesis

Mitochondrial biogenesis, together with fusion and fission that will be discussed later, is important for the maintenance of mitochondrial density and content and formation of new organelles. As it was mentioned previously, mitochondrial proteins and other components are produced by mtDNA and nDNA. This process is regulated by transcription factors that coordinate nuclear and mitochondrial gene expression.

One of these factors is the mitochondrial transcription factor A (TFAM). When TFAM binds to its recognition sites in mtDNA, it facilitates the interaction of other transcriptional components to the mitochondrial genes. Besides promoting the transcription of mtDNA, TFAM is also important for the stabilization and maintenance of the mitochondrial chromosome, having a histone-like architectural role in mtDNA.

Nuclear respiratory factors 1 and 2 (NRF-1 and NRF-2) represent a level of regulation of nuclear genes. NRF-1 binds to promoters involved in mitochondrial respiratory function. The genes activated by NRF-1 encode subunits of the respiratory complexes and enzymes of the heme biosynthetic pathway (a cofactor to the function of electron transport chain). The promoters of TFAM are also activated by NRF-1, as well as promoters of proteins required for mitochondrial protein import, like TOM20. NRF-2, like NRF-1, coordinates the expression of genes involved in respiratory function, including cytochrome oxidase and respiratory complex subunits, as well as genes for TFAM.

The peroxisome proliferator-activated receptor gamma coactivator-1alpha (PGC-1α) is considered to be the central regulator of mitochondrial biogenesis, as it can coordinate the expression and activity of NRF-1, NRF-2, and TFAM, besides other transcription factors. PGC-1α promotes the expression of proteins involved in thermogenic regulation, mitochondrial respiration, glucose uptake, glycolysis, and β-oxidation. It also links nuclear and mitochondrial events through activation of TFAM. The peroxisome proliferator-activated receptor gamma coactivator-1 beta (PGC-1β) is part of the same family of nuclear transcription factors as PGC-1α. PGC-1β shares sequence similarity with PGC-1α; however, they are functionally distinct. Despite that, both activate NRF-1 and induce mitochondrial biogenesis.

Besides TFAM, NRF-1, NRF-2, PCG-1α, and PGC-1β, there are other transcription factors that participate in mitochondrial biogenesis. The estrogen-related receptor alpha (ERRα) and peroxisome proliferator-activated receptors (PPARs) are examples of these other important transcription factors. They regulate genes involved in oxidative metabolism, such as the fatty acid oxidation enzymes.[9, 10]

1.1.4 Fusion and fission

Two other important processes that occur to assure mitochondrial health and function are fusion and fission. In fusion, inner and outer membranes from two mitochondria are united. This process can compensate for defects in mitochondria, as damaged components can be shared. In fission, one mitochondrion is divided into two mitochondria. This process is important to separate damaged mitochondria from mitochondrial network, to facilitate mitochondria trafficking in the cell, and to form new mitochondria to populate cells.[11]

There are key proteins that participate in mitochondrial fusion and fission. To fission, the inner and outer membrane lipid bilayers need to be divided and fuse. Dynamin-related/-like protein 1 (Drp1) is responsible for constricting the membranes. Drp1 is recruited by Mid49, Mid51, Fis1, and Mff. Dynamin2 (Dnm2) is responsible for the scission in the constricted membranes, leading to two mitochondria. In mitochondrial fusion, there is a participation of the mitofusins 1 and 2 (Mfn1 and Mfn2) and optic atrophy 1 (OPA1). Mfn1 and Mfn2 mediate fusion between mitochondrial outer membranes. OPA1 mediates fusion between mitochondrial inner membranes.[3]

1.1.5 Mitophagy

The selective degradation of damaged mitochondria by autophagy is called mitophagy. This process is also important to guarantee mitochondrial health and population, and could rescue the cell preventing its death. When mitochondria are damaged, the voltage-sensitive kinase, PTEN-induced putative kinase 1 (PINK1), accumulates in the mitochondrial outer membrane. Another protein, Parkin, an E3 ubiquitin ligase, is then translocated and recruited to mitochondria, and this process is facilitated by PINK1. Parkin ubiquitinates multiple proteins, leading to an accumulation of ubiquitin modifications that recruit p62 and conduct to mitochondrial degradation. Parkin can affect proteins of mitochondrial permeability transition pore opening, which can induce the release of cytochrome *c* and the consequent activation of apoptotic pathways. Mitophagy in red blood cells is mediated by another protein, the NIP3-like protein X (NIX), which binds to microtubule-associated protein 1A/1B-light chain 3 (LC3) and GABA receptor–associated protein, recruiting the autophagy machinery. This process needs to be tightly regulated to ensure cellular survival.[12, 13]

1.2 Mitochondria in brain

Our nervous system is formed by different cell types, including neurons and glial cells. Their functions guarantee the diverse activities performed by the nervous system, including control, regulation, and communication between all the other systems in the body. Mitochondria play an essential role in neurons and glial cells, for example, supplying their bioenergetic demand. During development, neurons proliferate and differentiate. Synapses are formed and neurotransmitters are released. Organelles need to be trafficked through a large axon. Neurons undergo programmed cell death. These and many other processes that take place in neurons and glial cells are modulated by mitochondrial dynamics. All this considered, it is not difficult to imagine what consequences mitochondrial abnormalities would bring to the nervous system.[14] In this section, we will overview some of the studies that highlight mitochondrial importance in brain.

After many years of controversy, now we know that our brain activity is primarily fueled by ATP from oxidative glucose consumption. The glutamatergic neurotransmission is one example of the high rates of oxidative metabolism used by neurons. About one molecule of glucose is spent to produce one molecule of glutamate that is released by neurons. This energy is supplied by glucose metabolism and citric acid cycle, being the last one fulfilled in mitochondria.[15] Not only do neurons use the energy provided by mitochondria, but the global energy budget is also spent by glial cells. A recent study pointed out that in astrocytes, the citric acid cycle is used as energy to perform glutamatergic activity, such as converting glutamate into glutamine.[16]

Mitochondria are also important for synaptic plasticity and morphogenesis. As synapses need a great amount of ATP, mitochondria are usually found in axonal terminals. Defective synapses were associated to mitochondrial loss in axonal terminals, supporting mitochondrial fundamental functions in neurons. Mitochondria are also found in dendritic spines, where they play an important role in synaptic activity. Experimental approaches revealed that the dendritic spine number and synaptic plasticity were increased after an addition of dendritic mitochondrial content. On the other hand, a reduction in dendritic mitochondrial content caused a loss of synapses and dendritic spines.[17] Presynaptic mitochondria are also involved in sequestering Ca^{2+}, enhancing synaptic transmission recovery. All this evidence indicates that mitochondria are crucial for neuronal and glial cell survival and function.[18] In the remainder of this chapter, it will be discussed how mitochondrial alterations could be related to aging and neurological disorders.

1.3 The role of mitochondria in aging

What regulates mortality and aging has always been one of the fundamental questions of life. Innumerous theories have been proposed to understand the process of aging. Some of the current theories involve the accumulation of undegradable byproducts of metabolism, impairment of regulatory pathways, telomere shortening, biological gene clocks, and mitochondria and ROS.[19] The free-radical theory of aging was proposed by Denham Harman in the 1950s. He proposed that degenerative changes associated with aging may be due to free-radical reactions that occur between ROS, likely coming from mitochondria, and proteins, lipids, and nucleic acids. This damage is not repaired and accumulate over the years, leading to aging and related desorders.[20]

This theory was debated along the years, and the current consensus is that the ROS generated by mitochondria are a contributor, but not solely, to aging and related disorders. Nevertheless, other aspects of mitochondria have grown in attention. Mutations in mtDNA increase with age. Despite more than 60% of mutations being needed to result in bioenergetic deficits in humans, animal models that can accumulate extremely high levels of mitochondrial mutations exhibit a shortened life span, indicating a possible relation between the mtDNA mutations and aging that needs to be explored. A decrease of mitochondrial respiratory chain activity was also reported in older subjects. It is interesting to note that studies observed that this reduction mainly takes place in complex I. The ability to regulate mitochondrial biogenesis is diminished in aging. There is also evidence that mitophagy declines in mammalian cells during aging. Thus, an increase in mitochondria damage associated with a decline in the capacity to remove and renew dysfunctional mitochondria could be regulating aging and age-related disorders.[21, 22]

2 Mitochondrial and neurological disorders

As we discussed previously, the brain is highly dependent on mitochondria. In this way, mitochondrial abnormalities could contribute to the pathogenesis of various neurological disorders. Some of the most common neurological diseases, such as Parkinson's, Alzheimer's, Huntington's diseases, amyotrophic lateral sclerosis, multiple sclerosis, and epilepsy, have mitochondrial dysfunctions as an important aspect (Fig. 2). Various investigations have been made to understand this involvement better and to use mitochondrial pathways as therapeutic targets for these diseases.

2.1 Parkinson's disease

Parkinson's disease (PD) is the second most common neurodegenerative disease among elderly. It was first described by James Parkinson in 1817, as Shaking Palsy.[23] PD has a prevalence of 1% in individuals over 60 years old and of 4% above 80 years old. The disease onset before 40 years old represents only 5% of the cases.[24] In addition, 5%–10% of cases are genetic, and they usually have an early onset.[25] Thus, sporadic cases are the most common, and there are still many debates on what would be the cause of this form of PD.

The main symptoms of PD are focused on the motor system. Patients present bradykinesia (slowness of movement), akinesia (absence of movement), rigidity (increased muscle tone), tremor, gait disturbances, and postural instability. An asymmetric onset spreading to become bilateral with time is also evident.[26] Recently, nonmotor symptoms have gained attention.

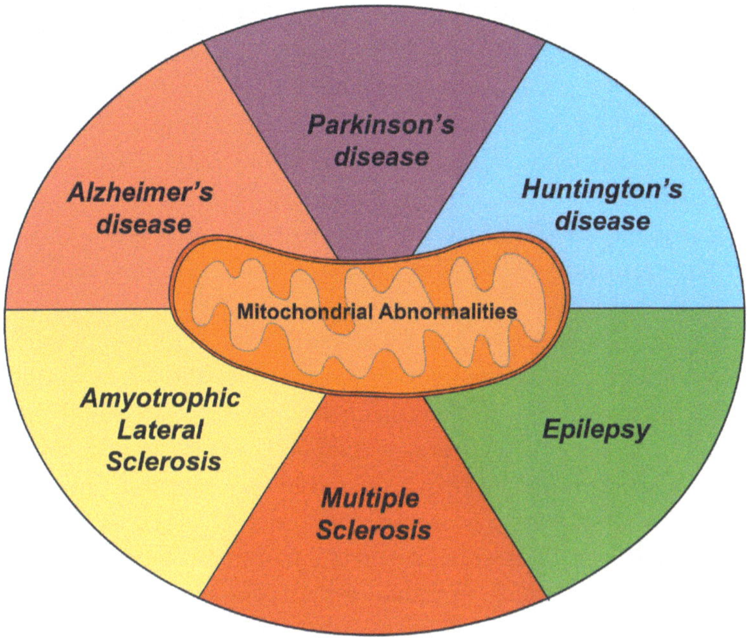

Fig. 2 Mitochondrial abnormalities in neurological disorders. All neurological disorders discussed in this chapter, Alzheimer's, Parkinson's, Huntington's diseases, amyotrophic lateral sclerosis, multiple sclerosis, and epilepsy, have mitochondrial deficits involved in their pathophysiological mechanism.

They usually appear before the motor symptoms and are still neglected in the clinic, which makes the diagnosis difficult. Among nonmotor symptoms, there are hyposmia, depression, sleep disturbances, cognitive decline, and pain.[27]

The involvement of basal nuclei in the pathogenesis of PD is well established. Basal nuclei, as the names say, are situated at the base of the forebrain. They are engaged in motor control, among other functions including emotion, behavior, and motor learning. The main components of basal nuclei are the striatum (or caudate-putamen), globus pallidus, subthalamic nucleus, and substantia nigra (SN) (divided into pars compacta and pars reticulata). In summary, the functioning of basal nuclei in motor control involves projections from cerebral cortex that activate striatum-pallido-thalamus circuits that close the motor loop with efferences back to motor cortex. Regulation of this system is able to facilitate or inhibit motor movements. Substantia nigra plays an important role in this regulation. Substantia nigra pars compacta (SNpc) releases dopamine in striatum. Nigrostriatal projections exert different effects on striatal neurons via two types of receptors. Either way, through dopamine receptor type 1 (D_1R) and dopamine receptor type 2 (D_2R), SNpc promotes movement facilitation.[28]

In PD, there is a degeneration of SNpc dopaminergic neurons. This results in severe deficiency in dopamine release in striatum, reducing movement initiation from cortical neuronal activation dependent on basal nuclei. As a consequence, dysfunctions on motor control are evident. They cause the cardinal features in PD patients, like bradykinesia and akinesia.[28] Besides SNpc degeneration, PD is also characterized by the presence of Lewy bodies. This structure was discovered by Fritz Jacob Heinrich Lewy in the 1900s, when he found inclusions in brains of patients with PD.[29] Lewy bodies are aggregates of α-synuclein in neuronal cell bodies, which alter their functional state. Interestingly, Lewy bodies are found not only in basal nuclei, but they are also present in olfactory bulb and lower brain stem, for example, of PD patients. This goes in accordance with the discovery that other areas are affected in PD, such as locus coeruleus, brain stem, raphe nucleus, dorsal motor nucleus of the vagus, amygdala, and hippocampus.[26]

About 200 years after its discovery, PD pathology remains unclear, PD's clinical diagnostic remains based on motor symptoms, and there is no cure for the disease. Many researchers have been studying PD, trying to answer these questions, and many theories have been raised. Some of them involve oxidative stress, mitochondrial dysfunction, altered proteolysis, genetic mutations, excitotoxic mechanisms, and neuroinflammation as possible pathogenic mechanisms for PD. The use of experimental models that mimic dopaminergic neuronal loss and Lewy bodies has been crucial to the advances in PD knowledge. However, there are still many gaps in the literature that need to be further investigated.

2.1.1 First reports of mitochondrial abnormalities in patients

The hypothesis that mitochondrial dysfunctions are associated with PD started to gain attention after studies published in the late 1970s and early 1980s. In 1979, a case was reported of an addicted man who presented muteness, severe rigidity, severe bradykinesia, generalized weakness, and tremor, among other symptoms. The man tried to synthetize 1-methyl-4-phenyl-4-propionoxypiperidine (MPPP) to use as an injected drug. He died after 2 years of the onset of the symptoms. Brain autopsy revealed the destruction of SN and structures similar to Lewy bodies.[30] In 1983, Langston and colleagues observed the presence of severe parkinsonism symptoms in four individuals, aged between 26 and

42 years, who used a synthetic form of heroin. The drug injected by the group contained varying amounts of 1-methyl-4-phenyl-1,2,3,6-tetrahydropyridine (MPTP) and MPPP. Patients revealed near total immobility, increase in tone, inability to speak comprehensibly, rigidity in the upper extremities, tremor at rest, and flexed posture, as observed in advanced cases of PD.[31] These cases were very similar to the one described in 1979.

Few years later, the mechanism of MPTP was discovered. In 1985, a study showed that MPP^+, the main metabolite of MPTP, is capable of inhibiting ADP-stimulated and uncoupled oxidation of NADH-linked substrates, thus inhibiting mitochondrial complex I.[32] This study provided a possible link between parkinsonism, the use of MPTP reported by the previous articles, and mitochondrial dysfunction. In 1989, mitochondrial alterations in PD patients were first described by Schapira and coworkers,[33] confirming the involvement of mitochondria in PD. This study reported the activity of mitochondrial respiratory chain in postmortem SN of nine PD patients. A reduction was found, when comparing complex I activity of PD patients with control individuals.

After Schapira's study, many other papers were published, trying to better understand the mitochondrial dysfunction role in PD. The complex I activity was found to be 50% less in platelets of PD patients than in controls.[34] In striatum of patients who died from PD, protein expression of three subunits of complex I was reduced.[35] A slower flux through the respiratory chain, and reduced activities of complexes I, II, and IV were observed in skeletal muscle samples from PD patients, when compared to controls.[36] Reduced complex I, IV, and IV activity was also reported in skeletal muscle of patients,[37] whereas no difference was observed in any mitochondrial complex of skeletal muscle by another other paper.[38] These studies did not agree on the presence of mitochondrial respiratory chain deficits, nor in which complexes are affected. Innumerous research groups have been studying mitochondria and PD association since then, trying to clarify these issues. In the next topic, the main results published until now will be discussed.

2.1.2 Genetic alterations

As mentioned previously, the majority of PD cases are sporadic, and familiar PD corresponds to a little percentage of total cases. However, it is important to mention what we know about the latter type of PD, since, as it will be shown later, an involvement of mitochondria is also observed in these cases. Mutations in different genes, with autosomal dominant or autosomal recessive inheritance, were related to PD familiar cases. Dominant mutations include SNCA (encoding for α-synuclein) and LRRK2 (leucine-rich repeat kinase 2) genes. Recessive mutations include Parkin (PARK2), PINK1 (PARK6), DJ-1 (PARK7), and cation-transporting ATPase 13A2 (ATP13A2) genes.[39]

In the dominant mutations, SNCA was the first gene reported in early onset PD. This gene encodes α-synuclein, and its mutation was first reported in one Italian and in three unrelated families of Greek origin with PD.[40] α-Synuclein was described to aggregate and form Lewy bodies. Mutations in SNCA cause an instability in α-synuclein, as the protein is not completely unfolded and loses its native conformation.[41] Studies have also pointed that increased levels of α-synuclein could induce mitochondrial fragmentation and ROS production,[42] cause disruption to Ca^{2+} homeostasis and to mitochondrial ATP production.[43] α-Synuclein has also been reported to reduce mitochondrial biogenesis, as it could downregulate PGC-1α.[44]

LRRK2 mutations are the most common causes of late-onset dominant PD. LRRK2 gene encodes a protein kinase that is capable of protein-protein interactions. Mutations reported in LRRK2 could alter the protein relationship with other proteins.[39] Increased vulnerability to mitochondrial toxins, defects in mitochondrial dynamics, and increased ROS production were also related to LRRK2 mutations. Other example of LRRK2 mutation deficit in mitochondria could be related to the fact that this protein is able to phosphorylate proteins involved in mitochondrial fission and fusion, such as Drp1, Mfn1/2, and OPA1. Recently, mutations in vacuolar protein sorting-associated protein 35 (VPS35), that seems to interact with mitochondrial fission and fusion proteins, and with *coiled-coil-helix-coiled-coil-helix domain containing 2* (*CHCHD2*), a mitochondrial intermembrane space protein that regulates complex IV, were identified in familial PD.[42]

In the recessive mutations, Parkin (PARK2) mutations are the most frequent causes of PD. Parkin ubiquitinates multiple proteins, playing an important role in mitophagy. Mutations in Parkin could bring abnormalities in mitochondrial morphology and function. Besides that, Parkin mediates the degradation of a repressor of PGC1-α. Thus, its loss leads to an accumulation of the PGC1-α repressor and to a decline in mitochondrial biogenesis. The other recessive mutation involves PINK1 that also participates in the mitophagy process. Loss of PINK1 function could impair mitochondrial degradation, morphology, and trafficking. ATP13A2 mutations were associated with reduced ATP production, increased mitochondrial fragmentation, and ROS production.[42] Lastly, mutations in DJ-1, which functions as a cellular sensor of oxidative stress, were also identified in familial PD.[39]

Mutations in mitochondrial genome have been associated with PD. Single-nucleotide polymorphisms (SNPs) in mtDNA were related to an increase or a decrease in risk to develop PD.[45] Individuals carrying SNP alleles 10398G and 13708A demonstrated decreased risk of PD. These SNPs cause modifications in NADH dehydrogenase 3 and NADH dehydrogenase 5 subunits of complex I. This protective effect could be explained by the fact that individuals who inherit these genotypes can have an increased complex I performance in brain and other tissues.[46] In addition, mtDNA deletions in individual SN neurons from individuals with PD were reported recently.[47] This study showed that SN neuron contains a large and heterogeneous pool of mtDNA deletions, which implies a susceptibility of dopaminergic neurons to accumulate mtDNA.

2.1.3 Mitochondrial dysfunction

After the first reports of altered complex I activity in PD by Schapira[33] and MPTP-induced parkinsonism by Langston,[31] compounds used to inhibit complex I activity gained attention. These compounds are able to reproduce parkinsonian features in primate and rodent experimental models. Innumerous animal studies adopt this strategy to study pathological mechanisms of PD. Nowadays, MPTP, 6-hydroxydopamine (6-OHDA), rotenone, and paraquat are commonly used to induce PD in animals. The MPTP model is induced systematically. In glial cells, MPTP is metabolized to MPP^+ that through the dopamine transporter and inhibits complex I of SN neurons. The 6-OHDA is also transported to dopaminergic neurons through the dopamine transporter, into the cell it is autoxidized, inhibits complex I, and causes oxidative stress. The pesticides rotenone and paraquat can affect mitochondria by inhibiting or interacting with complex I, respectively, oxidatively damaging the mitochondria. Despite some variation, the models can induce dopaminergic neuronal degenerations, altered behavior in animals, and Lewy bodies' formation.[48, 49] These factors highlight the association of complex I dysfunction and PD.

Recent human studies have reported abnormalities in mitochondrial complexes, which bring more support to mitochondrial hypothesis in PD pathology. Altered mitochondrial functions were shown in other brain areas of PD patients, such as frontal cortex and angular gyrus. Downregulation of genes encoding mitochondrial subunits and energy metabolism-related enzymes, decrease in the protein expression of mitochondrial subunits, and reduction in the activity of complexes I, II, III, and IV were among the dysfunctions reported.[50] Besides that, another study showed that complex I deficiency spreads along different brain areas, including in prefrontal cortex, hippocampus, striatum, cerebellar cortex, and dentate nucleus.[51]

Other alterations of energy were also associated with PD, including in glycolysis and oxidative decarboxylation of pyruvate.[52] One of the first studies measured the oxidative decarboxylation of pyruvate and succinate, important enzymes in citric acid cycle, in fibroblasts from patients with PD. They reported that pyruvate turnover was lower in PD fibroblasts.[53] In another study, dopaminergic cells were exposed to paraquat. Molecules involved in pentose phosphate pathway, a metabolic pathway parallel to glycolysis that also generates NADPH, were increased by the paraquat treatment. This effect was paralleled by the inhibition of enzymes involved in glycolysis and the citric acid cycle.[54] Thus, alterations in energy metabolism could also contribute to cell death progression, such as the one observed in PD.

2.1.4 Mitochondrial biogenesis

Analyses from 522 gene sets from PD patients were performed and revealed that the expression of genes in response to PGC-1α was decreased.[55] Another study showed that different PGC-1α SNPs present in PD patients were possibly associated with the risk of disease, and variants observed in controls were associated with longevity.[56] A reduction in PGC-1α expression and protein levels was observed in the SN of PD patients.[57] These results point to a possible role of mitochondrial biogenesis in the pathogenesis of PD. Further studies with an animal approach corroborate these ideas. Using cellular culture from human neurons, researchers compromised mitochondrial biogenesis through a knockdown of PGC-1α and observed accumulation of α-synuclein, as well as decreased levels of prosurvival proteins.[58] In α-synuclein transgenic mice, and in a cell culture model for α-synuclein oligomerization, both experimental models of PD, PGC-1α were downregulated. In the same study, they overexpressed and silenced PGC-1α; and observed reduction and accumulation of α-synuclein, respectively.[57] Using transgenic mice overexpressing PGC-1α in dopaminergic neurons, researchers observed resistance against cell degeneration induced by MPTP.[59]

Contributions of other factors involved in mitochondrial biogenesis have also been extensively studied. Some articles point to mitochondrial biogenesis deficiency in PD patients, while others explored this association using animal models. The postmortem tissue of SN from PD patients was submitted to histological assays that revealed a reduction in the numbers of subunits of complexes I and II, lower abundance of TFAM and TFB2M,[60] and decreased expression of TFAM.[61] In skin fibroblasts from PD familiar patients, the gene expression of NRF-1, NRF-2, and TFAM was reduced, as well as protein expression from complexes I, III, IV, and V.[62] Similar results were observed using animal models. In the MPTP PD model, a decreased expression of NRF-1 was found. In the 6-OHDA PD model, a reduction in NRF-1 expression was also observed, and this was accompanied with TFAM, PGC-1α, and complex I decreases.[63]

2.1.5 Fission and fusion

The involvement of disturbances in mitochondrial fission and fusion was reported after the studies with mitochondrial respiratory chain. One of the first studies to show mitochondrial dynamics impairment in neurodegeneration used neuronal cultures exposed to different toxins. It described fragmentation on mitochondria after exposure to nitric oxide, rotenone, MPP^+, and amyloid beta (Aβ) peptide. The fission observed was prior to neuronal death. Blocking fission with a negative mutant of Drp1 or expressing the protein involved with fusion, Mfn1, prevents mitochondrial fragmentation and neurodegeneration. These results suggest that a disbalance of fission may precede cell death, which could be related to what happens in PD.[64]

After that, other experimental studies reported mitochondrial dynamics impairment. Using 6-OHDA in cell culture, a profound mitochondrial fragmentation was observed. A similar effect was accomplished overexpressing Fis1, while silencing of Drp1 prevented 6-OHDA induced fragmentation.[65] More recently, using a PD model in which mitochondria from PD patient's platelets were transferred to cytoplasmic hybrid cells showed alteration in the morphology and distribution of the mitochondrial network, including a fragmented mitochondrion. A reduction in mitochondrial levels of the fusion protein OPA1 and an increase in Drp1 were also evident.[66]

2.1.6 Mitophagy

Two main proteins involved in mitophagy have been described as possible causes of familial PD, as mutations in their genes were reported in patients. These proteins are Parkin and PINK1. The importance of these proteins to PD was already described in this chapter. The discovery of PINK1/Parkin involvement in familial early onset of PD reinforced the quality control process of mitochondria as an essential process in PD. Besides the identification of genetic mutations in familial PD, studies have pointed that modifications in mitophagy are also a key player in sporadic form of PD.

Researchers hypothesized that the oxidative stress reported in sporadic PD can damage Parkin function through nitrosative and oxidative stress. Experimental approaches showed that cells treated with nitric oxid donor S-nitrosoglutathione[67] and α-synuclein[68] led to Parkin S-nitrosylation, which diminished Parkin's activity and function. In the MPTP model of PD, S-nitrosylation of Parkin was also observed.[67] Higher levels of nitrosative stress, followed by Parkin S-nitrosylation, were found in postmortem brains of PD patients.[67] Another hypothesis is that dopamine modifies Parkin in dopaminergic cells, which inactivates the protein function. Using postmortem tissue from sporadic PD patients, a study reported the presence of insoluble Parkin in striatum and SNpc, but not in other brain areas. Besides that, in cells expressing Parkin, treatment with dopamine leads to increased insoluble and aggregated Parkin, and loss of its ubiquitin ligase activity.[69] These results provide evidence for the association of Parkin dysfunction and sporadic PD.

Other proteins involved in mitophagy have also been linked with sporadic PD. LRRK2 gene mutations were reported in patients with sporadic PD that also showed nigral cell loss and Lewy bodies.[70] The same LRRK2 mutation was modeled in human-induced pluripotent stem cell–derived neurons from PD patients.[71] In the study, neurons that carried LRRK2 mutation presented impaired mitochondrial arrest and clearance. The authors proposed that LRRK2 forms a complex with Miro, a protein that anchors the microtubule motors, kinesin and dynein, to mitochondria, that binds to damage mitochondria, and that initiates mitophagy. In LRRK2 mutated cells, the complex is not formed, which consequently delays the onset of mitophagy, leading to the accumulation of damaged mitochondria.[71] Thus, deficient mitophagy is possibly present in both familial and sporadic PD.

2.2 Alzheimer's disease

Alzheimer's disease (AD) is a heterogeneous disease, acknowledged as a progressive multifactor neurodegenerative disorder.[72] AD is the most common cause of dementia, accounting for more than 80% of dementia cases worldwide in elderly people. An estimated 5.8 million Americans have Alzheimer's dementia and according to official death certificates, in 2017 more than 121,000 deaths from AD were recorded. This makes AD the sixth leading cause of death in the United States (USA) and the fifth leading cause of death among Americans aged 65 and older, and is becoming prevalent in many other countries. It is estimated that by 2050, one new case of AD is expected to develop every 33 s, or nearly a million new cases per year, and prevalence is projected to be of 13.8 million Americans affected with AD.[73]

AD was first described in 1907 by Alois Alzheimer, a German physician who examined a 51-year-old patient named Auguste Deter, who suffered with disorientation, cognitive and behavioral manifestations of dementia.[74,75] Upon her death in 1906, the autopsy performed by Alzheimer revealed diffuse cortical atrophy and "particular changes in cortical cell clusters" described as plaques and tangles in the cerebral cortex.[74] His discovery was followed by further research in the 1980s that revealed the presence of Aβ plaques and neurofibrillary tangles of tau protein in dementia patients.[74,75]

The neuropathology of AD consists of extracellular deposits of Aβ plaques and intracellular aggregations of neurofibrillary tangles (NFTs), composed of hyperphosphorylated tau.[72] These deposits are constituted chiefly of highly insoluble and proteolysis-resistant peptide fibrils produced by Aβ cleavage. Aβ peptides originate after altered cleavage of the amyloid precursor protein (APP) by the sequential enzymatic actions of beta-site amyloid precursor protein-cleaving enzyme 1 (BACE-1), a β-secretase, and γ-secretase, a protein complex with presenilin 1 (PSEN 1) at its catalytic core.[76,77] Nevertheless, Aβ is not formed if APP is firstly recognized and cleaved by the enzyme α-secretase, instead of β-secretase, in the nonamyloidogenic pathway. In the amyloidogenic pathway, there are two major isoforms of Aβ formed, soluble $A\beta_{40}$ and insoluble $A\beta_{42}$, and the latter peptide shows a greater propensity to aggregate.[78] Aβ aggregation is considered to be the primary reason for the neurotoxicity, and Aβ oligomers are the most neurotoxic form, diffusing into synaptic clefts and interfering with synaptic signaling. Moreover, the aggregation of Aβ might also promote free radicals such as ROS, local inflammation, oxidation, excitoxicity (excessive glutamate), and tau hyperphosphorylation and this process ends up leading to neuronal death.[74,76]

The NFTs are the result of tau protein hyperphosphorylation. Tau protein is a microtubule-associated protein that provides microtubule stability. Abnormally hyperphosphorylated tau forms insoluble fibrils and folds into intraneuronal tangles.[72,79] Consequently, it uncouples from microtubules, leading to an abnormal loss of communication between neurons, loss of signal processing, and finally apoptosis in neurons. It has been reported that soluble Aβ controls cleavage and phosphorylation of tau for NFT generation. Further, phosphorylation of tau is regulated by several kinases, including glycogen synthase kinase 3 (GSK3) and cyclin-dependent kinase 5 (CDK5) activated by extracellular Aβ.[74,80]

The vast majority of AD case occurrences are sporadic where there is no dominant genetic cause, but the genetics of AD should also be considered to play an influential role in the pathogenesis, and can present with very early age of onset and a more rapid rate of progression. The familial AD is the result of autosomal dominant mutations in APP, PSEN1, and PSEN2 genes, located on chromosomes 21, 14, and 1, respectively.[81] Moreover, allele ε4 of apolipoprotein E (ApoE) has been shown in recent studies as the strongest genetic risk factor related to the AD development process.[82] Clinical syndromes consistent with AD are defined by classical symptoms and cognitive profiles.[83] Typical amnestic cases are characterized by loss of basic and day-to-day functional thinking, remembering, reasoning, behavioral, and verbal abilities. Moreover, patients affected with AD may also present an early executive dysfunction or behavioral impairment to such an extent that it interferes with a person's daily life, including day-to-day activities, and renders them incapable of doing anything.[83,84]

Along the past years, Aβ peptides have been considered one of the most promising therapeutic targets for AD. However, based on the Aβ cascade hypothesis, many clinical studies failed. Moreover, some reports have shown weak correlations between amyloid burden and cognitive impairment in patients with AD, and the idea that Aβ pathology is not anymore the leading primary cause of AD has risen.[85,86] Clinical studies performed in post-mortem human brain samples by the National Alzheimer's Coordinating Center (NACC) demonstrated that nearly 30% of 919 subjects diagnosed with AD lacked the presence of neuritic plaques. In addition, these studies showed that nearly 40% of 279 subjects who were not clinically diagnosed with AD presented high levels of neuritic plaques. These findings indicate that amyloid pathology alone cannot be used to confirm a diagnosis of AD during early stages.[87] Instead, there is a need to understand the molecular AD pathogenesis and nowadays the belief that AD is a multifactorial disease is growing steadily.

In recent years, increasing lines of evidence have implicated mitochondrial bioenergetic defects as one of the factors that may actively contribute to the disease onset and progression of the neurodegenerative diseases.[88] More recently, studies have reported that neuroinflammation and oxidative stress reactions mediated by mitochondrial damage have also been implicated as key contributors to the pathophysiology of AD due to the fact that neurons have high and continuous energy demands to maintain their proper function.[88,89] Dysfunction of mitochondria has been linked to both normal aging and neurodegenerative disorders, with the latter occurring much more frequently at a higher age associated with elevated mutation load in mtDNA, defects in mitochondrial respiration and increased oxidative damage.[88,90] Specifically, genetic defects in mitochondria have been shown to accumulate during life, and certain mutations of mitochondrial genes have been implicated in the etiology of AD.[90]

2.2.1 First reports of mitochondrial abnormalities in patients

The impairment of energy metabolism and mitochondrial function in dementia and AD has been demonstrated since the 1970s. In one of these studies, Johnson and Blum, through electron microscopy pictures of frontal and temporal cortex obtained by biopsy from AD patients, revealed altered mitochondrial infrastructures.[91] In the early 1980s, studies evidenced a reduction in the rate of glucose utilization, using positron emission tomography (PET), evaluated in regions in brain from patients with senile dementia of the Alzheimer's type.[92,93] Ferris et al. also demonstrated that the degree of reduction in cerebral metabolic activity in AD patients was highly correlated with measures of degree of cognitive impairment,[93] contributing, in this way, to a better understanding of the pathology of AD. Subsequent investigations suggested that decrease of AD brain glucose utilization could be associated not only with decreased neuronal activity, but also with deficiencies in bioenergetic flux-related enzymes.[94] Corroborating with this information, Sorbi et al. showed that the activity of the pyruvate dehydrogenase complex (PDHC), an important mitochondria-localized enzyme, was reduced in affected areas of brain from patients with AD, reinforcing the hypothesis that the reduction in cerebral metabolic rate may relate to loss of tissue oxidative capacity.[95]

Few years later, Sims et al. performed the first detailed attempt to obtain in vitro estimates of oxygen uptake and evaluation of mitochondrial function carried out on neocortex samples from patients with dementia. In this study, the impairment of several parameters related to mitochondrial respiration was demonstrated. Furthermore, the authors speculated that if these changes persisted in vivo for long time periods, it could contribute to the damage and loss of neurons, which require a continuous supply of energy for normal function,[96] observed in AD. Around this time and shortly thereafter, studies have also shown reduction of cytochrome *c* oxidase activity (COX) in platelet mitochondria isolated from AD patients[97] and mitochondria isolated from brain tissue samples from AD patients.[98] After these findings, several lines of research demonstrated defects in the mitochondrial bioenergetic system in AD. In addition, Blass and Gibson proposed that metabolic stress due to mitochondrial impairment could contribute to the accumulation of senile plaques and NFTs in AD, suggesting that mitochondrial or other bioenergetic lesions could be an occurrence in the course of AD development.[99]

Indeed, given that mitochondria play a critical role in brain metabolism and function, mitochondrial perturbation may contribute significantly to AD pathophysiology. Furthermore, studies have shown that mtDNA deletions and point mutations accumulate with age, perhaps due to oxidative stress present in AD pathology. Years later, based on these new findings, Swerdlow and Khan proposed a new hypothesis to explain the onset of sporadic AD, the "mitochondrial cascade hypothesis." This hypothesis described that each human genetic heritage influences mitochondrial functions and it can influence the onset of AD pathology.[100] In short, this hypothesis states that the mitochondrial dysfunction is the primary process to trigger all the cascade of events that lead to sporadic late-onset AD.[101] And, in addition, the mitochondrial cascade hypothesis also allows overlap histopathologically between the sporadic, late-onset and autosomal dominant, and early onset forms of the disease.[100]

Currently, extensive research argues that mitochondria may contribute to a variety of AD pathologies, despite the fact that the validity of the mitochondrial cascade hypothesis has yet to be completely elucidated. In the next topics, we will consider studies and data that reinforce this hypothesis in AD research.

2.2.2 Genetic alterations

As mentioned previously, mitochondria contain their own DNA that encodes 13 mitochondrial core proteins of the ETC complexes and 2 rRNAs and 22 tRNAs necessary for mitochondrial protein synthesis. While mtDNA is critical to the proper function of mitochondria, it is susceptible to oxidative damage which gives rise to mutations. Mutations in mtDNA, whether through inheritance or through gradual somatic accumulation, propagate through clonal expansion, which may consequently gain important deleterious effects compromising mitochondrial function.[102]

Primary mtDNA mutations have been linked to several mitochondrial disorders that have psychiatric, neurological, and cognitive symptoms. Interestingly, many of these patients with mtDNA mutations evidenced cognitive deficits quite similar to those commonly seen in AD, which supports a critical role of mtDNA in proper cognitive function.[103] In addition, mtDNA changes are greater in postmortem brains from AD patients and aged subjects than from young subjects. In a study performed by Coskun et al., it was evidenced that 65% of the AD brains analyzed harbored the T414G mutation, whereas this mutation was absent from all control brains.[104] Although the T414G mutation was not associated with any obvious phenotypic effects, it occurred in a region that is crucial for mtDNA replication and transcription, thereby raising the possibility that they may have a detrimental effect on mitochondrial function within single cells.[105] This study also revealed, through cloning and sequencing of the mtDNA, that all AD brains had an average 63% increase in heteroplasmic mtDNA mutations in the control region and that AD brains from patients 80 years and older had a 130% increase in heteroplasmic mutations. These mutations can modulate known mtDNA regulatory elements and suppress mitochondrial transcription and replication, which may be responsible for some of the mitochondrial defects observed in AD.[104] Moreover, mtDNA mutations are positively associated with β-secretase activity, since the copy number of mtDNA was inversely associated with the levels of $A\beta_{40}$ and $A\beta_{42}$. Thus, somatic mtDNA mutations may not only be involved in aging phenotypes, but may also contribute to AD development.[106]

2.2.3 Mitochondrial dysfunction

Mitochondrial deficiency has been suggested to be a hallmark of AD, as the patients display early metabolic changes associated with deficiencies in bioenergetic flux-related enzymes that are prior to the emergence of any histopathological or clinical abnormalities. Multiple lines of evidence suggest that mitochondrial defects play a key role in the early stage of AD progression, partly due to oxidative stress. As ROS are a byproduct of mitochondrial respiration, any damage to mitochondria can result in interrupting the usual mechanisms by which ROS are destroyed, therefore increasing the number of ROS present in the organelle.[107] One of the main enzymes of the mitochondrial respiratory chain that has been reported to be decreased in postmortem AD brains is COX.[97] COX is responsible for reducing oxygen radicals and the commitment of its operation results in increase in radical production. This dysfunction forms a cycle in which damaged mitochondria produce more ROS and this in turn harms even more mitochondria leading to an increase in the number of ROS and so on. Excess of free radicals can cause various biochemical changes that are observed in neurodegeneration, such as lipid peroxidation.[107] Moreover, cumulative damage to nDNA and mtDNA can occur, which interferes with mitochondrial function, resulting in low rates of ATP production, increased dependence on anaerobic glycolysis and, therefore, a higher production of mitochondrial ROS.[102] Many studies reported in AD postmortem brains found increased free-radical production, lipid peroxidation, oxidative protein damage, decreased ATP production, and reduced cell viability.[106] Consequently, this results in cell damage that is responsible for some of the classic pathological changes observed in neurodegenerative diseases. In this way, the dysfunction of this organelle may be an important initiating or intermediate site for increased AD oxidative stress and neurodegeneration.[101]

Activities of the enzymes involved in mitochondrial energy production, such as complex IV, PDHC, mitochondrial isocitrate dehydrogenase, α-ketoglutarate dehydrogenase (αKGDH), and ATP synthase complex were found decreased in AD brains, which were associated with ROS overproduction and ATP underproduction.[95, 108, 109] More interestingly, all of the changes in ETC activities were correlated with the clinical state of AD patients.[109] Previous studies have also identified impaired mitochondrial oxidative phosphorylation including mitochondrial complexes I–IV,[110] as well as the defect of ATP synthase in AD individuals.[111] In addition, a genome-wide transcriptome study in laser-capture microdissected neurons demonstrated reduced expression of nuclear genes encoding mitochondrial ETC subunits in the posterior cingulate cortex, compared to those in the primary visual cortex, a region that is relatively spared metabolically in AD.[112] Moreover, using microarray analysis and quantitative RT-PCR, it was evidenced that 15 out of 51 members of the glycolytic, ETC, mitochondrial oxidative phosphorylation, and associated pathways were significantly downregulated in AD.[113]

Structural impairment of mitochondria has also been reported by Hirai et al. that demonstrated that mitochondria are significantly reduced in AD. In this study, through morphometric analysis performed in hippocampal, adjacent temporal cortex, cerebellum, and frontal cortex tissue from autopsy of patients with a diagnosis of AD, a significant decrease in mitochondria of vulnerable neurons in AD was observed, suggesting an intimate relationship between mitochondria and oxidative damage in AD.[114] In a more detailed analysis using human neuroblastoma M17 cells overexpressing APP mutant (APPswe), a study revealed mitochondrial fragmentation with altered size and number and reduced aspect ratio of mitochondria as well as abnormal distribution in these neurons, suggesting a potentially fragmented mitochondrial network in AD brain.[115] From a functional perspective, this study also demonstrated that APP overexpression affected mitochondria at multiple levels, including elevating ROS levels, decreasing mitochondrial membrane potential, reducing ATP production, and caused neuronal dysfunction,[115] all prominent features found in AD brain.

2.2.4 Mitochondrial biogenesis

Another key mitochondrial function is the mitochondrial biogenesis which plays an essential role in maintaining an adequate and functional neuronal mitochondrial mass. In this context, several studies have demonstrated mitochondrial biogenesis impaired in AD. The significant reduction in the number of mitochondria in hippocampal neurons in the brain from AD patients and in M17 cells overexpressing APPswe already suggests that the mitochondrial biogenesis is compromised.[114,115]

Qin et al. first demonstrated the reduced expression of PGC-1α in human postmortem brain samples from AD patients and primary neuronal cultures derived from Tg2576 transgenic mice model of AD, implicating an impaired mitochondrial biogenesis signaling in AD. In this study, it was evidenced that the decreased PGC-1α protein content in the hippocampal formation of the AD cases examined correlated with progression of amyloid neuropathology in AD. Moreover, they found that experimental hyperglycemic conditions in neuron cultures derived from Tg2576 may significantly inhibit PGC-1α expression coinciding with increased Aβ peptide content in the conditioned medium.[116] Mitochondrial biogenesis proteins have also been shown to be reduced in hippocampal tissues from AD patients,[117] in transgenic mice model of AD (Tg2576 line),[118] and in the mouse primary hippocampal (HT22) cells transfected with mutant APP complementary DNA.[119] In these studies, significantly reduced mRNA and protein levels of mitochondrial biogenesis genes, including PGC-1α, NRF-1, NRF-2, and TFAM, were found, indicating that reduced mitochondrial biogenesis is a typical AD feature. Sheng et al. also found that along with the reduced expression of PGC-1α, the level of cyclic AMP response element-binding protein (CREB) phosphorylated at Ser133 and cAMP-dependent protein kinase (PKA) was significantly decreased in APPswe M17 cells, suggesting that reduced PKA activity may be involved in the decreased CREB phosphorylation/activation and PGC-1α. They further demonstrated that reduced expression of phosphorylated CREB and PGC-1α in APPswe M17 cells could be rescued by cAMP in a dose-dependent manner.[117]

Indeed, the impaired mitochondrial biogenesis likely contributes to mitochondrial dysfunction in AD. More recently, Singulani et al. evidenced that impairment of PGC-1α-mediated mitochondrial biogenesis precedes mitochondrial dysfunction and AD pathology in the 3xTg mouse model of AD. In this study, it was demonstrated that essential factors for mitochondrial biogenesis, such as PKAC-α, CREB, PGC-1α, NRF-1, NRF-2, and TFAM, were all reduced in an age at which there were no significant

Aβ oligomer deposition and mitochondrial dysfunction, suggesting that reduced mitochondrial biogenesis causes mitochondrial dysfunction at later ages in 3xTg-AD mice.[120] Taken together, these observations point that impaired mitochondrial biogenesis is likely impaired in ages preceding the development of the AD pathology and contributes to mitochondrial dysfunction in AD.[120] Additionally, studies have shown that the expression of PGC-1α can contribute to pathological Aβ degradation, in part by promoting BACE1 ubiquitination and degradation, thus preventing neurodegeneration in AD.[121]

2.2.5 Fission and fusion

Mitochondrial fusion and fission are crucial for the maintenance of mitochondrial form and integrity. Therefore, deficits in either mitochondrial fusion or fission can also cause human neurological disorders, which accentuate the importance of a balance between mitochondrial fission and fusion in neuronal function and brain health.[102] Wang et al. evidenced that overexpression of APP (APPswe M17 cells) or Aβ-derived diffusible ligand treatment caused an imbalance of mitochondrial fission/fusion, assessed by levels of Drp1 (also referred to as DLP1), OPA1, and Fis1, leading to mitochondrial fragmentation and reduced mitochondrial coverage in neuronal processes in differentiated primary hippocampal neurons.[115] The impact of amyloid pathology on the impairment of mitochondrial fusion and fission has also been demonstrated in postmortem brain tissues from AD patients at different stages of disease progression,[122] and in mAPP-HT22 cells that express Swedish/Indiana mutations.[119] In both studies, the observations found strongly suggest that increased production of Aβ and the interaction of Aβ with Drp1 cause mitochondrial abnormalities and are crucial factors in mitochondrial fragmentation, abnormal mitochondrial dynamics, and synaptic damage, leading to neuronal dysfunction.[119, 122]

In another study, levels of Drp1, OPA1, Mfn1, and Mfn2 were significantly reduced, whereas levels of Fis1 were significantly increased in human neuroblastoma cell line and AD patient brains, suggesting that changes in the expression of these mitochondrial fission and fusion proteins likely underlie abnormal mitochondrial distribution in AD neurons.[123] Increased levels of Drp1 and Fis1 and reduced expression of Mfn1, Mfn2, and OPA1 were also detected in the mouse primary hippocampal (HT22) cells transfected with mutant APP complementary DNA,[119] in mouse models of AD,[124, 125] and in several other studies carried out on brain samples from patients with AD.[122, 123, 125] Moreover, an altered balance in mitochondrial fusion and fission could cause mitochondrial bioenergetic capacity deficits and enhance ROS generation,[126] as well as interfere with mitochondrial motility, thereby compromising mitochondrial trafficking and distribution in neurons, all prominent features found in AD brain.[106]

2.2.6 Mitophagy

Under physiological conditions, mitophagy, a selective autophagy for eliminating damaged mitochondria, plays an essential role in the basal mitochondrial turnover and maintenance. In this context, increasing evidence suggests autophagy/lysosome failure along with concurrent increases in oxidative stress levels as a contributor to an aggravating condition in the pathogenesis of AD. Among the various mitophagy pathways, PINK1-Parkin-dependent mitophagy has been the focus of current studies in AD.[127] Ye et al. provided the first line of evidence that chronic mitochondrial stress under AD-linked pathophysiological conditions effectively triggers Parkin-dependent mitophagy, valued in mutant hAPP neurons and AD patient brains.[128] Moreover, increased PINK1, Parkin were also evidenced in human fibroblasts and hippocampal biopsies from AD patients,[129] as well as in induced pluripotent stem cell–derived human AD neurons and in the APP/PS1 mouse model of AD.[130] In both studies, the findings indicated that mitophagy alterations are a pivotal event in AD pathogenesis and that mitophagy can be considered a new hallmark of AD, besides representing a potential site of therapeutic intervention.

2.3 Huntington's disease

Huntington's disease (HD) is an inherited, autosomal dominant, neurodegenerative disease. The mean age of onset of HD is 40 years, when patients present firsts motor symptoms. Usually, the disease leads to death in 15–20 years after the onset. There is another form of the disease, known as the juvenile form, which has its onset on childhood or adolescence. It is a less common form of HD; however, it progresses faster, leading to death in 10–15 years after the onset.[131] The worldwide prevalence is about 3 per 100,000 people. It has a higher prevalence in Europe, North American, and Australia, with about 6 per 100,000 people.[132] Interestingly, there are several areas with a high prevalence of HD, such as the Maracaibo region of Venezuela, in which the prevalence is about 700 per 100,000 people.[133]

In the 19th century, several doctors noted a hereditary nature of chorea (abnormal involuntary movement disorder). The first description of an inherited chorea was made by John Elliotson in 1832.[134] He observed hereditary chorea cases in adults, and he was probably facing a case of HD. In 1872, George Huntington, after whom the disease is named, lectured on a work called "On Chorea" that was published afterward.[135] He described the case of a patient with inherited chorea that he witnessed as a child. Two women, mother and daughter, were, as he described, "bowing, twisting, grimacing." After that, Huntington graduated in Medicine and focused on the study of the disease. The hereditary chorea, as he called, was described as a hereditary disease, with symptoms of an aggravated form of common chorea that manifests at adult life.[135]

The most common symptom in HD involves a motor disorder that is characterized by involuntary movements, impairment of voluntary movements, incoordination, dystonia, and bradykinesia. These alterations progress, leading to functional disability in patients. The condition is also characterized by cognitive impairment and neuropsychiatric alterations. Among the cognitive symptoms there are problems of attention, mental flexibility, planning and emotion recognition, cognitive slowing, and decrease in learning. Among the neuropsychiatric symptoms there are problems of apathy, depression, and irritability.[136]

In 1993, a genetic mutation was reported in HD patients.[137] An expanded CAG (cytosine, adenine, and guanine) trinucleotide repeat in the gene that encodes the protein later came to be known as huntingtin. The length of the trinucleotide repeat ranged from 42 to 66 copies in HD cases, which is considerably increased compared to healthy subjects that vary from 11 to 34 CAG copies.[137] This genetic alteration leads to an abnormal huntingtin that has an expanded polyQ segment near its amino terminus. Normal huntingtin plays important roles in nervous system development, cell adhesion, and production of neurotrophic factors. The mutated huntingtin has a toxic activity, as it can aggregate forming massive huntingtin-rich inclusions. The cytotoxic forms of the protein can lead to impaired functions, including disrupted intracellular signaling and transport, secretion, endocytic recycling, mitochondria, and synapses.[136]

It appears that there is a selective vulnerability to accumulate mutated huntingtin in striatum (caudate and putamen), globus pallidum, and cerebral cortex.[138] As the disease progresses, a degeneration in striatal neurons is observed. This degeneration is selective to the GABAergic medium spiny neurons that project to globus pallidus and the substantia nigra pars reticulata. As it was described previously, basal nuclei circuitry plays an important role in motor control. Thus, degeneration of these neurons impairs movements, which can explain the motor deficits observed in HD patients. Other brain areas are affected in HD with the disease progression, including hippocampus, the angular gyrus in the parietal lobe, the lateral tuberal nuclei of the hypothalamus, and cerebellum.[139]

2.3.1 Mitochondrial alterations

As like in PD and in AD, mitochondrial alterations have been implicated in the pathogenesis of HD. Between 1972 and 1989, it was reported in China cases of children with signs of a severe encephalopathy; dystonia was among the symptoms presented by them, and this occurred after the ingestion of sugar cane contaminated with the fungus *Phomopsis longicolla*. Computed tomography scan revealed bilateral lesions in putamen and globus pallidum. The toxin 3-nitropropionic acid (3-NP), known as the inhibitor of succinate dehydrogenase (complex II), was isolated from the fungus.[140] In 1993, 3-NP was systemically injected in rats. The researchers observed that a subacute systemic administration was able to produce bilateral striatal lesions that were confined to the striatum. They also performed a chronic administration and observed changes in striatal neurons that were similar to features of HD, suggesting an involvement of mitochondrial dysfunction in HD.[141]

Evidences of cellular metabolism impairment in HD patients were then reported in posterior studies, including alterations in lactate levels, *N*-acetyl aspartate, and creatinine.[142, 143] Reduced glucose metabolism was observed in caudate nucleus and putamen of HD patients, which could be a reflection of mitochondrial abnormalities.[144] The confirmation of mitochondrial impairment in HD was supported after postmortem studies on the brain tissue. Activities from complexes II, III, and IV were found to be reduced in the caudate nucleus of HD patients.[145] A decrease in the activity of complex IV, and respiratory activity, measured by maximal rates of oxygen consumption, were observed in the caudate nucleus of patients.[146] In addition, a decrease in complex I activity in platelet mitochondria from HD patients was also evident.[147] Besides that, structural alterations in mitochondria were observed in the frontal cortex of HD patients, including an enlarged mitochondrion, with dense granular matrix and very few cristae, and unusual crystalline fibrillar arrays within the matrix.[148]

Moreover, it appears that mutant huntingtin has diverse effects on mitochondria. This protein can interact with mitochondria and alter its function. A study demonstrated that mutant huntingtin interacts with outer mitochondrial membrane and increases susceptibility to mitochondrial permeability transition pore opening, which can result in the release of cytochrome *c* and activation of apoptotic pathways.[149] Mutant huntingtin was shown to interfere with mitochondrial trafficking. When mutant huntingtin is associated to mitochondria, it reduces the association of trafficking proteins to mitochondria, and this leads to an impairment in mitochondrial transport and distribution in neurons. This results in the loss of the energetic capacity in neurons and reduction of ATP levels in the axonal terminal.[150] Mitochondrial transcriptional genes can also be affected by mutant huntingtin. For example, mutant huntingtin binds to p53 that regulates genes associated to mitochondria, upregulating nuclear p53 levels and transcription activity. Consequently, p53 induces the expression of proteins involved in mitochondrial depolarization and apoptosis, causing damage.[151]

Disturbances in mitochondrial biogenesis are another alteration that is evident in HD. It seems that PGC-1α levels are downregulated in HD, and this was corroborated by many studies with HD patients, animal models, and cell culture. A 30% decrease of PGC-1α gene expression was reported in the caudate nucleus of striatum from postmortem brain sample of HD patients. This was accomplished with decreased gene expression of other mitochondrial genes, such as subunits of mitochondrial complexes and enzymes of citric acid cycle. Besides that, the authors showed that a decrease in PGC-1α gene expression was observed in the caudate of an animal model of HD, in which mutant huntingtin is expressed. Using the cellular culture of striatal neurons, it was observed that mutant huntingtin modulates the expression of the PGC-1α gene, repressing PGC-1α transcription. Interestingly, authors found that the mutant huntingtin occupies the PGC-1α promoter region, which decreased the transcription of PGC-1α gene as the transcriptional machinery could not bind to the gene.[152] Other studies also reported PGC-1α as an important player in pathogenesis of HD.[153, 154]

Mitochondrial biogenesis alterations go beyond PGC-1α repression. In another study, analyses of caudate from HD patients showed that protein levels of TFAM were reduced, as well as levels of PGC-1α. This reduction was directly proportional to the severity of the disease, meaning that a greater reduction was found in more severe cases of HD.[155] In addition, SNPs in TFAM and NRF-1 genes showed associations with the age onset of HD in patients with the disease. The researchers also performed a combined analysis with the discovered polymorphisms and previously described modifications in PGC-1α and mitochondrial respiratory chain genes, reporting a joint effect of these factors on the age onset of HD.[156] These results reinforce the presence of an impaired mitochondrial function among the mechanisms of HD.

As it was mentioned previously, mitochondrial structural and functional changes are evident in HD. These alterations can be regulated by mitochondrial dynamics, such as the process of fission and fusion. This hypothesis was supported by the discovery of an increased gene and protein expression of fission proteins Drp1 and Fis1, and decreased gene and protein expression of fusion proteins Mfn1, Mfn2, and OPA1 in the frontal cortex of HD patients. It appears that these alterations are restricted to specific brain areas, as more fission proteins and less fission proteins were observed in the striatum and frontal cortex, respectively, when compared to the cerebellum in HD patients.[157] The role of mutant huntingtin as a link between HD and mitochondrial abnormalities is evident in fission and fusion as well. Mutant huntingtin can interact with Drp1, which enhances Drp1 enzymatic activity, increasing mitochondrial fragmentation. This effect seems to result in defective anterograde mitochondrial movement and abnormal mitochondria distribution.[158]

2.4 Amyotrophic lateral sclerosis

Amyotrophic lateral sclerosis (ALS) is an idiopathic, fatal neurodegenerative disease of the human motor system. It is a progressive disorder characterized by progressive muscular paralysis reflecting degeneration of motor neurons in the primary motor cortex, corticospinal tracts, brain stem, and spinal cord.[159] Disease progression is generally rapid with a prognosis of 2–5 years from onset to fatality. The worldwide prevalence of ALS is nearly 6 cases per 100,000 individuals, while annual incidence varies worldwide from nearly 1 to 2.6 cases per 100,000 individuals. Regarding geographic distribution, the occurrence is higher in populations of European than Asian origin.[160] The origin of geographic difference in ALS incidence is still unknown. Perhaps, this is partly due to genes (C9ORF72) and partly due to environmental risk factors.[161]

The ALS was first described by a French physician Jean-Martin Charcot in the 1870s, employing the clinic anatomic method that he planned.[162] Charcot was the one who presumed the association between the clinical signs and the findings at autopsy. He also identified patients with muscle weakness without sensory difficulties, impaired intellect, epilepsies, or other involuntary movements.[163] In clinical and anatomical neurology, Charcot's contributions were numerous. One of his contributions was the differentiation in the clinical picture of multiple sclerosis from PD, two conditions predominated by tremor and heretofore diagnostically confused.[162]

Approximately 90% of cases are sporadic ALS (sALS), and the mean age of onset varies between 55 and 65 years. However, juvenile sporadic (jALS) onset cases are being increasingly recognized, affecting 5% of cases that have an onset before the age of 30 years. Although most cases of ALS are sporadic, approximately 5% to 10% of cases of ALS are familial (FALS), with a Mendelian pattern of inheritance.[164, 165] To date, mutations in more than two dozen genes have been identified. About 20% of familial cases are caused by mutations in Cu/Zn superoxide dismutase 1 (SOD1), which encodes for copper/zinc ion–binding superoxide dismutase. The remaining 80% are caused by mutations in other genes, such as Fused in Sarcoma (FUS), C9orf72, CHCHD10, TAR DNA-binding protein 43 (TDP-43), Sqstm1/p62, ANG (encodes angiogenin, ribonuclease, RNase A family, 5), and OPTN (encodes optineurin).[165, 166] Mutations in SOD1 induce cell toxicity, since this enzyme is responsible for destroying free superoxide radicals in the body. About more than 170 mutations have now been identified in this gene. The characteristics of mutated SOD1 that lead to the selective death of motor neurons are unknown. Current knowledge of neurodegeneration in ALS suggests a complex relationship between multiple genetic, physiological, and environmental factors, which may contribute to the phenotypic unpredictability associated with both sALS and fALS.[166]

ALS is characterized by the progressive degeneration of both upper motor neurons (UMN), in the motor cortex, and lower motor neurons (LMN), in the spinal cord and brain stem. The main clinical features of ALS include gradual muscle denervation that leads to weakness, atrophy, and paralysis, which consequently results in lethal respiratory failure.[164] Other less common clinical features are the weight loss, cramps, and fasciculations in the absence of muscle weakness, emotional lability, and frontal lobe-type cognitive dysfunction. The exact molecular pathway implicated in the pathophysiology of ALS is unknown. It is believed to be a complex interplay between multiple pathogenic cellular mechanisms, including the numerous genetic mutations described so far, excitotoxicity induced by stimulation of the glutamate receptors, oxidative stress, impaired axonal transport, abnormal assembly with accumulation of neurofilaments, inflammatory processes, and mitochondrial dysfunction, among others.[164, 167]

2.4.1 Mitochondrial alterations

Mitochondrial dysfunction has been evidenced as playing a role in motor neuron death in ALS. Abnormalities in mitochondrial morphology are well described in sALS. Using an electron microscopic examination of the mitochondria in dorsal root ganglion cells from sporadic ALS patients, Sasaki et al. evidenced inclusion bodies observed in the mitochondria in the somata of the ganglion cells. These inclusions were found in the cristae and/or intermembrane space, along with stubby mitochondria, suggesting that mitochondrial abnormalities may be involved in the degenerative processes in the dorsal root ganglion cells in sporadic ALS.[168] Abnormal accumulation of mitochondria was also reported in motor axons of mouse expressing the ALS-linked SOD1-G93A mutation.[169]

Using a cellular model overexpressing mutant SOD1, Menzies et al. demonstrated that the presence of mutant SOD1 results in the development of abnormally swollen and pale staining mitochondria. Moreover, these morphological changes are accompanied by biochemical abnormalities with specific decreases in the activities of complexes II and IV of the mitochondrial ETC.[170] Deficiency in the energy metabolism and a decrease in the activity of the ETC were also described in animal models of ALS as well as in patients with either sporadic or familial forms of ALS.[171–173] Reduced activity of mitochondrial complex I, along with a lower mitochondrial membrane potential and reduced levels of cytosolic ATP, was evidenced in mouse neuroblastoma cells (N2a) transfected with mutant human SOD1 (G37R).[174] In addition, using peripheral blood mononuclear cells from ALS patients, Araujo et al. showed an impairment in several mitochondrial functional parameters such as a lower mitochondrial Ca^{2+} uptake/retention, mitochondrial depolarization, and redox homeostasis unbalance, decrease in ATP and pyruvate levels, and the augmented lactate levels observed in patient cells, suggesting significant mitochondrial dysfunction in the pathology of ALS.[175]

Since PGC-1α functions as a key regulator of mitochondrial biogenesis and its commitment is involved in triggering important neurodegenerative diseases, studies have shown a diminished expression of PGC-1α in the spinal cord and muscle tissues of SOD-G93A ALS mice and in ALS patients,[176] as well as, in peripheral blood mononuclear cells from ALS patients.[175] These studies also demonstrated reduction in the expression of important factors related to mitochondrial biogenesis, such as PGC-1α, NRF-1, NRF-2, and TFAM, suggesting deficit in mitochondrial biogenesis in neurodegeneration in ALS.[175, 176] Moreover, PGC-1α plays a protective role in ALS by using a double transgenic mouse model where PGC-1α is overexpressed in an SOD1 transgenic mouse (TgSOD1-G93A/PGC-1α).[177, 178] In these studies, it was evidenced that PGC-1α significantly improves motor function and survival of SOD1-G93A mice,[177] and that improved motor activity was also associated with a decreased loss of motor neurons and less degeneration of neuromuscular junctions.[178] Taken together, these studies suggest that a decrease in mitochondrial biogenesis may be a common feature in ALS contributing to a reduction in cell survival, and that the modulation of PGC-1α may be a potential therapeutic target for ALS.

Another important contributing factor to ALS pathology–related mitochondrial dysfunction is mitochondrial fragmentation, mediated by Drp1 hyperactivation. Joshi et al. demonstrated in their study that ALS patient–derived fibroblasts show altered mitochondrial function and mitochondrial fragmentation associated with Drp1 hyperactivation. Furthermore, using P110, a selective peptide inhibitor of Drp1/Fis1 interaction, it was verified to have reduced mitochondrial fission in ALS patient-derived fibroblasts, motor neurons expressing G93A mutation, and an SOD1G93A mouse model, suggesting that inhibition of Drp1 hyperactivation may be useful in preventing or slowing the progression of ALS in humans.[179] Additionally, in a study conducted in transgenic mice that express human SOD1 carrying the pathogenic G93A mutation (SOD1-G93A), it was evidenced that the degradation of dysfunctional mitochondria is activated due to the recruitment of the autophagy receptor p62. Furthermore, Parkin, a ubiquitin ligase, Miro1 and Mfn2, which are mitochondrial dynamics proteins, and mitochondrial biogenesis regulator PGC-1α were depleted in these mice models. Together, these results suggest that Parkin depletion, along with decreased PGC-1α levels, causes a decline in the pool of mitochondrial proteins, which can accelerate ALS neurodegeneration.[180] Another study also demonstrated an augmentation in mitochondrial fission, which was verified by an increase in dynamin 1-like (DNM1L) and Fis-1 expression, along with a reduction in mitochondrial degradation signaling, which was verified by a significant decrease in Parkin expression, in peripheral blood mononuclear cells from ALS patients. Taken together, these results suggest that mitophagy, the main machinery to remove dysfunctional mitochondria, could be impaired in peripheral blood mononuclear cells from individuals with ALS.[175]

In fact, many of the event-related mitochondrial dysfunctions in ALS can be the cause for and consequence of each other, resulting in a vicious cycle that results in motor axon denervation and consequently motor neuron degeneration in ALS.

2.5 Multiple sclerosis

Multiple sclerosis (MS) is a chronic, inflammatory, and demyelinating disease that affects central nervous system and can progress to a neuroaxonal neurodegeneration.[181] It is the most common inflammatory neurological disease among young adults. The disease starts before the age of 40 years, being the diagnosis age approximately at 30 years. About 2 million people worldwide had multiple sclerosis in 2016. An increase of about 10% in the prevalence of MS was estimated between 1990 and 2016. There is also a sex influence on the prevalence of MS, with females more affected than males. There is a report in 2016 of 18,932 deaths due to MS worldwide.[182] MS patients were found to have a life expectancy 7 years shorter and a mortality rate three times higher than the general population.[183]

There are reports of people with symptoms similar to those of MS since the 18th century. One of the best-known descriptions is from the diary of Sir Augustus d'Este (1794–1848). The notes from his diary enable a postmortem diagnosis of MS. Some of the symptoms described are now known as typical MS characteristics, such as retrobulbar neuritis, progressive weakness, numbness, difficulty in walking, painful spasms, and depression.[184] Later, in the 19th century, the first illustration of the pathological anatomy of MS was made by Robert Carswell in 1838.[185] It was in 1868 that Jean-Martin Charcot described the disease now known as MS.[186] He detailed clinical signs that are known as Charcot's triad: intention tremor, nystagmus, and scanning speech. He also reported that brain and spinal cord are affected by sclerotic plaques, and that there is a destruction of myelin. He named the disease as "*le sclerose en plaques*," and was able to differentiate it from other diseases such as PD and ALS.[187, 188]

The clinical manifestation of MS is quite heterogeneous, mainly because it depends on the location of the affected brain area with demyelination. In about 80% of the patients, disease onset starts with an acute episode (usually last for days or weeks) that affects one site, for instance, the optic nerve, spinal cord, brain stem, cerebellum, or cerebral hemispheres. Demyelination in these areas is responsible for some of the neurological dysfunctions observed, including optic neuritis (loss of vision with a blind spot in the visual field and color vision loss), paresthesia, Lhermitte sign (characterized by a transient electrical sensation that radiates down the spine or limbs on neck flexion), Uhthoff phenomenon (worsening of symptoms when body temperature increases), impairment in ocular movements, ataxia, gait imbalance, paresis, spasticity, cognitive deficits, and fatigue. This acute episode is known as the clinically isolated syndrome (CIS).[189, 190] If a second clinical episode occurs, the disease is diagnosed as relapse-remitting MS (RRMS), with phases of relapse and remission. A progressive worsening of the symptoms from the onset can also happen; in this case, a primary progressive MS (PPMS) is the diagnostic. Finally, a secondary progressive MS (SPMS) is characterized when after episodes of relapse and remission, permanent neurological deficits with a steady progression become evident.[190]

The presence of demyelinated lesions that form scars in the brain and the spinal cord represents the cardinal feature of the disease. Besides that, MS pathology involves the breakdown of the blood-brain barrier (BBB), production of proinflammatory cytokines and chemokines, and the activation of T cells, B cells, and macrophages that can reach the central nervous system through the damaged BBB. This leads to an inflammatory microenvironment, astrocyte activation, microglia activation, and gliosis. Microglia, astrocytes, and immune cells interact with oligodendrocytes, which are responsible for the production of myelin and contribute to oligodendroglial cell injury, for example, B cells produce myelin-specific antibodies against cells of the oligodendrocyte lineage and myelin. A neuroaxonal neurodegeneration, probably due to the neurotoxic inflammatory mediator production, is also observed.[190]

Different studies have reported environmental and genetic factors involved in the susceptibility to MS. Among the environmental factors, there are smoking, lack of sunlight exposure, obesity, and deficiency of vitamin D, among others.[189] The Epstein-Barr virus infection is one of the well-known environmental risk factors for MS. It was reported that infection with Epstein-Barr virus in adolescence and early adulthood increased the risk of developing MS.[191] Later, it was discovered that Epstein-Barr virus and a myelin protein are recognized by the same receptor in T cells. They hypothesize that the immune response to the virus cross-reacts with myelin, which could be causing the demyelination.[192] There were also identified genetic variants as risk for developing MS. One of the most relevant polymorphisms is found in genes encoding human leukocyte antigen (HLA), which is involved in the immune system. Other polymorphisms were reported in genes related to T cell activation and proliferation, and tumor necrosis factor (TNF), suggesting that alterations in immune processes mediate risk to MS.[193]

2.5.1 Mitochondrial alterations

The activation of immune system (lymphocytes, macrophage, and microglia) and release of proinflammatory cytokines in MS lead to an inflammatory environment with the production of ROS and RNS, and a secondary damage of mitochondria.[194] Oxidative damage has been reported in MS patients' brains. Oxidized DNA and lipids were observed in brain areas affected by MS, and these damages were associated with areas that also presented inflammation markers, like microglia activation and T cells. In addition, lipid peroxidation and oxidized DNA were mainly reported to be localized in oligodendrocyte cells, suggesting that oxidative damage plays an important role in MS pathology.[195] In accordance with this idea, it was shown that oxidative stress disrupts oligodendrocyte differentiation in oligodendrocyte lineage cells and leads to loss of mature myelinating oligodendrocytes, which could interfere with the myelination process.[196]

As it has already been described along this chapter, mitochondria are one of the main production sites of ROS. In this way, it is reasonable to associate mitochondria and MS. In fact, several studies have reported the involvement of mitochondria in the demyelination process of MS, pointing that mitochondrial function is essential for proper oligodendrocyte differentiation and myelination. One of the processes required for myelination is the cholesterol biosynthesis, which requires acetyl-CoA produced by mitochondria. Cholesterologenic and mitochondrial transcripts are induced by oligodendroglial differentiation. In addition, low doses of the mitochondrial inhibitor rotenone are able to stop the differentiation of oligodendroglia, highlighting the importance of mitochondrial function in this cell lineage differentiation.[197] In animal models, the involvement of mitochondria in MS has also been reported. For example, by inducing breaks in mtDNA in a mouse model, impairment of locomotor function, chronic demyelination, glial activation, and axonal degeneration, characteristic of MS, were observed.[198]

Mitochondrial dysfunction has been demonstrated in MS patients. One study that compared the gene expression in postmortem motor cortex of MS and control brains revealed that activities of mitochondrial complexes I and III were decreased in MS brains. They also observed a decrease in 488 and an increase in 67 transcripts of mtDNA, and a decrease of 26 nuclear-encoded mitochondrial genes, demonstrating that neurons in MS patients probably have an impaired mitochondrial function.[199] In accordance with these results, others have reported loss of function and protein expression of subunits of complex IV, and reduction of mitochondrial density in postmortem brain from MS patients.[200,201]

Reduction of mitochondrial protein expression in MS brains was later associated with impaired mitochondrial biogenesis. Witte and coauthors analyzed cingulate gyrus and frontal cortex samples of MS patients and observed, besides a reduced expression of mitochondrial respiratory complexes, a decrease of PGC-1α mRNA and protein levels in both areas, and a decrease of NRF-1 in frontal cortex.[202] In another study, a decrease in the binding activity of NRF-2 was identified in nuclear extracts isolated from MS cortex, along with a reduction in mitochondrial complex I, III, and V gene expression.[203] Another paper used an approach that promoted a tissue-specific deletion of TFAM, an important mitochondrial biogenesis factor, in Schwann cells of mice. The authors reported that TFAM deletion causes a shift in lipid metabolism that resulted in the depletion of important myelin lipid components. These abnormalities followed by mitochondrial biogenesis deficits could also be relevant to the pathogenesis of MS.[204]

Regarding mitochondrial dynamics, alterations in fusion, fission, and mitophagy have also been involved in MS. Pathogenic genetic mutations in OPA1, protein involved in mitochondrial fusion, were identified in patients who fulfilled criteria for the diagnosis of MS.[205] Recently, another study corroborates this finding by showing a deregulation of OPA1 in peripheral blood mononuclear cells (lymphocytes, monocytes, and dendritic cells) in patients with MS. OPA1 is physiologically processed into an uncleaved long OPA1 (L-OPA1) in a cleaved short OPA1 (S-OPA1) form. The authors reported alterations in OPA1 processing in MS patients, as S-OPA1 was formed at a higher molecular weight. In addition, proteins involved in OPA1 processing and stabilization were also altered.[206] Concerning mitophagy, blood samples from MS patients were analyzed and serum concentration of Parkin, protein involved in mitophagy, and autophagy-related protein 5 (ATG5), protein involved in autophagy, were increased in MS. These results point to an impairment of mitophagy and autophagy in MS, and provide novel serum biomarkers that could be applied in the diagnosis of the disease.[207]

2.6 Epilepsy

Epilepsy is a neurological disorder characterized by enduring predisposition to unprovoked epileptic seizures, in which the person can become unconscious, present uncontrolled shaking and stiffening movements, or face momentary loss of awareness. Epileptic seizures are caused by an abnormally excessive or synchronous neuronal firing in the brain.[208] It is one of the most common neurological diseases, affecting around 50 million people worldwide.[209] It has a prevalence of about 7 cases per 1000 persons.[210] The disease carries a great burden as, for example, the risk of premature death is three times higher in people with epilepsy. Furthermore, epilepsy patients suffer from stigma and discrimination. These people are more likely to have lower self-esteem, quality of life, and psychological health, as well as they tend to be socially isolated, unemployed, and have a delay to search for treatment.[209]

The first description of an epileptic seizure dates from 2000 BC, in the region of Mesopotamia, where it described a man who presented uncontrolled shaking movements of neck, hands, and feet, and loss of consciousness. The condition was diagnosed as a divine origin. The word "epilepsy" has a Greek origin, and it means "to seize, possess, or afflict." The characterization of a disease as a divine punishment for sinner perpetuates for many years. Hippocrates, the father of medicine, was the first to question the divine origin of epilepsy. He suggested that epilepsy is a brain disease that is inherited, and he also introduced the ideas of posttraumatic epilepsy and focal epileptic seizures.[211] However, it was in the 18th and 19th centuries that great advances were made to the understanding of epilepsy. Through the work of many scientists, it was proven that epilepsy derives from the brain, localized lesions on cortex are involved in epileptic convulsions, abnormal electric discharges occurs during seizures, among other findings. These studies contributed to a better comprehension of the disease pathology, with improved diagnose methods and therapeutic approaches.[212]

Generally, in epilepsy the symptoms include uncontrollable jerking movements of the arms and legs and loss of consciousness or awareness. However, the epileptic seizure can be classified in three categories: focal, generalized, and unknown onset. The focal onset is when the seizures originate from an abnormal neuronal activity that arises in a localized brain region in one of the hemispheres. The generalized onset is characterized by an abnormal neuronal activity that arises in many brain areas in both hemispheres. Lastly, the unknown onset is identified, when it is not possible to determine the onset. These types can be identified and diagnosed through an electroencephalogram and neuroimaging. Besides the seizure itself, other complications usually follow the disease. For instance, epileptic seizures can directly bring damage and risk of falling, drowning, and having a car accident, or indirectly increase psychological problems (depression and anxiety) and medication side effects (obesity and cardiovascular effects).[208]

An epileptic seizure occurs when there is a disbalance between excitation and inhibition in the brain, which conduct to the hypersynchronization of a neuronal population. The onset of epilepsy includes several factors, such as brain tumors, stroke, brain infections, and morphological, genetic, immune, and metabolic alterations.[208] For example, aberrant hippocampal neurogenesis, increased proliferation of neural progenitors, and neuronal hypertrophy were all observed in epilepsy. The suppression of adult neurogenesis in a rat model was able to reduce chronic seizure frequency.[213] Another cause of seizure could be related to astrogliosis (astrocyte activation) in epileptic brain, as in a mouse model of chronic astrogliosis spontaneous seizures and neuronal hyperexcitability were observed in the first 6 weeks of animal's life.[214]

Some of the epilepsy cases have a genetic origin. For instance, mutations in sodium channels are responsible for several cases of epilepsy. Two gene mutations in neuronal voltage-gated sodium channel were observed in patients with epilepsy. These mutations may reduce the inactivation of the channel, increasing sodium influx, which contribute to an enhanced excitability of neurons and increase seizure susceptibility.[215] Epigenetic mechanisms are also reported in epileptic patients. For example, the promotor gene of reelin, a protein involved in the development and plasticity in the adult brain, was found to be more methylated in a hippocampal DNA sample from epileptic patients than control subjects. This epigenetic modulation may contribute to reduced protein levels of reelin, and a consequent alteration in most brain regions that enhance the development of epileptic seizures.[216]

2.6.1 Mitochondrial alterations

There is evidence that associated mitochondrial dysfunction with epilepsy arises from genetic alteration in mitochondrial DNA. The most frequent mutation is the mitochondrial $tRNA^{Lys}$ (MT-TK) gene, which is identified in patients with myoclonic epilepsy with ragged red fibers (MERRF) syndrome. Since mtDNA encodes several subunits of the respiratory chain, a decrease in mitochondrial translation by mutations in tRNA can result in reduced activity of mitochondrial complexes and oxidative phosphorylation deficiencies.[217] Another epileptic syndrome, the mitochondrial encephalomyopathy with lactic acidosis and stroke-like episodes (MELAS), was associated with mutation in $tRNA^{Lys}$.[218] Moreover, mutations in $tRNA^{Phe}$ (MT-TF) gene have been reported in maternally inherited severe epilepsy,[219] and in MERRF.[220] Mutations in the mitochondrial DNA polymerase γ (POLG1) were also identified in epileptic individuals.[221] These studies highlight the involvement of mitochondria in epilepsy pathology.

Animal models have been of great relevance to the study of epilepsy. One of the first evidences showed that the administration of the mitochondrial toxin aminooxyacetic acid triggered seizures in mice.[222] This toxin is a potent inhibitor of an enzyme involved in the malate-aspartate shunt, which is important in ATP synthesis, and has been known to impair oxidation and reduce ATP production.[223] The mitochondrial toxin 3-NP, when applied systematically, induced convulsions in mice that resemble those evoked by aminooxyacetic acid.[224] These reports reinforce the crucial role of mitochondria in epilepsy. After that, innumerous researchers characterized mitochondrial abnormalities in epileptic phenotype. Recently, a group of scientists have developed a model of mitochondrial epilepsy. They validated a brain slice model of epilepsy by application of mitochondrial respiratory inhibitors and an inhibitor of astrocytic aconitase, an enzyme of the citric acid cycle. This model adds strong evidences that mitochondrial dysfunction is important of the pathogenesis of epilepsy.[225]

Deficits in mitochondrial complexes have been reported in epileptic patients. The most common form of focal epilepsy, the temporal lobe epilepsy (TLE), has been extensively correlated with mitochondrial dysfunctions. A study performed analyses of brain sections from TLE patients with different onset epileptic focuses, they being hippocampus or parahippocampal gyrus. Interestingly, mitochondrial alterations were specific to the epileptic focus, as lower complex I activity was observed in CA3 hippocampal area of patients with hippocampus focus, and decreases in complex activity were found in the parahippocampal gyrus but not in the hippocampus of patients with parahippocampal gyrus focus.[226] It appears that mitochondrial deficits are not restricted to neurons, as the astrocytic mitochondria showed decreased expression of subunits of complexes I and IV in patients with mitochondrial disease who had epilepsy and patients with TLE.[225] Animal models of epilepsy support this idea as reduction in mitochondrial complex I activity was reported in rat brain areas of genetic and kainic acid–induced experimental temporal lobe epilepsy.[227, 228]

In addition to respiratory deficits, loss of neurons in areas of epileptogenesis is observed in epileptic patients. For instance, neuronal loss in CA1 of hippocampus and cell damage occurring after a prolonged seizure in hippocampus, amygdala, and cortex were reported in patients.[229] A study that investigated patients with mitochondrial disease revealed that inhibitory interneurons (GABAergic) are particularly vulnerable to mitochondrial defects. These neurons presented altered complexes I and IV, and were reduced in cortical regions of patients, with increased cell loss and increased respiratory chain deficiency trend to positively correlate. The reduced inhibitory interneurons lead to lower amount of inhibitory activity in the brain.[230] A model was proposed by Chan and coworkers linking mitochondrial abnormalities as genetic mutations in mtDNA leads to disruptions in transcription and translation of mitochondrial genome. Impairment of oxidative phosphorylation, especially complex I, might be the following stage. Finally, these deficits lead to inhibitory interneuron death and disbalance between excitation and inhibition in the brain that could contribute to epilepsy pathology.[225]

Mitochondrial biogenesis, fusion, and fission have all been suggested to play an important role in epilepsy. Using an animal model of rats with chronic seizures, it was demonstrated that mtDNA content and mtDNA-encoded protein were reduced, which may indicate an impairment in mitochondrial biogenesis. Besides that, mRNA and protein levels of PGC-1α, NRF-1, and TFAM were elevated in rats with chronic seizures. They also observed reduced DNA-binding activity of TFAM, which can contribute to impairment of mitochondrial biogenesis.[231] It was suggested that this increase in PGC-1α, NRF-1, and TFAM may be a protective approach in epilepsy.[232, 233]

The activation of fission has been reported after prolonged seizure activity in rats. Sustained epileptic seizures promoted an increase in protein levels of Drp1, as well as smaller size, altered shape, and increased numbers of mitochondria, which are features of mitochondrial fission. These features were accompanied by activation of apoptosis pathway (involving cytochrome *c* and caspase-3) and neuronal cell death.[234] In addition, abnormal mitochondrial fission and fusion can cause distinct cell death patterns in different neuron populations, which may be a reflection of the variety of epilepsy types (for example, the epileptic focus). For instance, while an excessive mitochondrial fission leads to an aggravated cell loss in parvalbumin cells, an abnormal mitochondrial elongation induced CA1 hippocampal neuronal death.[235]

3 Conclusion

Mitochondria are essential organelles for cellular function not only due to energy production but also for allowing a vast range of reactions necessary for cellular fate and life. This includes processes such as cellular differentiation, regulation of the cellular growth cycle, and cell death. Recent discoveries have greatly expanded our understanding of the role of mitochondria in the pathogenesis of neurological diseases. Impairment of mitochondrial function is observed in several neurological diseases including Parkinson's disease (PD), Alzheimer's disease (AD), Huntington's disease (HD) amyotrophic lateral sclerosis (ALS), multiple sclerosis (MS), and epilepsy. Among the main factors related to mitochondrial dysfunction, the oxidative stress, impairment of energy metabolism, mitochondrial fragmentation, reduction of important factors related to mitochondrial biogenesis, and impairment mitophagy are widely reported and may be a common feature in all of the diseases. Most interesting, several proteins related to neurodegenerative diseases, such as APP, Aβ, presenilin, α-synuclein, parkin, DJ-1, PINK1,

LRRK2, HTRA2, SOD1, and huntingtin, have interactions with the mitochondrial biology, resulting in a vicious cycle of mitochondrial dysfunction, neuronal death, and disease progression, as discussed in this chapter (Fig. 3). Thus, drugs that improve mitochondrial function bring exciting new possibilities for therapeutic targets that may have the potential to treat

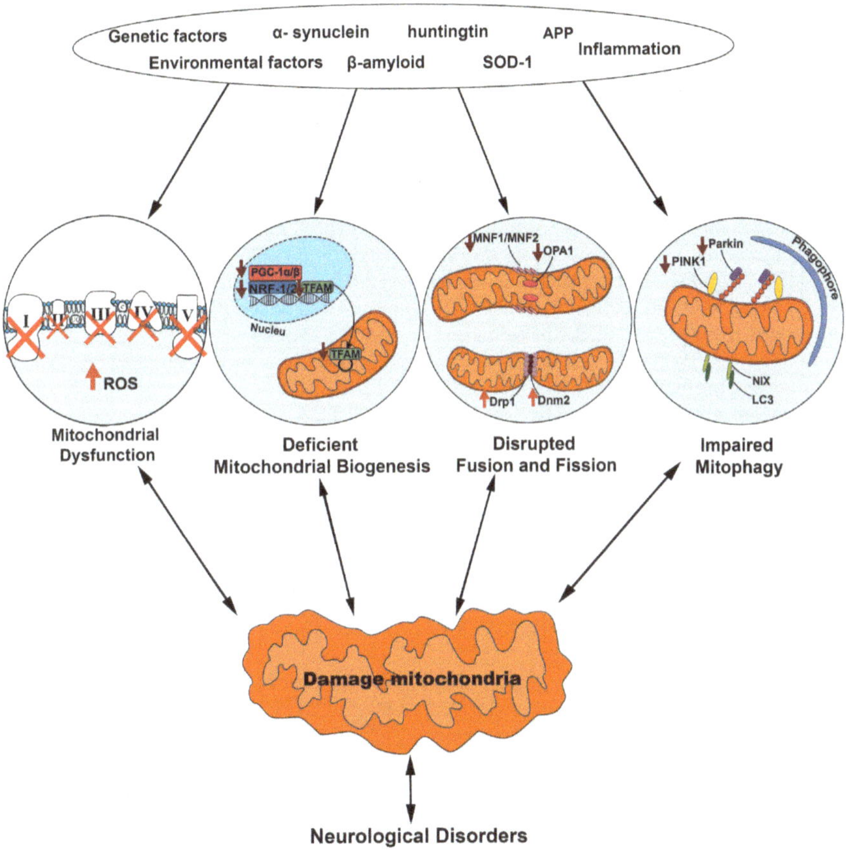

Fig. 3 The main mitochondrial aspects altered in neurological disorders. It is still unclear whether mitochondrial abnormalities are the cause or consequence of several neurological disorders. Nevertheless, literature has provided evidence for the involvement of factors related to neurological disorders, such as genetic and environmental factors, α-synuclein, β-amyloid, huntingtin, SOD1, APP, and inflammation, which can directly or indirectly interact with diverse mitochondrial aspects. These factors can lead to a mitochondrial dysfunction (by impairment of mitochondrial complexes and increase of ROS production), deficient mitochondrial biogenesis (by reduction of PGC-1α/β, NRF-1/2, and TFAM), disrupted fusion and fission (by increase of Drp1 and Dnm2, and reduction of MNF1, MNF2, and OPA2), and impaired mitophagy (by reduction of Parkin and PINK1). All these alterations result in mitochondrial damage, which can consequently bring more mitochondrial dysfunctions, creating a vicious cycle of damage. These factors can contribute to the pathological mechanisms of neurological disorders.

these neurological diseases. However, pharmacological agents that enhance mitochondrial integrity, scavenge the excessive ROS, or normalize the autophagic flux still face many challenges to be developed. Despite that, mitochondrial involvement in neurological diseases holds a great potential for enhancing our understanding of the pathological mechanism and the subsequent development of treatments for these devastating diseases.

References

1. van der Bliek AM, Sedensky MM, Morgan PG. Cell biology of the mitochondrion. *Genetics.* 2017;207(3):843–871.
2. Kühlbrandt W. Structure and function of mitochondrial membrane protein complexes. *BMC Biol.* 2015;13(1):89.
3. Tilokani L, Nagashima S, Paupe V, Prudent J. Mitochondrial dynamics: overview of molecular mechanisms. *Essays Biochem.* 2018;62(3):341–360.
4. Nicholls DG. *Bioenergetics.* Elsevier Science; 2013.
5. Brand MD, Orr AL, Perevoshchikova IV, Quinlan CL. The role of mitochondrial function and cellular bioenergetics in ageing and disease. *Br J Dermatol.* 2013;169(Suppl. 2):1–8.
6. Spinelli JB, Haigis MC. The multifaceted contributions of mitochondria to cellular metabolism. *Nat Cell Biol.* 2018;20(7):745–754.
7. Martínez-Reyes I, Chandel NS. Mitochondrial TCA cycle metabolites control physiology and disease. *Nat Commun.* 2020;11(1):102.
8. Adam-Vizi V, Chinopoulos C. Bioenergetics and the formation of mitochondrial reactive oxygen species. *Trends Pharmacol Sci.* 2006;27(12):639–645.
9. Procaccio V, Bris C, de la Chao Barca JM, et al. Perspectives of drug-based neuroprotection targeting mitochondria. *Rev Neurol.* 2014;170(5):390–400.
10. Scarpulla RC. Transcriptional paradigms in mammalian mitochondrial biogenesis and function. *Physiol Rev.* 2008;88(2):611–638.
11. Youle RJ, van der Bliek AM. Mitochondrial fission, fusion, and stress. *Science.* 2012;337(6098):1062–1065.
12. Rambold AS, Lippincott-Schwartz J. Mechanisms of mitochondria and autophagy crosstalk. *Cell Cycle.* 2011;10(23):4032–4038.
13. Youle RJ, Narendra DP. Mechanisms of mitophagy. *Nat Rev Mol Cell Biol.* 2011;12(1):9–14.
14. Belenguer P, Duarte JMN, Schuck PF, Ferreira GC. Mitochondria and the brain: bioenergetics and beyond. *Neurotox Res.* 2019;36(2):219–238.
15. Hyder F, Rothman DL. Quantitative fMRI and oxidative neuroenergetics. *Neuroimage.* 2012;62(2):985–994.
16. Sonnay S, Poirot J, Just N, et al. Astrocytic and neuronal oxidative metabolism are coupled to the rate of glutamate-glutamine cycle in the tree shrew visual cortex. *Glia.* 2018;66(3):477–491.
17. Li Z, Okamoto K, Hayashi Y, Sheng M. The importance of dendritic mitochondria in the morphogenesis and plasticity of spines and synapses. *Cell.* 2004;119(6):873–887.
18. Mattson MP, Gleichmann M, Cheng A. Mitochondria in neuroplasticity and neurological disorders. *Neuron.* 2008;60(5):748–766.
19. Sergiev PV, Dontsova OA, Berezkin GV. Theories of aging: an ever-evolving field. *Acta Naturae.* 2015;7(1):9–18.
20. Harman D. The biologic clock: the mitochondria? *J Am Geriatr Soc.* 1972;20(4):145–147.

21. Gureev AP, Shaforostova EA, Popov VN. Regulation of mitochondrial biogenesis as a way for active longevity: interaction between the Nrf2 and PGC-1α signaling pathways. *Front Genet.* 2019;10:435.
22. Jang JY, Blum A, Liu J, Finkel T. The role of mitochondria in aging. *J Clin Invest.* 2018;128(9):3662–3670.
23. Parkinson J. An essay on the shaking palsy. 1817. *J Neuropsychiatry Clin Neurosci.* 2002;14(2):223–236 [discussion 222].
24. Pringsheim T, Jette N, Frolkis A, Steeves TD. The prevalence of Parkinson's disease: a systematic review and meta-analysis. *Mov Disord.* 2014;29(13):1583–1590.
25. Tysnes OB, Storstein A. Epidemiology of Parkinson's disease. *J Neural Transm.* 2017;124(8):901–905.
26. Dexter DT, Jenner P. Parkinson disease: from pathology to molecular disease mechanisms. *Free Radic Biol Med.* 2013;62:132–144.
27. WHO. *Neurological Disorders: Public Health Challenges.* World Healthy Organization; 2006:232.
28. Lanciego JL, Luquin N, Obeso JA. Functional neuroanatomy of the basal ganglia. *Cold Spring Harb Perspect Med.* 2012;2(12):a009621.
29. Lewy FH. Paralysis Agitans. I. Pathologische Anatomie. *Handbuch der Neurologie, Dritter Band, Spezielle Neurologie I.* Berlin: Julius Springer; 1912.
30. Davis GC, Williams AC, Markey SP, et al. Chronic parkinsonism secondary to intravenous injection of meperidine analogues. *Psychiatry Res.* 1979;1(3):249–254.
31. Langston JW, Ballard P, Tetrud JW, Irwin I. Chronic Parkinsonism in humans due to a product of meperidine-analog synthesis. *Science.* 1983;219(4587):979–980.
32. Nicklas WJ, Vyas I, Heikkila RE. Inhibition of NADH-linked oxidation in brain mitochondria by 1-methyl-4-phenyl-pyridine, a metabolite of the neurotoxin, 1-methyl-4-phenyl-1,2,5,6-tetrahydropyridine. *Life Sci.* 1985;36(26):2503–2508.
33. Schapira AH, Cooper JM, Dexter D, Jenner P, Clark JB, Marsden CD. Mitochondrial complex I deficiency in Parkinson's disease. *Lancet.* 1989;1(8649):1269.
34. Parker Jr WD, Boyson SJ, Parks JK. Abnormalities of the electron transport chain in idiopathic Parkinson's disease. *Ann Neurol.* 1989;26(6):719–723.
35. Mizuno Y, Ohta S, Tanaka M, et al. Deficiencies in complex I subunits of the respiratory chain in Parkinson's disease. *Biochem Biophys Res Commun.* 1989;163(3):1450–1455.
36. Bindoff LA, Birch-Machin M, Cartlidge NEF, Parker WD, Turnbull DM. Mitochondrial function in Parkinson's disease. *Lancet.* 1989;334(8653):49.
37. Cardellach F, Martí MJ, Fernández-Solá J, et al. Mitochondrial respiratory chain activity in skeletal muscle from patients with Parkinson's disease. *Neurology.* 1993;43(11):2258–2262.
38. DiDonato S, Zeviani M, Giovannini P, et al. Respiratory chain and mitochondrial DNA in muscle and brain in Parkinson's disease patients. *Neurology.* 1993;43(11):2262–2268.
39. Klein C, Westenberger A. Genetics of Parkinson's disease. *Cold Spring Harb Perspect Med.* 2012;2(1):a008888.
40. Polymeropoulos MH, Lavedan C, Leroy E, et al. Mutation in the alpha-synuclein gene identified in families with Parkinson's disease. *Science.* 1997;276(5321):2045–2047.
41. Bertoncini CW, Fernandez CO, Griesinger C, Jovin TM, Zweckstetter M. Familial mutants of alpha-synuclein with increased neurotoxicity have a destabilized conformation. *J Biol Chem.* 2005;280(35):30649–30652.
42. Park JS, Davis RL, Sue CM. Mitochondrial dysfunction in Parkinson's disease: new mechanistic insights and therapeutic perspectives. *Curr Neurol Neurosci Rep.* 2018;18(5):21.
43. Paillusson S, Gomez-Suaga P, Stoica R, et al. α-Synuclein binds to the ER-mitochondria tethering protein VAPB to disrupt Ca(2+) homeostasis and mitochondrial ATP production. *Acta Neuropathol.* 2017;134(1):129–149.

44. Ryan SD, Dolatabadi N, Chan SF, et al. Isogenic human iPSC Parkinson's model shows nitrosative stress-induced dysfunction in MEF2-PGC1α transcription. *Cell.* 2013;155(6):1351–1364.
45. Martín-Jiménez R, Lurette O, Hebert-Chatelain E. Damage in mitochondrial DNA associated with Parkinson's disease. *DNA Cell Biol.* 2020;39(8):1421–1430.
46. van der Walt JM, Nicodemus KK, Martin ER, et al. Mitochondrial polymorphisms significantly reduce the risk of Parkinson disease. *Am J Hum Genet.* 2003;72(4):804–811.
47. Nido GS, Dölle C, Flønes I, et al. Ultradeep mapping of neuronal mitochondrial deletions in Parkinson's disease. *Neurobiol Aging.* 2018;63:120–127.
48. Giannoccaro MP, La Morgia C, Rizzo G, Carelli V. Mitochondrial DNA and primary mitochondrial dysfunction in Parkinson's disease. *Mov Disord.* 2017;32(3):346–363.
49. Meredith GE, Sonsalla PK, Chesselet M-F. Animal models of Parkinson's disease progression. *Acta Neuropathol.* 2008;115(4):385–398.
50. Garcia-Esparcia P, Koneti A, Rodríguez-Oroz MC, Gago B, Del Rio JA, Ferrer I. Mitochondrial activity in the frontal cortex area 8 and angular gyrus in Parkinson's disease and Parkinson's disease with dementia. *Brain Pathol.* 2018;28(1):43–57.
51. Flønes IH, Fernandez-Vizarra E, Lykouri M, et al. Neuronal complex I deficiency occurs throughout the Parkinson's disease brain, but is not associated with neurodegeneration or mitochondrial DNA damage. *Acta Neuropathol.* 2018;135(3):409–425.
52. Duarte JM, Schuck PF, Wenk GL, Ferreira GC. Metabolic disturbances in diseases with neurological involvement. *Aging Dis.* 2014;5(4):238–255.
53. Mytilineou C, Werner P, Molinari S, Di Rocco A, Cohen G, Yahr MD. Impaired oxidative decarboxylation of pyruvate in fibroblasts from patients with Parkinson's disease. *J Neural Transm Park Dis Dement Sect.* 1994;8(3):223–228.
54. Lei S, Zavala-Flores L, Garcia-Garcia A, et al. Alterations in energy/redox metabolism induced by mitochondrial and environmental toxins: a specific role for glucose-6-phosphate-dehydrogenase and the pentose phosphate pathway in paraquat toxicity. *ACS Chem Biol.* 2014;9(9):2032–2048.
55. Zheng B, Liao Z, Locascio JJ, et al. PGC-1α, a potential therapeutic target for early intervention in Parkinson's disease. *Sci Transl Med.* 2010;2(52):52ra73.
56. Clark J, Reddy S, Zheng K, Betensky RA, Simon DK. Association of PGC-1alpha polymorphisms with age of onset and risk of Parkinson's disease. *BMC Med Genet.* 2011;12:69.
57. Eschbach J, von Einem B, Müller K, et al. Mutual exacerbation of peroxisome proliferator-activated receptor γ coactivator 1α deregulation and α-synuclein oligomerization. *Ann Neurol.* 2015;77(1):15–32.
58. Ebrahim AS, Ko LW, Yen SH. Reduced expression of peroxisome-proliferator activated receptor gamma coactivator-1alpha enhances alpha-synuclein oligomerization and down regulates AKT/GSK3beta signaling pathway in human neuronal cells that inducibly express alpha-synuclein. *Neurosci Lett.* 2010;473(2):120–125.
59. Mudò G, Mäkelä J, Di Liberto V, et al. Transgenic expression and activation of PGC-1α protect dopaminergic neurons in the MPTP mouse model of Parkinson's disease. *Cell Mol Life Sci.* 2012;69(7):1153–1165.
60. Grünewald A, Rygiel KA, Hepplewhite PD, Morris CM, Picard M, Turnbull DM. Mitochondrial DNA depletion in respiratory chain-deficient Parkinson disease neurons. *Ann Neurol.* 2016;79(3):366–378.
61. Chen C, Vincent AE, Blain AP, Smith AL, Turnbull DM, Reeve AK. Investigation of mitochondrial biogenesis defects in single substantia nigra neurons using post-mortem human tissues. *Neurobiol Dis.* 2020;134:104631.
62. Pacelli C, De Rasmo D, Signorile A, et al. Mitochondrial defect and PGC-1α dysfunction in parkin-associated familial Parkinson's disease. *Biochim Biophys Acta (BBA) Mol Basis Dis.* 2011;1812(8):1041–1053.

63. Ferreira AFF, Binda KH, Singulani MP, et al. Physical exercise protects against mitochondria alterations in the 6-hidroxydopamine rat model of Parkinson's disease. *Behav Brain Res.* 2020;387:112607.
64. Barsoum MJ, Yuan H, Gerencser AA, et al. Nitric oxide-induced mitochondrial fission is regulated by dynamin-related GTPases in neurons. *EMBO J.* 2006;25(16):3900–3911.
65. Gomez-Lazaro M, Bonekamp NA, Galindo MF, Jordán J, Schrader M. 6-Hydroxydopamine (6-OHDA) induces Drp1-dependent mitochondrial fragmentation in SH-SY5Y cells. *Free Radic Biol Med.* 2008;44(11):1960–1969.
66. Santos D, Esteves AR, Silva DF, Januário C, Cardoso SM. The impact of mitochondrial fusion and fission modulation in sporadic Parkinson's disease. *Mol Neurobiol.* 2015;52(1):573–586.
67. Chung KK, Thomas B, Li X, et al. S-nitrosylation of parkin regulates ubiquitination and compromises parkin's protective function. *Science.* 2004;304(5675):1328–1331.
68. Wilkaniec A, Lenkiewicz AM, Czapski GA, et al. Extracellular alpha-synuclein oligomers induce parkin S-nitrosylation: relevance to sporadic Parkinson's disease etiopathology. *Mol Neurobiol.* 2019;56(1):125–140.
69. LaVoie MJ, Ostaszewski BL, Weihofen A, Schlossmacher MG, Selkoe DJ. Dopamine covalently modifies and functionally inactivates parkin. *Nat Med.* 2005;11(11):1214–1221.
70. Gilks WP, Abou-Sleiman PM, Gandhi S, et al. A common LRRK2 mutation in idiopathic Parkinson's disease. *Lancet.* 2005;365(9457):415–416.
71. Hsieh C-H, Shaltouki A, Gonzalez AE, et al. Functional impairment in miro degradation and mitophagy is a shared feature in familial and sporadic Parkinson's disease. *Cell Stem Cell.* 2016;19(6):709–724.
72. Querfurth HW, LaFerla FM. Alzheimer's disease. *N Engl J Med.* 2010;362(4):329–344.
73. Alzheimer's disease facts and figures. *Alzheimers Dement.* 2020;2020.
74. Tiwari S, Atluri V, Kaushik A, Yndart A, Nair M. Alzheimer's disease: pathogenesis, diagnostics, and therapeutics. *Int J Nanomedicine.* 2019;14:5541–5554.
75. Bondi MW, Edmonds EC, Salmon DP. Alzheimer's disease: past, present, and future. *J Int Neuropsychol Soc.* 2017;23(9–10):818–831.
76. Sun X, Chen WD, Wang YD. β-Amyloid: the key peptide in the pathogenesis of Alzheimer's disease. *Front Pharmacol.* 2015;6:221.
77. O'Brien RJ, Wong PC. Amyloid precursor protein processing and Alzheimer's disease. *Annu Rev Neurosci.* 2011;34:185–204.
78. Wang J, Dickson DW, Trojanowski JQ, Lee VM. The levels of soluble versus insoluble brain Abeta distinguish Alzheimer's disease from normal and pathologic aging. *Exp Neurol.* 1999;158(2):328–337.
79. Iqbal K, Alonso AEC, Chen S, et al. Tau pathology in Alzheimer disease and other tauopathies. *Biochim Biophys Acta.* 2005;1739(2–3):198–210.
80. Engmann O, Giese KP. Crosstalk between Cdk5 and GSK3beta: implications for Alzheimer's disease. *Front Mol Neurosci.* 2009;2:2.
81. Van Cauwenberghe C, Van Broeckhoven C, Sleegers K. The genetic landscape of Alzheimer disease: clinical implications and perspectives. *Genet Med.* 2016;18(5):421–430.
82. Stocker H, Möllers T, Perna L, Brenner H. The genetic risk of Alzheimer's disease beyond APOE ε4: systematic review of Alzheimer's genetic risk scores. *Transl Psychiatry.* 2018;8(1):166.
83. McKhann GM, Knopman DS, Chertkow H, et al. The diagnosis of dementia due to Alzheimer's disease: recommendations from the National Institute on Aging-Alzheimer's Association workgroups on diagnostic guidelines for Alzheimer's disease. *Alzheimers Dement.* 2011;7(3):263–269.

84. Long JM, Holtzman DM. Alzheimer disease: an update on pathobiology and treatment strategies. *Cell*. 2019;179(2):312–339.
85. Terry RD, Masliah E, Salmon DP, et al. Physical basis of cognitive alterations in Alzheimer's disease: synapse loss is the major correlate of cognitive impairment. *Ann Neurol*. 1991;30(4):572–580.
86. Jack CR, Lowe VJ, Weigand SD, et al. Serial PIB and MRI in normal, mild cognitive impairment and Alzheimer's disease: implications for sequence of pathological events in Alzheimer's disease. *Brain*. 2009;132(Pt 5):1355–1365.
87. Beach TG, Monsell SE, Phillips LE, Kukull W. Accuracy of the clinical diagnosis of Alzheimer disease at National Institute on Aging Alzheimer Disease Centers, 2005–2010. *J Neuropathol Exp Neurol*. 2012;71(4):266–273.
88. Swerdlow RH. Mitochondria and mitochondrial cascades in Alzheimer's disease. *J Alzheimers Dis*. 2018;62(3):1403–1416.
89. Swerdlow RH, Burns JM, Khan SM. The Alzheimer's disease mitochondrial cascade hypothesis. *J Alzheimers Dis*. 2010;20(Suppl 2):S265–S279.
90. Fukui H, Moraes CT. The mitochondrial impairment, oxidative stress and neurodegeneration connection: reality or just an attractive hypothesis? *Trends Neurosci*. 2008;31(5):251–256.
91. Johnson AB, Blum NR. Nucleoside phosphatase activities associated with the tangles and plaques of Alzheimer's disease: a histochemical study of natural and experimental neurofibrillary tangles. *J Neuropathol Exp Neurol*. 1970;29(3):463–478.
92. Friedland RP, Budinger TF, Ganz E, et al. Regional cerebral metabolic alterations in dementia of the Alzheimer type: positron emission tomography with [18F]fluorodeoxyglucose. *J Comput Assist Tomogr*. 1983;7(4):590–598.
93. Ferris SH, de Leon MJ, Wolf AP, et al. Positron emission tomography in the study of aging and senile dementia. *Neurobiol Aging*. 1980;1(2):127–131.
94. Blass JP, Zemcov A. Alzheimer's disease. A metabolic systems degeneration? *Neurochem Pathol*. 1984;2(2):103–114.
95. Sorbi S, Bird ED, Blass JP. Decreased pyruvate dehydrogenase complex activity in Huntington and Alzheimer brain. *Ann Neurol*. 1983;13(1):72–78.
96. Sims NR, Finegan JM, Blass JP, Bowen DM, Neary D. Mitochondrial function in brain tissue in primary degenerative dementia. *Brain Res*. 1987;436(1):30–38.
97. Parker WD, Filley CM, Parks JK. Cytochrome oxidase deficiency in Alzheimer's disease. *Neurology*. 1990;40(8):1302–1303.
98. Parker WD, Parks J, Filley CM, Kleinschmidt-DeMasters BK. Electron transport chain defects in Alzheimer's disease brain. *Neurology*. 1994;44(6):1090–1096.
99. Blass JP, Gibson GE. The role of oxidative abnormalities in the pathophysiology of Alzheimer's disease. *Rev Neurol*. 1991;147(6–7):513–525.
100. Swerdlow RH, Khan SM. A "mitochondrial cascade hypothesis" for sporadic Alzheimer's disease. *Med Hypotheses*. 2004;63(1):8–20.
101. Cenini G, Voos W. Mitochondria as potential targets in Alzheimer disease therapy: an update. *Front Pharmacol*. 2019;10:902.
102. Wang W, Zhao F, Ma X, Perry G, Zhu X. Mitochondria dysfunction in the pathogenesis of Alzheimer's disease: recent advances. *Mol Neurodegener*. 2020;15(1):30.
103. Inczedy-Farkas G, Trampush JW, Perczel Forintos D, et al. Mitochondrial DNA mutations and cognition: a case-series report. *Arch Clin Neuropsychol*. 2014;29(4):315–321.
104. Coskun PE, Beal MF, Wallace DC. Alzheimer's brains harbor somatic mtDNA control-region mutations that suppress mitochondrial transcription and replication. *Proc Natl Acad Sci USA*. 2004;101(29):10726–10731.
105. Chinnery PF, Taylor GA, Howell N, Brown DT, Parsons TJ, Turnbull DM. Point mutations of the mtDNA control region in normal and neurodegenerative human brains. *Am J Hum Genet*. 2001;68(2):529–532.
106. Cai Q, Tammineni P. Mitochondrial aspects of synaptic dysfunction in Alzheimer's disease. *J Alzheimers Dis*. 2017;57(4):1087–1103.

107. Huang WJ, Zhang X, Chen WW. Role of oxidative stress in Alzheimer's disease. *Biomed Rep.* 2016;4(5):519–522.
108. Ojaimi J, Masters CL, McLean C, Opeskin K, McKelvie P, Byrne E. Irregular distribution of cytochrome c oxidase protein subunits in aging and Alzheimer's disease. *Ann Neurol.* 1999;46(4):656–660.
109. Bubber P, Haroutunian V, Fisch G, Blass JP, Gibson GE. Mitochondrial abnormalities in Alzheimer brain: mechanistic implications. *Ann Neurol.* 2005;57(5):695–703.
110. Manczak M, Park BS, Jung Y, Reddy PH. Differential expression of oxidative phosphorylation genes in patients with Alzheimer's disease: implications for early mitochondrial dysfunction and oxidative damage. *Neuromolecular Med.* 2004;5(2):147–162.
111. Beck SJ, Guo L, Phensy A, et al. Deregulation of mitochondrial F1FO-ATP synthase via OSCP in Alzheimer's disease. *Nat Commun.* 2016;7:11483.
112. Liang WS, Reiman EM, Valla J, et al. Alzheimer's disease is associated with reduced expression of energy metabolism genes in posterior cingulate neurons. *Proc Natl Acad Sci USA.* 2008;105(11):4441–4446.
113. Brooks WM, Lynch PJ, Ingle CC, et al. Gene expression profiles of metabolic enzyme transcripts in Alzheimer's disease. *Brain Res.* 2007;1127(1):127–135.
114. Hirai K, Aliev G, Nunomura A, et al. Mitochondrial abnormalities in Alzheimer's disease. *J Neurosci.* 2001;21(9):3017–3023.
115. Wang X, Su B, Siedlak SL, et al. Amyloid-beta overproduction causes abnormal mitochondrial dynamics via differential modulation of mitochondrial fission/fusion proteins. *Proc Natl Acad Sci USA.* 2008;105(49):19318–19323.
116. Qin W, Haroutunian V, Katsel P, et al. PGC-1alpha expression decreases in the Alzheimer disease brain as a function of dementia. *Arch Neurol.* 2009;66(3):352–361.
117. Sheng B, Wang X, Su B, et al. Impaired mitochondrial biogenesis contributes to mitochondrial dysfunction in Alzheimer's disease. *J Neurochem.* 2012;120(3):419–429.
118. Manczak M, Kandimalla R, Yin X, Reddy PH. Hippocampal mutant APP and amyloid beta-induced cognitive decline, dendritic spine loss, defective autophagy, mitophagy and mitochondrial abnormalities in a mouse model of Alzheimer's disease. *Hum Mol Genet.* 2018;27(8):1332–1342.
119. Reddy PH, Yin X, Manczak M, et al. Mutant APP and amyloid beta-induced defective autophagy, mitophagy, mitochondrial structural and functional changes and synaptic damage in hippocampal neurons from Alzheimer's disease. *Hum Mol Genet.* 2018;27(14):2502–2516.
120. Singulani MP, Pereira CPM, Ferreira AFF, et al. Impairment of PGC-1α-mediated mitochondrial biogenesis precedes mitochondrial dysfunction and Alzheimer's pathology in the 3xTg mouse model of Alzheimer's disease. *Exp Gerontol.* 2020;133:110882.
121. Gong B, Pan Y, Vempati P, et al. Nicotinamide riboside restores cognition through an upregulation of proliferator-activated receptor-γ coactivator 1α regulated β-secretase 1 degradation and mitochondrial gene expression in Alzheimer's mouse models. *Neurobiol Aging.* 2013;34(6):1581–1588.
122. Manczak M, Calkins MJ, Reddy PH. Impaired mitochondrial dynamics and abnormal interaction of amyloid beta with mitochondrial protein Drp1 in neurons from patients with Alzheimer's disease: implications for neuronal damage. *Hum Mol Genet.* 2011;20(13):2495–2509.
123. Wang X, Su B, Lee HG, et al. Impaired balance of mitochondrial fission and fusion in Alzheimer's disease. *J Neurosci.* 2009;29(28):9090–9103.
124. Calkins MJ, Manczak M, Mao P, Shirendeb U, Reddy PH. Impaired mitochondrial biogenesis, defective axonal transport of mitochondria, abnormal mitochondrial dynamics and synaptic degeneration in a mouse model of Alzheimer's disease. *Hum Mol Genet.* 2011;20(23):4515–4529.
125. Manczak M, Reddy PH. Abnormal interaction between the mitochondrial fission protein Drp1 and hyperphosphorylated tau in Alzheimer's disease neurons: implications for

mitochondrial dysfunction and neuronal damage. *Hum Mol Genet.* 2012;21(11):2538–2547.
126. Westermann B. Bioenergetic role of mitochondrial fusion and fission. *Biochim Biophys Acta.* 2012;1817(10):1833–1838.
127. Cai Q, Jeong YY. Mitophagy in Alzheimer's disease and other age-related neurodegenerative diseases. *Cells.* 2020;9(1).
128. Ye X, Sun X, Starovoytov V, Cai Q. Parkin-mediated mitophagy in mutant hAPP neurons and Alzheimer's disease patient brains. *Hum Mol Genet.* 2015;24(10):2938–2951.
129. Martín-Maestro P, Gargini R, Perry G, Avila J, García-Escudero V. PARK2 enhancement is able to compensate mitophagy alterations found in sporadic Alzheimer's disease. *Hum Mol Genet.* 2016;25(4):792–806.
130. Fang EF, Hou Y, Palikaras K, et al. Mitophagy inhibits amyloid-β and tau pathology and reverses cognitive deficits in models of Alzheimer's disease. *Nat Neurosci.* 2019;22(3):401–412.
131. Ross CA, Tabrizi SJ. Huntington's disease: from molecular pathogenesis to clinical treatment. *Lancet Neurol.* 2011;10(1):83–98.
132. Pringsheim T, Wiltshire K, Day L, Dykeman J, Steeves T, Jette N. The incidence and prevalence of Huntington's disease: a systematic review and meta-analysis. *Mov Disord.* 2012;27(9):1083–1091.
133. Avila-Giron RJAN. Medical and social aspects of Huntington's chorea in the state of Zulia, Venezuela. *Adv Neurol.* 1973;1:261–266.
134. Elliotson J. Clinical lecture. *Lancet.* 1832;18(467):589–593.
135. Huntington G. *On Chorea.* Medical and Surgical Reporter of Philadelphia; 1872.
136. Bates GP, Dorsey R, Gusella JF, et al. Huntington disease. *Nat Rev Dis Primers.* 2015;1(1):15005.
137. MacDonald ME, Ambrose CM, Duyao MP, et al. A novel gene containing a trinucleotide repeat that is expanded and unstable on Huntington's disease chromosomes. *Cell.* 1993;72(6):971–983.
138. Mann DM, Oliver R, Snowden JS. The topographic distribution of brain atrophy in Huntington's disease and progressive supranuclear palsy. *Acta Neuropathol.* 1993;85(5):553–559.
139. Han I, You Y, Kordower JH, Brady ST, Morfini GA. Differential vulnerability of neurons in Huntington's disease: the role of cell type-specific features. *J Neurochem.* 2010;113(5):1073–1091.
140. Ludolph AC, He F, Spencer PS, Hammerstad J, Sabri M. 3-Nitropropionic acid-exogenous animal neurotoxin and possible human striatal toxin. *Can J Neurol Sci.* 1991;18(4):492–498.
141. Beal MF, Brouillet E, Jenkins BG, et al. Neurochemical and histologic characterization of striatal excitotoxic lesions produced by the mitochondrial toxin 3-nitropropionic acid. *J Neurosci.* 1993;13(10):4181–4192.
142. Jenkins BG, Koroshetz WJ, Beal MF, Rosen BR. Evidence for impairment of energy metabolism in vivo in Huntington's disease using localized 1H NMR spectroscopy. *Neurology.* 1993;43(12):2689–2695.
143. Sánchez-Pernaute R, García-Segura JM, del Barrio AA, Viaño J, de Yébenes JG. Clinical correlation of striatal 1H MRS changes in Huntington's disease. *Neurology.* 1999;53(4):806–812.
144. Antonini A, Leenders KL, Spiegel R, et al. Striatal glucose metabolism and dopamine D2 receptor binding in asymptomatic gene carriers and patients with Huntington's disease. *Brain.* 1996;119(Pt. 6):2085–2095.
145. Gu M, Gash MT, Mann VM, Javoy-Agid F, Cooper JM, Schapira AH. Mitochondrial defect in Huntington's disease caudate nucleus. *Ann Neurol.* 1996;39(3):385–389.

146. Brennan Jr WA, Bird ED, Aprille JR. Regional mitochondrial respiratory activity in Huntington's disease brain. *J Neurochem*. 1985;44(6):1948–1950.
147. Parker Jr WD, Boyson SJ, Luder AS, Parks JK. Evidence for a defect in NADH: ubiquinone oxidoreductase (complex I) in Huntington's disease. *Neurology*. 1990;40(8):1231–1234.
148. Tellez-Nagel I, Johnson AB, Terry RD. Studies on brain biopsies of patients with Huntington's chorea. *J Neuropathol Exp Neurol*. 1974;33(2):308–332.
149. Choo YS, Johnson GVW, MacDonald M, Detloff PJ, Lesort M. Mutant huntingtin directly increases susceptibility of mitochondria to the calcium-induced permeability transition and cytochrome c release. *Hum Mol Genet*. 2004;13(14):1407–1420.
150. Orr AL, Li S, Wang C-E, et al. N-Terminal mutant huntingtin associates with mitochondria and impairs mitochondrial trafficking. *J Neurosci*. 2008;28(11):2783.
151. Bae BI, Xu H, Igarashi S, et al. p53 mediates cellular dysfunction and behavioral abnormalities in Huntington's disease. *Neuron*. 2005;47(1):29–41.
152. Cui L, Jeong H, Borovecki F, Parkhurst CN, Tanese N, Krainc D. Transcriptional repression of PGC-1alpha by mutant huntingtin leads to mitochondrial dysfunction and neurodegeneration. *Cell*. 2006;127(1):59–69.
153. Johri A, Starkov AA, Chandra A, et al. Truncated peroxisome proliferator-activated receptor-γ coactivator 1α splice variant is severely altered in Huntington's disease. *Neurodegener Dis*. 2011;8(6):496–503.
154. Weydt P, Soyal SM, Landwehrmeyer GB, Patsch W. A single nucleotide polymorphism in the coding region of PGC-1α is a male-specific modifier of Huntington disease age-at-onset in a large European cohort. *BMC Neurol*. 2014;14:1.
155. Kim J, Moody JP, Edgerly CK, et al. Mitochondrial loss, dysfunction and altered dynamics in Huntington's disease. *Hum Mol Genet*. 2010;19(20):3919–3935.
156. Taherzadeh-Fard E, Saft C, Akkad DA, et al. PGC-1alpha downstream transcription factors NRF-1 and TFAM are genetic modifiers of Huntington disease. *Mol Neurodegener*. 2011;6(1):32.
157. Shirendeb U, Reddy AP, Manczak M, et al. Abnormal mitochondrial dynamics, mitochondrial loss and mutant huntingtin oligomers in Huntington's disease: implications for selective neuronal damage. *Hum Mol Genet*. 2011;20(7):1438–1455.
158. Shirendeb UP, Calkins MJ, Manczak M, et al. Mutant huntingtin's interaction with mitochondrial protein Drp1 impairs mitochondrial biogenesis and causes defective axonal transport and synaptic degeneration in Huntington's disease. *Hum Mol Genet*. 2012;21(2):406–420.
159. Rowland LP, Shneider NA. Amyotrophic lateral sclerosis. *N Engl J Med*. 2001;344(22):1688–1700.
160. Filippini T, Fiore M, Tesauro M, et al. Clinical and lifestyle factors and risk of amyotrophic lateral sclerosis: a population-based case-control study. *Int J Environ Res Public Health*. 2020;17(3).
161. Logroscino G, Piccininni M. Amyotrophic lateral sclerosis descriptive epidemiology: the origin of geographic difference. *Neuroepidemiology*. 2019;52(1–2):93–103.
162. Goetz CG. Amyotrophic lateral sclerosis: early contributions of Jean-Martin Charcot. *Muscle Nerve*. 2000;23(3):336–343.
163. Katz JS, Dimachkie MM, Barohn RJ. Amyotrophic lateral sclerosis: a historical perspective. *Neurol Clin*. 2015;33(4):727–734.
164. Ferraiuolo L, Kirby J, Grierson AJ, Sendtner M, Shaw PJ. Molecular pathways of motor neuron injury in amyotrophic lateral sclerosis. *Nat Rev Neurol*. 2011;7(11):616–630.
165. Maruyama R, Yokota T. Molecular diagnosis and novel therapies for neuromuscular diseases. *J Pers Med*. 2020;10(3).
166. Calió ML, Henriques E, Siena A, Bertoncini CRA, Gil-Mohapel J, Rosenstock TR. Mitochondrial dysfunction, neurogenesis, and epigenetics: putative implications

for amyotrophic lateral sclerosis neurodegeneration and treatment. *Front Neurosci.* 2020;14:679.
167. Wijesekera LC, Leigh PN. Amyotrophic lateral sclerosis. *Orphanet J Rare Dis.* 2009;4:3.
168. Sasaki S, Horie Y, Iwata M. Mitochondrial alterations in dorsal root ganglion cells in sporadic amyotrophic lateral sclerosis. *Acta Neuropathol.* 2007;114(6):633–639.
169. Sotelo-Silveira JR, Lepanto P, Elizondo V, et al. Axonal mitochondrial clusters containing mutant SOD1 in transgenic models of ALS. *Antioxid Redox Signal.* 2009;11(7):1535–1545.
170. Menzies FM, Cookson MR, Taylor RW, et al. Mitochondrial dysfunction in a cell culture model of familial amyotrophic lateral sclerosis. *Brain.* 2002;125(Pt. 7):1522–1533.
171. Browne SE, Yang L, DiMauro JP, Fuller SW, Licata SC, Beal MF. Bioenergetic abnormalities in discrete cerebral motor pathways presage spinal cord pathology in the G93A SOD1 mouse model of ALS. *Neurobiol Dis.* 2006;22(3):599–610.
172. Bowling AC, Schulz JB, Brown RH, Beal MF. Superoxide dismutase activity, oxidative damage, and mitochondrial energy metabolism in familial and sporadic amyotrophic lateral sclerosis. *J Neurochem.* 1993;61(6):2322–2325.
173. Crugnola V, Lamperti C, Lucchini V, et al. Mitochondrial respiratory chain dysfunction in muscle from patients with amyotrophic lateral sclerosis. *Arch Neurol.* 2010;67(7):849–854.
174. Coussee E, De Smet P, Bogaert E, et al. G37R SOD1 mutant alters mitochondrial complex I activity, Ca(2+) uptake and ATP production. *Cell Calcium.* 2011;49(4):217–225.
175. Araujo BG, Silva LF SE, de Barros Torresi JL, et al. Decreased mitochondrial function, biogenesis, and degradation in peripheral blood mononuclear cells from amyotrophic lateral sclerosis patients as a potential tool for biomarker research. *Mol Neurobiol.* 2020;57(12):5084–5102.
176. Thau N, Knippenberg S, Körner S, Rath KJ, Dengler R, Petri S. Decreased mRNA expression of PGC-1α and PGC-1α-regulated factors in the SOD1G93A ALS mouse model and in human sporadic ALS. *J Neuropathol Exp Neurol.* 2012;71(12):1064–1074.
177. Zhao W, Varghese M, Yemul S, et al. Peroxisome proliferator activator receptor gamma coactivator-1alpha (PGC-1α) improves motor performance and survival in a mouse model of amyotrophic lateral sclerosis. *Mol Neurodegener.* 2011;6(1):51.
178. Liang H, Ward WF, Jang YC, et al. PGC-1α protects neurons and alters disease progression in an amyotrophic lateral sclerosis mouse model. *Muscle Nerve.* 2011;44(6):947–956.
179. Joshi AU, Saw NL, Vogel H, Cunnigham AD, Shamloo M, Mochly-Rosen D. Inhibition of Drp1/Fis1 interaction slows progression of amyotrophic lateral sclerosis. *EMBO Mol Med.* 2018;10(3).
180. Palomo GM, Granatiero V, Kawamata H, et al. Parkin is a disease modifier in the mutant SOD1 mouse model of ALS. *EMBO Mol Med.* 2018;10(10).
181. Friese MA, Schattling B, Fugger L. Mechanisms of neurodegeneration and axonal dysfunction in multiple sclerosis. *Nat Rev Neurol.* 2014;10(4):225–238.
182. Wallin MT, Culpepper WJ, Nichols E, et al. Global, regional, and national burden of multiple sclerosis 1990–2016: a systematic analysis for the Global Burden of Disease Study 2016. *Lancet Neurol.* 2019;18(3):269–285.
183. Lunde HMB, Assmus J, Myhr K-M, Bø L, Grytten N. Survival and cause of death in multiple sclerosis: a 60-year longitudinal population study. *J Neurol Neurosurg Psychiatry.* 2017;88(8):621.
184. Firth D. The case of Augustus d'Este (1794–1848): the first account of disseminated sclerosis: (section of the history of medicine). *Proc R Soc Med.* 1941;34(7):381–384.
185. Carswell R. *Illustrations of the Elementary Forms of Disease.* Longman; 1838.
186. Charcot M. Histologie de la sclerose en plaque. *Gaz Hôp.* 1868;41:554–556.
187. Pearce JMS. Historical descriptions of multiple sclerosis. *Eur Neurol.* 2005;54(1):49–53.

188. Talley C. Multiple sclerosis: the history of a disease (review). *Bull Hist Med.* 2006;80:180–182.
189. Compston A, Coles A. Multiple sclerosis. *Lancet.* 2008;372(9648):1502–1517.
190. Filippi M, Bar-Or A, Piehl F, et al. Multiple sclerosis. *Nat Rev Dis Primers.* 2018;4(1):43.
191. Levin LI, Munger KL, Rubertone MV, et al. Multiple sclerosis and Epstein-Barr virus. *JAMA.* 2003;289(12):1533–1536.
192. Lang HL, Jacobsen H, Ikemizu S, et al. A functional and structural basis for TCR cross-reactivity in multiple sclerosis. *Nat Immunol.* 2002;3(10):940–943.
193. Cotsapas C, Mitrovic M. Genome-wide association studies of multiple sclerosis. *Clin Transl Immunol.* 2018;7(6):e1018.
194. IPd B, Troxell RM, Graves JS. Mitochondrial dysfunction and multiple sclerosis. *Biology.* 2019;8(2):37.
195. Haider L, Fischer MT, Frischer JM, et al. Oxidative damage in multiple sclerosis lesions. *Brain.* 2011;134(Pt. 7):1914–1924.
196. French HM, Reid M, Mamontov P, Simmons RA, Grinspan JB. Oxidative stress disrupts oligodendrocyte maturation. *J Neurosci Res.* 2009;87(14):3076–3087.
197. Schoenfeld R, Wong A, Silva J, et al. Oligodendroglial differentiation induces mitochondrial genes and inhibition of mitochondrial function represses oligodendroglial differentiation. *Mitochondrion.* 2010;10(2):143–150.
198. Madsen PM, Pinto M, Patel S, et al. Mitochondrial DNA double-strand breaks in oligodendrocytes cause demyelination, axonal injury, and CNS inflammation. *J Neurosci.* 2017;37(42):10185–10199.
199. Dutta R, McDonough J, Yin X, et al. Mitochondrial dysfunction as a cause of axonal degeneration in multiple sclerosis patients. *Ann Neurol.* 2006;59(3):478–489.
200. Mahad D, Ziabreva I, Lassmann H, Turnbull D. Mitochondrial defects in acute multiple sclerosis lesions. *Brain.* 2008;131(Pt. 7):1722–1735.
201. Campbell GR, Ziabreva I, Reeve AK, et al. Mitochondrial DNA deletions and neurodegeneration in multiple sclerosis. *Ann Neurol.* 2011;69(3):481–492.
202. Witte ME, Nijland PG, Drexhage JA, et al. Reduced expression of PGC-1α partly underlies mitochondrial changes and correlates with neuronal loss in multiple sclerosis cortex. *Acta Neuropathol.* 2013;125(2):231–243.
203. Pandit A, Vadnal J, Houston S, Freeman E, McDonough J. Impaired regulation of electron transport chain subunit genes by nuclear respiratory factor 2 in multiple sclerosis. *J Neurol Sci.* 2009;279(1–2):14–20.
204. Viader A, Sasaki Y, Kim S, et al. Aberrant Schwann cell lipid metabolism linked to mitochondrial deficits leads to axon degeneration and neuropathy. *Neuron.* 2013;77(5):886–898.
205. Yu-Wai-Man P, Spyropoulos A, Duncan HJ, Guadagno JV, Chinnery PF. A multiple sclerosis-like disorder in patients with OPA1 mutations. *Ann Clin Transl Neurol.* 2016;3(9):723–729.
206. De Rasmo D, Ferretta A, Russo S, et al. PBMC of multiple sclerosis patients show deregulation of OPA1 processing associated with increased ROS and PHB2 protein levels. *Biomedicines.* 2020;8(4):85.
207. Joodi Khanghah O, Nourazarian A, Khaki-Khatibi F, et al. Evaluation of the diagnostic and predictive value of serum levels of ANT1, ATG5, and parkin in multiple sclerosis. *Clin Neurol Neurosurg.* 2020;197:106197.
208. Devinsky O, Vezzani A, O'Brien TJ, et al. Epilepsy. *Nat Rev Dis Primers.* 2018;4(1):18024.
209. World Health Organization. *Epilepsy: A Public Health Imperative.* Geneva: World Health Organization; 2019.
210. Fiest KM, Sauro KM, Wiebe S, et al. Prevalence and incidence of epilepsy: a systematic review and meta-analysis of international studies. *Neurology.* 2017;88(3):296–303.

211. Magiorkinis E, Sidiropoulou K, Diamantis A. Hallmarks in the history of epilepsy: epilepsy in antiquity. *Epilepsy Behav.* 2010;17(1):103–108.
212. Magiorkinis E, Diamantis A, Sidiropoulou K, Panteliadis C. Highights in the history of epilepsy: the last 200 years. *Epilepsy Res Treat.* 2014;2014:582039.
213. Cho KO, Lybrand ZR, Ito N, et al. Aberrant hippocampal neurogenesis contributes to epilepsy and associated cognitive decline. *Nat Commun.* 2015;6:6606.
214. Robel S, Buckingham SC, Boni JL, et al. Reactive astrogliosis causes the development of spontaneous seizures. *J Neurosci.* 2015;35(8):3330–3345.
215. Escayg A, MacDonald BT, Meisler MH, et al. Mutations of SCN1A, encoding a neuronal sodium channel, in two families with GEFS+2. *Nat Genet.* 2000;24(4):343–345.
216. Kobow K, Jeske I, Hildebrandt M, et al. Increased reelin promoter methylation is associated with granule cell dispersion in human temporal lobe epilepsy. *J Neuropathol Exp Neurol.* 2009;68(4):356–364.
217. Shoffner JM, Lott MT, Lezza AMS, Seibel P, Ballinger SW, Wallace DC. Myoclonic epilepsy and ragged-red fiber disease (MERRF) is associated with a mitochondrial DNA tRNALys mutation. *Cell.* 1990;61(6):931–937.
218. Zeviani M, Muntoni F, Savarese N, et al. A MERRF/MELAS overlap syndrome associated with a new point mutation in the mitochondrial DNA tRNALys gene. *Eur J Hum Genet.* 1993;1(1):80–87.
219. Zsurka G, Hampel KG, Nelson I, et al. Severe epilepsy as the major symptom of new mutations in the mitochondrial tRNA(Phe) gene. *Neurology.* 2010;74(6):507.
220. Mancuso M, Filosto M, Mootha VK, et al. A novel mitochondrial tRNAPhe mutation causes MERRF syndrome. *Neurology.* 2004;62(11):2119.
221. Zsurka G, Baron M, Stewart JD, et al. Clonally expanded mitochondrial DNA mutations in epileptic individuals with mutated DNA polymerase γ. *J Neuropathol Exp Neurol.* 2008;67(9):857–866.
222. Turski WA, Dziki M, Urbanska E, Calderazzo-Filho LS, Cavalheiro EA. Seizures induced by aminooxyacetic acid in mice: pharmacolgical characteristics. *Synapse.* 1991;7(3):173–180.
223. Kauppinen RA, Sihra TS, Nicholls DG. Aminooxyacetic acid inhibits the malate-aspartate shuttle in isolated nerve terminals and prevents the mitochondria from utilizing glycolytic substrates. *Biochim Biophys Acta.* 1987;930(2):173–178.
224. Urbanska EM, Blaszczak P, Saran T, Kleinrok Z, Turski WA. Mitochondrial toxin 3-nitropropionic acid evokes seizures in mice. *Eur J Pharmacol.* 1998;359(1):55–58.
225. Chan F, Lax NZ, Voss CM, et al. The role of astrocytes in seizure generation: insights from a novel in vitro seizure model based on mitochondrial dysfunction. *Brain.* 2019;142(2):391–411.
226. Kunz WS, Kudin AP, Vielhaber S, et al. Mitochondrial complex I deficiency in the epileptic focus of patients with temporal lobe epilepsy. *Ann Neurol.* 2000;48(5):766–773.
227. Chuang YC, Chen SD, Liou CW, et al. Contribution of nitric oxide, superoxide anion, and peroxynitrite to activation of mitochondrial apoptotic signaling in hippocampal CA3 subfield following experimental temporal lobe status epilepticus. *Epilepsia.* 2009;50(4):731–746.
228. Simeone KA, Matthews SA, Samson KK, Simeone TA. Targeting deficiencies in mitochondrial respiratory complex I and functional uncoupling exerts anti-seizure effects in a genetic model of temporal lobe epilepsy and in a model of acute temporal lobe seizures. *Exp Neurol.* 2014;251:84–90.
229. Meldrum BS. Excitotoxicity and selective neuronal loss in epilepsy. *Brain Pathol.* 1993;3(4):405–412.
230. Lax NZ, Grady J, Laude A, et al. Extensive respiratory chain defects in inhibitory interneurones in patients with mitochondrial disease. *Neuropathol Appl Neurobiol.* 2016;42(2):180–193.

231. Han Y, Lin Y, Xie N, et al. Impaired mitochondrial biogenesis in hippocampi of rats with chronic seizures. *Neuroscience.* 2011;194:234–240.
232. Chuang YC, Chen SD, Jou SB, et al. Sirtuin 1 regulates mitochondrial biogenesis and provides an endogenous neuroprotective mechanism against seizure-induced neuronal cell death in the hippocampus following status epilepticus. *Int J Mol Sci.* 2019;20(14).
233. Chuang YC, Chen SD, Hsu CY, Chen SF, Chen NC, Jou SB. Resveratrol promotes mitochondrial biogenesis and protects against seizure-induced neuronal cell damage in the hippocampus following status epilepticus by activation of the PGC-1α signaling pathway. *Int J Mol Sci.* 2019;20(4).
234. Chen SD, Zhen YY, Lin JW, et al. Dynamin-related protein 1 promotes mitochondrial fission and contributes to the hippocampal neuronal cell death following experimental status epilepticus. *CNS Neurosci Ther.* 2016;22(12):988–999.
235. Kim JE, Kang TC. Differential roles of mitochondrial translocation of active caspase-3 and HMGB1 in neuronal death induced by status epilepticus. *Front Cell Neurosci.* 2018;12:301.

CHAPTER 10

Mitochondrial abnormalities in psychological disorders

Taraneh Mousavi and Mohammad Abdollahi
Toxicology and Diseases Group (TDG), Pharmaceutical Sciences Research Center (PSRC), The Institute of Pharmaceutical Sciences (TIPS), Tehran University of Medical Sciences, Tehran, Iran
Department of Toxicology and Pharmacology, School of Pharmacy, Tehran University of Medical Sciences, Tehran, Iran

Contents

1 Introduction

Being the leading cause of years lost due to disability (YLDs),[1] mental disorders are nowadays regarded as the most significant challenge to affected individuals, their relatives, health care systems, and governments, which as well accounted for 183.9 million disability-adjusted life years (DALYs) and 8.6 million years of life lost (YLLs), according to the Global Burden of Disease (GBD) 2010 report.[1] The enormous burden of mental disorders carried global costs of USD 2.5 trillion (USD 1.7 trillion, and USD 0.8 trillion for indirect and direct expenses, respectively); both are estimated to be doubled by 2030.[2] The cumulative loss of economic output is forecasted to reach USD 16.3 trillion worldwide by the indicated period,[2] highlighting the increasing impact of psychiatric disorders.

According to the Diagnostic and Statistical Manual of Mental Disorders, Fifth Edition (DSM-5),[3] a principal diagnostic criterion in the United States

Mitochondrial Metabolism
https://doi.org/10.1016/B978-0-12-822416-8.00013-0

published by the American Psychiatric Association (APA), mental disorders can be classified into major groups of neurodevelopmental disorders, schizophrenia spectrum and other psychiatric disorders, bipolar and related disorders, depressive disorders, anxiety disorders, obsessive-compulsive and related disorders, trauma and stress-related disorders, dissociative disorders, somatic symptoms and related disorders, feeding and eating disorders, elimination disorders, sleep-wake disorders, sexual dysfunctions, gender dysphoria, disruptive, impulse-control and conduct disorders, substance-related and addictive disorders, neurocognitive disorders, and personality disorders, paraphilic disorders. Both intrinsic (mainly, genetic, inflammation, and mitochondrial abnormalities) and extrinsic (mostly, maternal and pregnancy-related factors, lifestyle, physiological diseases, toxicants, medications, and past experiences) are responsible for the incidence of psychiatric disorders. Nonetheless, almost all the mentioned disorders' exact pathophysiology remains unknown, justifying the lack of specific diagnostic tests and precise medication in this regard. Mental illnesses are usually diagnosed based on clinical and self-reported evaluations without measuring any particular biomarkers. As well, available pharmacological medications, i.e., antidepressants, mood-stabilizers, antipsychotics, stimulants, and antianxiety agents, are nonspecifically approved for different mental disorders classes prescribed the primary therapy or as an adjuvant to nonpharmacological interventions, mainly psychotherapy and brain stimulation. What exacerbates the situation is that two or more mental illnesses can go together, providing additional emotional and economic burdens.

Nevertheless, the point here is that noncommunicable diseases, specifically mental disorders, have been neglected in most global regions, especially in low- and middle-income countries.[4,5] However, they hold a large share of YLDs and DALYs (see Fig. 1) comparable to other communicable and noncommunicable diseases.[1] For instance, it has been estimated that in low, middle, and high-income countries, about 76%, 85%, and 42% of people with severe mental illnesses do not receive any treatment to improve their condition.[6] Therefore, as the World Health Organization (WHO) stated,[4,7] it is crucial to employ significant health and government sectors to improve diagnostic systems, treatments and medication access, and developing policies. One future approach could be introducing biomarkers and precise medications targeting specific sites/pathways within each psychiatric disorder. Since people with mitochondrial diseases are at a higher risk of developing mental illnesses, it seems that mitochondrion could be a promising target. Thus, investigating abnormalities because of

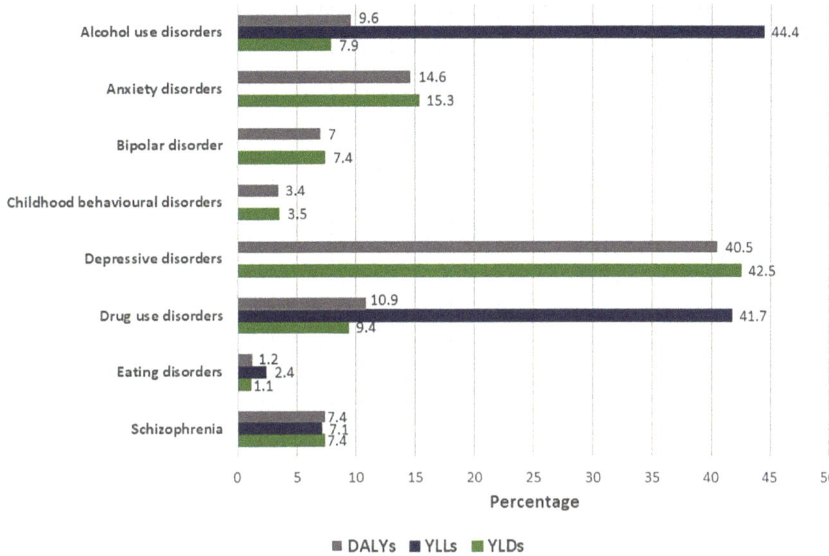

Fig. 1 Percentage of years lost due to disability (YLDs), disability-adjusted life years (DALYs), and years of life lost (YLLs) within some mental disorders, according to Global Burden of Disease (GBD) 2010 report. YLLs of anxiety disorders, bipolar disorder, childhood behavioral disorders, and depressive disorders were not reported.

mitochondrial DNA (mtDNA), biogenesis, calcium metabolism, respiration, oxidative stress, and apoptotic pathways within psychiatric disorders may shine the light upon remarkable progress in the context of psychology. In this chapter, we comprehensively discuss mitochondrial abnormalities in psychological disorders, based on the DSM-5 classification.

2 Mitochondrial abnormalities

Mitochondria are one of the essential organelles, particularly in neuronal cells as the brain's primary energy source, supplying 92% of overall energy demand.[8–10] They are found to be at play in different stages of neuronal development, including neuronal plasticity (i.e., the ability of neurons to adapt to changing conditions and environments), synaptic plasticity,[8, 11] synaptic connectivity, neurogenesis[8] and neuronal differentiation[11] as well as being responsible for proper learning, memory[11] and behavioral adaption.[8] They are mainly accumulated in the gray matter, especially in neurons' presynaptic terminals and Ranvier nodes in axons, with the highest energy demand than other components.[12] However, mitochondria are dynamic structures,[13] which can be translocated through the neurons, where there is a need for calcium buffering or a higher amount of energy.[12] Mitochondrial dynamics are implemented via microtubules and other factors,[13] specifically disrupted-in-schizophrenia 1 (DISC1), which plays a vital role within mental disorders.[14] Nonetheless, it should be noted that mitochondrial function within each part of the neuron could be different; that is, dendrites mitochondria have more active metabolism while axons mitochondria are more motile.[10] Aside from this, mitochondria can join together (fission) or separate from each other (fusion), the disruption of which reduces cell function and leads to apoptosis and cell death.[11] Mitochondria are composed of two bi-layer phospholipid membranes, with the inner membrane being folded and creates cristae. There is a small space between the intracellular spaces between two membranes and the mitochondrial matrix is covered by the inner membrane.[13] Mitochondria are the only organelles possessing their genetic content (1% of total cellular DNA) in the shape of circular and double-strand DNA; each has 16,569 base pairs.[15] There are 37 genes in the mtDNA inherited from mothers, 13 of which express polypeptides in the respiratory chain, 22 genes responsible for transfer RNA (tRNA), and the last 2 code ribosomal RNA (rRNA).[9, 15] The number of mtDNA may vary between 1 and 10 based on the cell type[15]; however, the higher quantity of mtDNA and mtDNA copy numbers can incur heteroplasmy risk, i.e., the existence of both intact and mutant genetic contents within the cell.[15] Put it differently, since mtDNA lacks telomerase and intron,[16] and mtDNA repair is rare,[8] it is more prone to internal and external damages resulting in mutations such as base substitutions, deletions, and insertions which change the homoplasmic state to heteroplasmic. The issue of heteroplasmy is more pronounced within neurons as mutations can be more quickly transferred to daughter cells due to the rapid neuronal growth.[15] Ultimately, the existence of massive mutant genetic contents in cells/tissues leads to the development of various physiological and psychological disorders. Mutations in mtDNA and mitochondrial-related genes (Tables 1 and 2) are contributory factors in the majority of mental disorders, which will be discussed in detail in the following sections.

There are two significant systems within mitochondria called respiratory chain and Krebs or tricarboxylic acid (TCA) or citric acid cycle to supply the energy demand. The glycolysis pathway in the cytoplasm provides molecules needed for the activation of the TCA cycle. Then, the energy-carrying molecules, nicotinamide adenine dinucleotide phosphate (NADPH), and flavin adenine dinucleotide ($FADH_2$) are being converted to an enormous amount of adenosine triphosphate (ATP) through oxidative phosphorylation

Table 1 Mitochondrial-related genes regulating glycolysis, TCA cycle, OXPHOS pathways, and mitochondrial dynamics.

	Glycolysis	TCA cycle	Respiratory chain					Fission	Fusion	Transport
			COM I	COM II	COM III	COM IV	COM V			
mtDNA genes			MT-ND1/2		MT-CYB	MT-CO1	MT-ATP6			
			MT-ND3/4			MT-CO2	MT-ATP8			
			MT-ND4L			MT-CO3				
			MT-ND5/6							
Nuclear genes	ALDOA	ACO2	ACAD9	SDHA	CYC1	COX4I1/2	ATP5A1	BNIP3	AFG3L2	ATPIF1
	ENO1	CS	ATP5SL	SDHB	CYCS	COX5A	ATP5B	COX10	BAK1	BNIP3
	GAPDH	FH	BCS1L	SDHC	UQCR10	COX5B	ATP5C1	DHODH	BAX	BNIP3L
	GPI	IDH2	COA1	SDHD	UQCRB	COX6A1/2	ATP5D	DNM1L	BNIP3	CISD2
	HK1	MDH1	ECSIT	SDHAF1	UQCRC1/2	COX7B	ATP5E	DNM2/3	CHCHD3	DNM1L
	PFKL	OGDH	FOXRED1	SDHAF2	UQCRFS1	COX7C	ATP5F1	FIS1	DNM1L	FIS1
	PGAM1	SDHA	NDUFA1-13	SDHAF3	UQCRH	COX8A	ATP5G1	GDAP1	FIS1	FUNDC1
	PGK1	SUCLA2	NDUFAB1		UQCRQ	COX8C	ATP5G2	LRRK2	GDAP1	FUNDC2
	PKM		NDUFAF1-8		UQCRC1/2	BCS1L	ATP5H	MARCH5	MFF	HK2
	TPI1		NDUFB1-11		UQCRFS1	CHCHD5	ATP5I	MFF	MFN1/2	HTRA2
			NDUFC1/2		UQCRH	COA1/3/4/5	ATP5J	MIEF1/2	MIEF1/2	MFN2
			NDUFS1-8		UQCRQ	COX10/14	ATP5J2	MTFP1	MUL1	MUL1
			NDUFV1-3		BCS1L	COX15/17	ATP5L	MTFR1	OMA1	PARK7
			NUBPL		LYRM7	COX19/20	ATP5O	MTFR1L	OPA1	PINK1
			OXA1L		PMPCB	PET100/117	ATP5S	MTFR2	PLD6	SQSTM1
			PARK7		SLC25A33	SCO1/2	ATPIF1	MUL1	PRKN	TOMM7
			TIMM21		TTC19	SURF1	ATPAF1/2	MYO19	STOML2	TSPO
			TIMMDC1		UQCC1/2	TIMM21	OXA1L	OPA1	USP30	VDAC1
			TMEM126B		UQCR10/11		PRIF	PINK1		
							TMEM70	PRKN		
							USMG5	SLC25A46		

COM, complex; *mtDNA*, mitochondrial DNA; *OXPHOS*, oxidative phosphorylation; *TCA*, tricarboxylic acid.

(OXPHOS) in the respiratory chain or electron transport chain (ETC).[17] As is depicted in Fig. 2, there are five multisubunit complexes responsible for producing ATP. These are named ubiquinone oxidoreductase (complex 1), succinate dehydrogenase (complex II), ubiquinol-cytochrome *c* oxidoreductase (complex III), cytochrome *c* oxidase (complex IV), and ATP synthase (complex V). Complexes I, III, and IV provide proton gradient,[11] and complex V is directly responsible for converting adenosine monophosphate (ADP) and inorganic phosphate to ATP. Respiratory chain and altered expression/mutations of its related genes could be another promising target in psychiatric illnesses since several deregulations have been reported in this regard.

The respiratory chain, although being essential for glucose metabolism and energy production, its components, mainly complex I,[11] is responsible for producing highly active radical oxygen species (ROS). As with ROS, reactive nitrogen species (RNS) may also be made by NADPH oxidases (NOXs), mainly NOX2 and NOX4, which are more prevalent and active with different parts of the brain cells.[18] Under normal conditions, cells can take out excessive intracellular oxygen and eliminate damaged cellular components. As well, an endogenous antioxidant system composed of enzymatic,

Table 2 Mitochondrial-related genetic alternations in psychiatric disorders, according to previous clinical studies.

Psychiatric disorders	Mitochondrial-related genes (except ETC) with altered expression	Electron transport complex (ETC) genes with altered expression	Mitochondrial/Nuclear DNA variants/polymorphisms correlated with disorder
Neurodevelopmental disorders			
ASD	**Decreased:** AIP, DNAJC19, DNM1L, FTH1, GCLM, LRPPRC, MFN1, MFN2, MIPEP, MTX2, NCF2, NEFL, OPA1, RHOT2, SLC25A12, SLC25A14, SLC25A15, SLC25A22, SLC25A24, SLC25A27, SLC25A3, SLC25A4, SLC25A5, TIMM17B, TIMM17A, TIMM23, TIMM50, TIMM8A, TIMM9, TOMM20, TOMM22, TOMM34, TOMM70A, TXN/**Increased:** BCL2, DRP1, FIS1, HSP90AA1, SLC25A25, SLC25A37, TIMM44, TP53, TSPO	**Decreased:** ATP5A1, ATP5B, ATP5F1, ATP5G3, ATP5H, ATP5J, ATP6VA2, COX5A, COC6A1, COX7A2, COX7B, COX8A, CYC1, NDUFA2, NDUFA4, NDUFA5, NDUFA8, NDUFAB1, NDUFB3, NDUFB4, NDUFB5, NDUFB6, NDUFS2, NDUFS4, UQCRC1, UQCRC2, UQCRFS1, UQCRH,	ATP6, ATP8, BCKDK, CO1, CO2, CO3, CYB, NC7, ND1, ND2, ND3, ND4, ND5, RNR2, SLC25A12, TL1, T1, TQ
Schizophrenia spectrum and other psychotic disorders			
SCZ	**Decreased:** ACAT1, CBR3, CNP, ETFA, GOT2, GSTM3, GSTP1, GSTTL, IDH3A, IDH3B, MDH1, MDH2, SDHA, SDHB OPA1, PDHA1, PDHB, PPA2, PRDX1, SCO1, SDHA, SLC25A4, TIMM17A, UCP2, TF/ **Contradictory:** ACO2 (↑or↓), ALDOC (↑or↓), ATP6V1A (↑or↓), BR1 (↑or↓), CA2 (↑or↓), CKB (↑or↓), DLD (↑or↓), ENO2 (↑or↓), GAPDH (↑or↓), GLUL (↑or↓), PGAM1 (↑or↓), PRDX2 (↑or↓), QDPR (↑or↓), TPI1 (↑or↓)/**Increased:** HK1, LARS2, ME3, LETM2, PGK1, PRDX6, TKT	**Decreased:** ATP5B, ATP5C1, ATP5F1, ATP5H, ATP5L, MT-CO2, COX7A1, COX7B, COX7C, COX8A, NDUFA2, NDUFS4, NDUFA6, NDUFA8, NDUFA11, NDUFB2, NDUFB3, NDUFB5, NDUFB6, NDUFB7, NDUFB9, NDUFC1, NDUFS1, NDUFS2, NDUFS3, NDUFS4, NDUFS6, UQCRFS1, UQCRQ, /**Contradictory:** ATP5A1 (↑or↓), NDUFB9 (↑or↓), NDUFV1 (↑or↓), NDUFV2 (↑or↓),	ACN9, ACO1, ACSL4, ACSL5, ALAS1, ALDH1A1, ALDH1A2, ALDHSA1, ATP5A1, ATP5J, ATP6, BCAT1, BDH1, CBR4, CERK, CLPX, CO2, COX4I1, CYB5B, DISC1, EARS2, FOXRED1, G72, GCLM, GOT2, GSTA1, GSTM1, GTPBP3, GTPBP5, GSTT1, GSTT2, HSDL1, KYNU, LARS2, LONP1, MDH2, MnSOD, MRPL27, MRPL40, ND1, ND2, ND4, ND5, NDUFV2, NOS1, NOS1AP, PRDX1, PRDX2, PRODH, RNMTL1, RNR1, SLC25A1, SLC25A12, SLC25A23, SND1, TDRKH, TUFM, TW, TXNRD2, VDAC1
Depressive disorders			
MDD	**Decreased:** BCL-2, MnSOD	-	BCL-2, CAT, PON1, SOD
Bipolar and related disorders			
BD	APAF (↑), BAD (↑), BAK (↑), BAX (↑), BCL2 (↓), CAT (↓), DISC1 (↓), DRP1 (±), GP_x2 (↑), GP_x1 (↓), GP_x4 (↓) GSTA4 (↓), GSTA2 (↑), GSTM5 (↑), HO1 (↑), IDH3A (↓), IDH3B (↓), MFN1 (±), MGST3 (↓), NOS (↑), OPA1 (±), SOD1 (↓), UCP2 (↓)	**Decreased:** ATP5C1, ATP5G3, ATP5H, ATP5J2, ATP5O, COX15, COX5A, COX6C, COX7A2, COX7B, NDUFAB1, NDUF1, NDUF2, NDUFS1, NDUFS3, NDUSF7, NDUFS8, NDUFV2, UQCRC2,	CACNA1C, CYB, DISC1, G72, ND1, ND2, ND3, ND5, NDUFV2, TRPM2
Eating disorders			
AN	**Decreased:** NOX5, PTGS2/**Increased:** DUOX2, GP_X1, GP_X3, PRDX2, PRDX6	-	-
Obsessive-compulsive and related disorders			
OCD	-	-	UCP2 and SOD
Stress-related disorders			
PTSD	-	-	ALOX12, ALOX15, RORA

AN, anorexia nervosa; *ASD*, autism spectrum disorder; *BD*, bipolar disorder; *MDD*, major depressive disorder; *OCD*, obsessive-compulsive disorder; *PTSD*, posttraumatic stress disorder; *SCZ*, schizophrenia.

e.g., glutathione reductase (GR), glutathione S-transferase (GST), glutathione peroxidase (GP_x), superoxide dismutase (SOD), paraoxonase (PON), and catalase (CAT) as well as nonenzymatic antioxidants, e.g., vitamin C and E, and glutathione (GSH) can sufficiently neutralize ROS and avoid further complications.[19] However, when accompanied by external and internal factors[7, 14, 19] like dietary deficiencies, toxins, pollutants, polymorphic changes, inflammation, fatty acid oxidation, nucleic acid metabolism, and reactive species overcome antioxidants capability developing oxidative stress. Following this, reactive species can inflict damages to macromolecules, intracellular

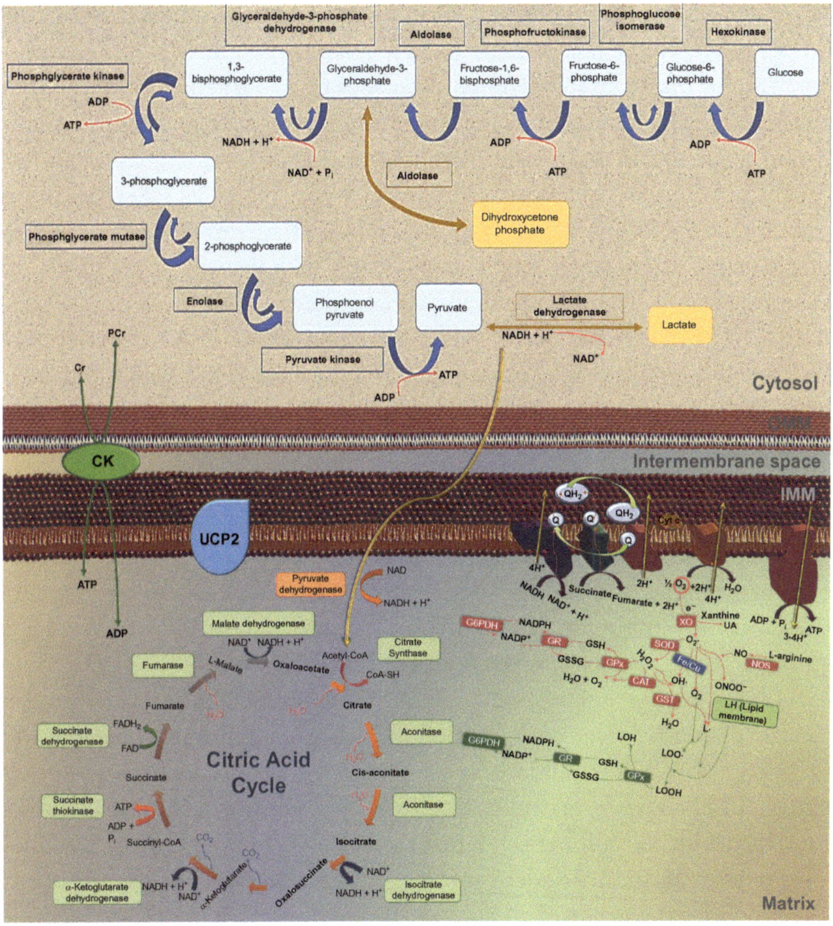

Fig. 2 Pathways of glycolysis (in the cytosol), citric acid cycle (in the matrix), oxidative phosphorylation (in the inner mitochondrial membrane; from left to right: complex I, II, III, IV, and V), and the related oxidative stress (in the matrix) were illustrated in this figure. *ADP*, adenosine diphosphate; *ATP*, adenosine triphosphate; *CAT*, catalase; *CK*, creatine kinase; *CO_2*, carbon dioxide; *CoA-SH*, coenzyme A with sulfhydryl functional group; *Cr*, creatine; *Cu*, copper; *Cyt c*, cytochrome complex; *FAD*, flavin adenine dinucleotide; *$FADH_2$*, reduced FAD; *Fe*, iron; *G6PDH*, glucose 6-phosphate dehydrogenase; *GP_x*, glutathione peroxidase; *GSH*, glutathione; *GSSG*, glutathione disulfide; *GST*, glutathione S-transferase; *H_2O*, dihydrogen oxide; *H_2O_2*, hydrogen peroxide; *IMM*, inner mitochondrial membrane; *L*, oxidized lipid; *LOH*, hydroxy lipid; *LOO*, lipid peroxyl radical; *LOOH*, lipid hydroperoxide; *NADH*, nicotinamide adenine dinucleotide; *NADPH*, nicotinamide adenine dinucleotide phosphate; *NO*, nitric oxide; *NOS*, nitric oxide synthase; *O_2*, oxygen; *O_2*, superoxide radical; *OH*, hydroxyl radical; *OMM*, outer mitochondrial membrane; *$ONOO^-$*, peroxynitrite; *PCr*, phosphocreatine; *P_i*, inorganic phosphate; *Q*, ubiquinone; *QH2*, ubiquinol; *SOD*, superoxide dismutase; *UA*, uric acid; *UCP2*, uncoupling protein 2; *XO*, xanthine oxidase.

pathways, and, more importantly, morphological changes in neuronal cells and the brain. For instance, it can oxidize guanine base in DNA,[19] intensify overall aging, telomere shortening,[18] and apoptosis,[13] and disrupt mitochondrial biogenesis and function. The mentioned destructions are more exacerbated in the brain due to the large oxygen utility, lower antioxidants, and higher lipid content. When occurred in the brain, oxidative stress causes increased permeability of blood-brain barrier (BBB), neuroinflammation, synaptic damage, and morphologic changes along with disrupting neurogenesis, neuronal growth, synaptic plasticity, neuroprotection, and neuronal regeneration.[18, 20] Noteworthy, poor sleeping is considerable risk factors putting the brain at a higher risk of oxidative stress. Since most psychiatric disorders are associated with sleep disturbances, negative impacts of oxidative stress are more pronounced within such illnesses.[20, 21] Another interesting point is that some psychiatric medications can alter mitochondrial function, mainly through affecting the balance of oxidants/antioxidants and OXPHOS.[22–25]

Besides the mentioned aspects, excessive intracellular calcium concentration and hemostasis are also being at play in mental disorders. Calcium is essential for mitochondrial function, synaptic plasticity, and neuronal signaling,[11, 15, 26] which can be quickly taken up by mitochondria from the condition's cytosol of high energy demand and excessive cytoplasm calcium content.[11] The basal mitochondrial level is low and required to regulate TCA enzymes and complex V;hence, ATP production. However, even slight changes in matrix calcium concentration can initiate oxidative stress, recurrent mitochondrial depolarization, and mitochondrial dysfunctions; provoking mitochondrial swelling and cellular death through opening mitochondrial permeability transition pore (mPTP), releasing apoptotic factors to the cytoplasm, activating caspases, and inducing apoptosis.[10, 11, 13] Apoptosis or programmed cell death can occur either through intrinsic or extrinsic pathways. Although mitochondria can be involved in both,[10] its role is dominant within the intrinsic way. To briefly explain the intrinsic apoptosis pathway, under the stress condition, B-cell lymphoma 2 (Bcl-2) family pro-apoptotic proteins, namely Bcl-2 associated X protein (Bax) and Bcl-2 homologous antagonist killer (Bak) affect the mitochondrial membrane, resulting in the release of cytochrome *c* into the cytoplasm. Following this, cytochrome *c* generates a complex (azotosome) with apoptotic protease-activating factor 1 (Apaf-1) and pro-caspase 9, which activates caspase 9. The activated caspase 9 activates caspases 3, 6, and 7; hence, inducing apoptosis and cell death.[15]

Mitochondrial abnormalities can be mainly categorized into mtDNA and mitochondrial-related genes impairments, OXPHOS disruptions, oxidative stress, calcium hemostasis dysregulation, and apoptosis. In the next section, the association of the mentioned abnormalities within each mental disorder will be discussed.

3 Mitochondrial abnormalities and psychological disorders

Insufficient attention of the health sector to mental disorders is also reflected in the existing literature. Unfortunately, the previous preclinical and clinical studies mainly focused on autism spectrum disorders (ASDs), attention-deficit/hyperactivity disorder (ADHD), schizophrenia, depressive disorders, bipolar and related disorder, eating disorders, obsessive-compulsive disorder (OCD), and stress-related disorders. This is while there are almost no clinical studies on the other categories of psychiatric illnesses. The existing related sources are limited to a few animal studies with controversial results, making it entirely unfeasible to extrapolate the reports to the real-world data. Thus, the following sections discuss mental disorders by providing reliable evidence from both animals and human investigations. As obtaining accurate data from living human brain regions usually requires invasive strategies, the majority of former reports are from postmortem assessment, where postmortem factors may also influence the brain. This implies two crucial issues: first, it is urgent to find specific biomarkers existing in the blood, plasma, urine, and blood cells to more simply diagnose mental disorders and rule out similar conditions; second, implementing concurrent noninvasive techniques is required to confirm postmortem findings. As briefly depicted in Fig. 3, oxidative stress has been the central focus of recent studies on all mental disorders. This may primarily be due to the noninvasive nature of in vivo and clinical tests to measure oxidants/antioxidants' level through body fluids, mainly blood. Regarding oxidative stress biomarkers in different psychiatric illnesses, alternations in malondialdehyde (MDA), and GSH, which are considered the most reliable markers, are coordinated in almost all studies. Therefore, due to such methods' availability and simplicity, one should carefully identify related differences among mental disorders; hence, attempting to design and develop new molecules to modulate oxidative stress through following drug discovery strategies.[27]

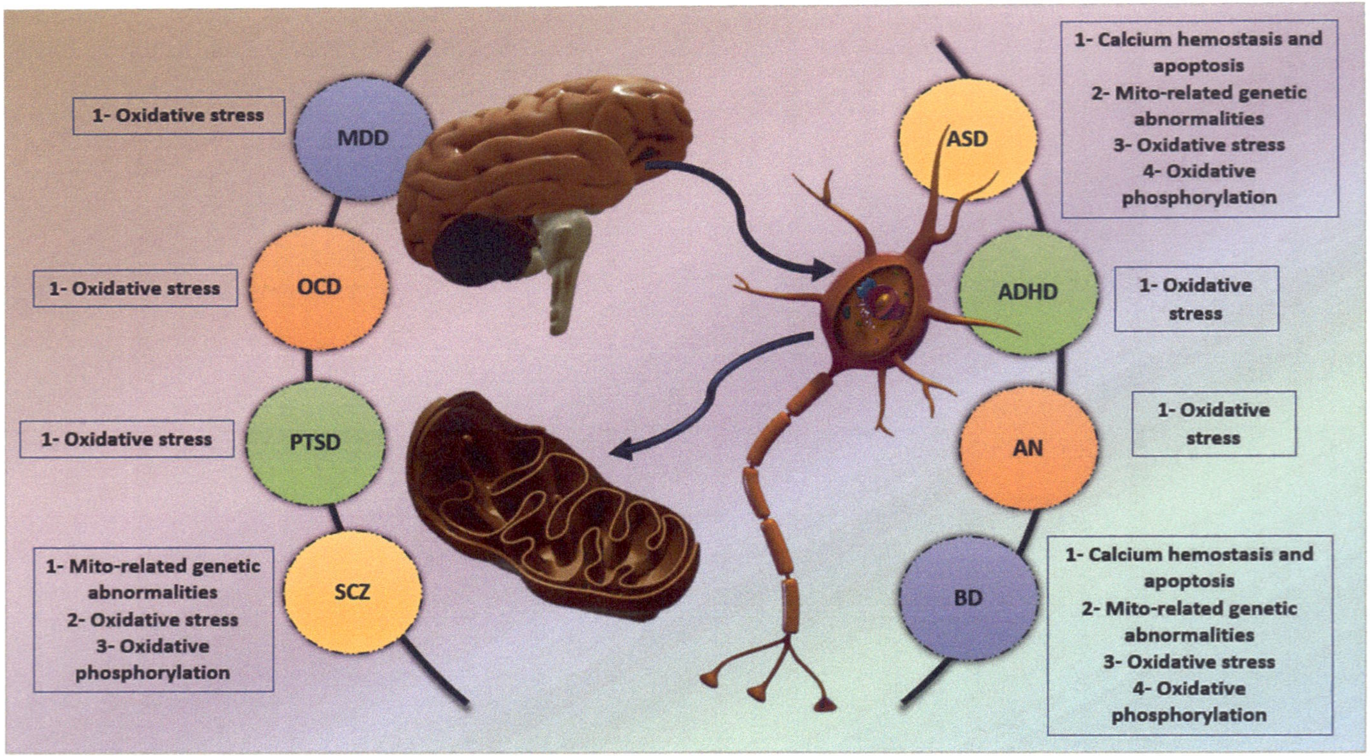

Fig. 3 Mitochondrial abnormalities within each mental disorder were the focus of attention in former studies. *ADHD*, attention-deficit/hyperactivity disorder; *AN*, anorexia nervosa; *ASD*, autism spectrum disorder; *BD*, bipolar disorder; *Mito*, mitochondria; *MDD*, major depressive disorder; *OCD*, obsessive-compulsive disorder; *PTSD*, posttraumatic stress disorder; *SCZ*, schizophrenia.

3.1 Neurodevelopmental disorders

3.1.1 Autism spectrum disorder

According to the Centers for Disease Control and Prevention's (CDC) Autism and Developmental Disabilities Monitoring (ADDM) network, the prevalence of ASD was growing from 1 in 150 children in 2000 to 1 in 54 cases (about 2%) in 2016, in the United States.[9,28] Although ASD can occur at any age, it is classified as developmental disorders since it is common in early childhood, particularly among boys. Indeed, ASD covers a broad set of symptoms in the first developmental period. Patients experience low social communication/interactions and limited and repetitive patterns of interests and behaviors, causing further educational, occupational, and social complications.[29,30] Considering the pathogenesis of the disease, ASD can be divided into two major subgroups of "genetic disorders," including fragile X syndrome, Rett syndrome and tuberous sclerosis, and "idiopathic disorders" like Asperger syndrome, autistic disorder, and pervasive developmental disorder-not otherwise specified (PDD-NOS).[31] Although the exact nature of the disease has not been revealed yet, the role of several environmental and internal factors like prenatal and postpartum complications,[32] rubella infection during pregnancy,[33] toxicants exposure,[34] and familial and genetic issues (i.e., parental age, having a sibling with ASD)[15] has been proved. These can increase the incidence of ASD by reducing the level of cerebral gamma-aminobutyric acid (GABA) and *N*-acetyl aspartate (NAA) and causing neuroinflammation and cellular abnormalities in different brain regions, metabolic anomalies (i.e., dysfunction of serotonin, acetylcholine, and glutamate), oxidative stress, and ultimately mitochondrial dysfunction. Considering an extensive literature on the association of mitochondrial abnormalities with ASD since 1985,[28] this topic has paved the way for a better understanding of the disease etiology and management. ASD-related mitochondrial abnormalities can be generally categorized into oxidative stress, mtDNA abnormalities, mitochondrial-related metabolite abnormalities, OXPHOS dysfunction, and mitochondrial-related genes with altered expression. It has been repeatedly claimed that the level of both GSH and total GSH in the plasma, whole blood, and blood mononuclear cells reduced during ASD, which, when accompanied with the elevated level of oxidized glutathione (GSSG), led to the lower GSH/GSSG ratio in the plasma, whole blood, cerebellum, and temporal cortex samples. There was also an increase in the activity of GST among ASD patients compared to the control group.[30] Despite these, results are controversial regarding enzymatic antioxidants' level/activity, namely SOD, GP_x, and CAT, though they exhibited a dropping pattern in general.[9,15,29,35] Dysregulation of these enzymes may be related to the lower level of elements, zincand iron, which are cofactors of the mentioned enzymes.[36] The mean concentration of vitamin blood levels was also controversial regarding vitamin B9, vitamin B12, vitamin C, vitamin D3, and carotene while reduced vitamin A and vitamin E levels and elevation in the level of vitamin B6 were repored. Likewise, the blood level of transmethylation and transsulfuration metabolites, including methionine, homocysteine, cystathionine, cysteine, and cysteinylglycine were not concurrently altered among ASD patients across previous studies.[37] Nonetheless, downregulation of genes responsible for the ROS metabolism (i.e., FTH1, NCF2, TXN, PTGS2, PRNP, and GCLM)[29] along with the alleviation in the concentration of oxidative stress biomarkers, including nitric oxide (NO),[37] 8-isoprostane,[15] MDA, and 8-hydroxy-2′-deoxyguanosine (8-OHdG),[29] are of the other evidences for oxidative/antioxidative imbalance in ASD. Besides oxidative stress, wide-ranging reports on mtDNA sequence anomalies such as increased copy number, duplication, point and pathogenic mutations, depletion, and deletion justify metabolite abnormalities and ETC dysfunction in ASD.[9,30,38–40] Such irregularities mostly appeared as elevations in lactate, pyruvate, alanine, creatine kinase, and ammonia blood levels of ASD patients.[28] Other mitochondrial plasma markers with abnormal levels were lactate dehydrogenase, α-keto acid dehydrogenase, and caspase 7[29]; the latter was also the most significant marker discrimination of ASD patients from healthy controls.

Aside from altering mitochondrial plasma markers, mtDNA genetic abnormalities can disrupt respiratory chain (RC) activity through suppressing the activity of ETC complexes, i.e., complexes I, II, III, IV, and V in the muscle, lymphocyte, buccal, and brain region biopsies.[9,30,38,39] Indeed, mtDNA sequence anomalies may directly downregulate ETC genes, including some subgroups of NDUF (complex II), UQCR (complex III), CYC (complex III), COX (complex IV), ATP5, and ATP6 (both for RCV) genes (Table 2).[39] Here, another valuble point is the association of ETC-related genetic variants with the incidence of ASD, providing the opportunity to establish personalized medicine for this disorder. As mentioned in the previous section, the impairment of OXPHOS may also be owing to the calcium homeostasis dysregulation in favor of matrix calcium accumulation. This itself can provoke inflammation and oxidative stress, which further causes neuronal damage and cell death. The candidate genes thought to maintain calcium homeostasis in ASD patients are SLC2512, ATP13A4, CAPS2, IL1RAPL1, CACNA1C, ATP2B2, and DISC1.[30,41]

As summarized in Table 2, the expression of some mitochondria-related genes may also be altered during ASD.[39] The most crucial ones are MTX2, NEFL, and SLC25 family, which contribute to the importation of proteins to mitochondrion/metabolism of proteins, intracellular transport to axons, and dendrites the production of mitochondrial aspartate/glutamate carriers,[30] respectively. Finally, mitochondrial dynamics can also be disrupted during ASD; thereby, undermining mitochondrion ability to attenuate stress. High levels of FIS1 and DRP1 (responsible for mitochondrial fission) and reduced levels of MFN1, MFN2, and OPA1 (controlling mitochondrial fusion) in postmortem studies of ASD patients are proofs of this claim.[9]

To sum up, there have been by far neither a diagnostic method nor a specific treatment indicated in ASD.[28,29] Put it differently, management of ASD is only restricted to the use of nonpharmacological therapies or symptoms relief through the administration of antipsychotics, stimulants, anxiolytics, anticonvulsants, and antidepressants.[42] On the other hand, mitochondrial abnormalities, particularly OXPHOS deficiencies, have been closely associated with the incidence and propagation of ASD since they are reported to be present in about 80% of ASD-affected children.[30] Thus, there is an ample opportunity to develop mitochondrial-targeted medications or diagnostic methods, considering the biomarkers mentioned above and genetic anomalies.

3.1.2 Attention-deficit/hyperactivity disorder

Despite ASD, the contribution of mitochondrial abnormalities in the pathogenesis of ADHD has not been comprehensively investigated, and available data are mostly limited to the role of oxidative stress. ADHD is the most prevalent neurobehavioral disorder affecting about 7% of children and 4% of adults.[19,43] According to the DSM-5, patients experience different chronic states of inattention and hyperactivity/impulsivity,[44,45] which in some cases, may remain until the end of life. This exerts occupational, educational, and social problems and further provokes other psychological and physiological issues such as depression, substance use disorders, sleep apnea, ASD, anxiety, antisocial behavior, and oppositional defiant disorder (ODD).[19,44,45] As with ASD, the pathology of ADHD has not been clear yet. However, dopaminergic and noradrenergic systems were reported to play the dominant role, i.e., their related neurotransmitters are lower in some areas of the brain, including the caudate nucleus, cerebellum, and prefrontal cerebral cortex.[43,44] Such defects can be associated with genetic alterations, including cytokines polymorphisms, variations in active dopamine transporters, dopamine receptors D1,2,4,5, norepinephrine transporters, alpha-2A-adrenergic receptor genes, etc.[45]

Additionally, several environmental factors[43, 45, 46] (e.g., prenatal smoking and alcohol exposure, malnutrition, lead, pesticides, toxins exposures, etc.) are being at play here, ultimately developing oxidative/nitrosative stress. Indeed, the increased amount of ROS and RNS is the primary cause of structural damage in the brain's mentioned parts, dopamine neurons, and transporters,[42, 44] which suppresses neurotransmitters synthesis by inhibiting 5,6,7,8-tetrahydrobiopterin and dopamine release.[43] Reactive species reduce brain-derived neurotrophin, an essential factor for neurogenesis, neuroprotection, synaptic plasticity, learning, and memory.[43] These events are more exacerbated in patients with lower levels of zinc, vitamin D, omega-3, and iron storage (ferritin)[46]; hence, the dietary regimen should be the center of attention in ADHD patients.[43] Another possible reason for oxidative stress propagation is the dopamine molecule itself, which can be converted to its highly active metabolite, dopamine quinone, and further oxidative stress.[43] Interestingly, conventional ADHD treatments, e.g., methylphenidate and atomoxetine, were also proved to have the potential for generating oxidative stress in preclinical models.[19] It is difficult to extrapolate these data to real-world experiments due to the existence of multivariate factors like treatment duration, dosage, patients' age, affected region of the brain, types of measured biomarkers, and so on.[43]

In line with these, the elevated concentration of oxidative stress biomarkers, including MDA, NO, nitric oxide synthase (NOS), and xanthine oxidase (XO), were detected in the urine, saliva, and mainly plasma/serum of ADHD patients in almost all clinical studies.[19, 43, 45] In contrast, results regarding alternations in antioxidant biomarkers level, i.e., CAT, GST, are controversial. This is while the SOD and PON1 were decreased or remained unchanged.[19, 46] What is clear is that ADHD patients had a higher ratio of total oxidant capacity (TOC) to total antioxidant capacity (TAC), highlighting the insufficiency of the antioxidant system to combat the elevated level of reactive species.[46] Therefore, targeting mitochondria in view of maintaining the balance in oxidative-antioxidative ratio could become a possible approach toward managing ADHD though little effort has been dedicated to this scope. More importantly, the genetic profile of ADHD patients regarding mitochondrial respiratory chain and OXPHOS should be further investigated.

3.2 Schizophrenia spectrum and other psychotic disorders

3.2.1 Schizophrenia (SCZ)

SCZ, with affecting about 1% of the population worldwide,[12] is one of the mental disorders with considerable impacts on belief, cognition, speech, memory, and cognition of patients, especially in their late-adolescence; highlighting the possibility for developing preventive approaches.[12, 47] SCZ patients may suffer from four states of symptoms, i.e., positive symptoms with psychotic signs, mainly delusions (suspensions and false beliefs) and hallucinations (seeing or hearing something that does not exist), negative symptoms with lack of speech, interests, and suppressed feeling, cognitive symptoms with neurocognitive deficits and alternate mood symptoms. It has been demonstrated that glutaminergic, GABAergic, and dopaminergic systems may be deregulated[48]; however, types of disruptions may vary across different symptoms.[49] For instance, the dopaminergic system's overactivity happens within positive symptoms, while this system's downregulation accompanies negative states. Regarding risk factors, genetics, prenatal, and early childhood issues are the most important, the effects of which can be varied among different individuals.[17] Therefore, the disease's heterogeneity makes it even tougher to find a diagnostic marker or an effective treatment.[48, 50] Nonetheless, the intensity of mitochondrial defects, particularly the level of ATP and phosphocreatine (PCr), is reported to be directly associated with the severity of symptoms.[50, 51] Moreover, imaging studies revealed that the number of enlarged mitochondria with destroyed inner membrane increases within the disease's longer duration.[52] Thus, mitochondria can be regarded as promising targets. The lower ratio of ATP to PCr can be correlated with the dopaminergic system's hyperactivity and a higher level of dopamine, specifically during positive states. This is because the excessive uptake of dopamine by mitochondria can inhibit complex I activity and reduce ATP production; hence, various neuronal aspects are disrupted, resulting in abnormal behavior and cognition.[48] Dopamine can also cause mitochondrial dysfunction in another way. The metabolism of dopamine by monoamine oxidase produces hydrogen peroxide, one of the initiators of oxidative stress.[51] Then, it is not surprising that total ROS, lipid peroxidation (MDA, pentane, and isoprostanes), protein oxidation (ethane, 3-nitrotyrosine, carbonyls, 4-hydroxynonenal, and pentosidine), DNA oxidation (8-OHdG) markers levels as well as NOS-related biomarkers (i.e., NOS activity, $NO/NO_2^-/NO_2$concentration) were elevated in body fluids, blood cells and breath of SCZ patients and animal models.[53] In line with this, dehydroepiandrosterone-sulfate, lipid hydroperoxide (LOOH), MDA, and homocysteine were also increased in first-episode psychoses (FEP) patients.[54] The latter is taught to be an indicator of early cognitive deficits in these patients. Contrary to these, a decline in the levels of GSH, SOD, vitamin A, vitamin B6, vitamin C, uric acid, and TAC pointed to the insufficiency of the antioxidant system in SCZ.[29] Likewise, CAT, GSH, SOD, TAC, and uric acid concentrations decreased in the FEP population.[54] However, the elevated levels of uric acid, vitamin A, and vitamin E in patients with early-onset psychoses in another meta-analysis indicated possible compensation.[55] Despite these, results regarding CAT and GPx activity were inconsistent across studies.[53] Levels of trace metals could also be changed in SCZ. Reports suggested that SCZ patients had lower selenium and iron concentrations and higher concentrations of lead, chromium, and cadmium. This is while, there might be intrapatient variations regarding copper and zinc levels.[56] Mutations in oxidative/antioxidative genes, namely GCLM, GSTM1, GSTT1, GSTT2, GSTA1, MnSOD, NOS1, and NOS1AP[53] in different populations can be responsible for the incidence and progress of SCZ.

Another useful finding to properly implement preventive approaches or initiate medications is the alternation in energy metabolism within SCZ, which is usually the indicator of the disease's early stages.[47] Dysregulation in ATP productions can occur in each step responsible for ATP synthesis, including glycolysis, TCA, and OXPHOS.[57] This is due to the large number of related genetic variations in previous clinical studies (Table 2).[17, 48, 52, 58–60] Despite ASD where the activity of all respiratory chain complexes is reduced, in SCZ, risk factors mainly affected complex I/II and IV activity in the temporal cortex/basal ganglia and frontal cortex/temporal cortex of SCZ patients, respectively.[17, 50, 51, 61] For instance, deregulations in NDUFV1, NDUFV1, and NDUFS1 activity were obtained in different hippocampus, parietal-occipital cortex, prefrontal cortex, and somatic cells of SCZ patients' brain.[48] Interestingly, the mentioned alternations were also seen in bipolar disorder and to a lesser extent in patients with depressive disorders. However, they are distributed in different neuroanatomical sites creating specific patterns for each disorder; albeit sometimes with shared symptoms.[47] Rather than respiratory chain complexes, lower expression of OPA1 responsible for mitochondrial fusion in postmortem evaluations[52] can bring about dynamicity issues, further leading to mitochondrial dysfunction. As well, AS3MT, ATPAF2, C2orf47, DISC1, DNAJA3, DRG2, FOXO3, G72, HSPA9, HSPE1-MOB4, IMMP2L, NDUFA6, PCCB, SMDT1, UCP2, and UCP4 were found to be gene candidates highly associated with SCZ.[48, 58, 62, 63] All mentioned factors together can justify mitochondrial-related metabolite abnormalities in SCZ, which are about a reduction in NAA,[17, 51, 61] citrate, α-ketoglutarate, and pyruvate[47, 61] levels along with aconitase, succinate thiokinase, and pyruvate dehydrogenase activity[47] in various brain regions.[61] More importantly, TCA/OXPHOS dysfunction is associated with a shift toward the glycolysis pathway, further leading to an increased lactate concentration indicative of lower cognition and functional capacity in SCZ patients.[47] Elevated lactate concentration reduces pH, similar to what could be seen within bipolar disorder (discussed later).[47]

Regarding numerous studies on the connotation of mitochondrial abnormalities in STZ, what is important now is the paramount need to establish investigations to develop more effective and specific medications as well as preventive approaches. Additionally, mitochondrial calcium hemostasis and related risk genes should be more precisely identified as the literature is almost lacking in this field.

3.3 Depressive disorders

3.3.1 Major depressive disorder (MDD)

According to the GBD 2010 report, among mental illnesses, depressive disorders are responsible for the largest share of DALYs (40.5%) and YLDs (42.5%) worldwide, which have been growing over recent years.[1, 64] Depression mostly affects females in their 40–59 years of age. Patients should have at least five of the following symptoms during at least 2 weeks, one of which must be anhedonia or reduced mood. The symptoms are depressed mood, reduced interest, significant dietary pattern or weight change, sleep disturbance, psychomotor agitation or retardation, fatigue or loss of energy, feelings of worthlessness, decreased ability to concentrate and think, and recurrent thoughts of suicide, death, or suicide attempt.[17, 65] As the symptoms markedly vary among individuals and patients usually refuse to accept their disorder, screening and diagnostic tests do not efficiently work. They are just limited to a few questionnaires and diagnostic scales. That is, there have been by far no available laboratory tests in this regard. Many depression symptoms overlap with other mental disorders; hence, it would be difficult to differentiate it. What is responsible for the significant emotional and economic burden of depression,[65] is its multifactorial nature, i.e., depression has the highest comorbidities with other physiological/psychological conditions and is linked to many internal and environmental factors. The specific pathology of depression has not been revealed yet though the role of some neurotransmitters, including serotonin, norepinephrine, dopamine, glutamate, and brain-derived neurotrophic factor (BDNF) has been clarified.[66, 67] In line with these, selective serotonin reuptake inhibitors, serotonin/norepinephrine reuptake inhibitors, atypical antidepressants, tricyclic antidepressants, monoamine oxidase inhibitors, and *N*-methyl-D-aspartate receptor antagonists have been introduced besides nonpharmacological managements. However, about 30% of patients become treatment-resistant,[68] and relapses occur in half of the treated patients.[66] Considering the life-lasting and recurrent nature of depression, the indicated issue highlights the need to overcome medication resistance and develop treatments preventing relapses. To this end, mitochondrial dysfunction has got little attention over the last decades. While there has been close attention to oxidative stress,[69–71] other aspects, particularly mitochondrial-related genetic abnormalities, have been out of focus, with conducting very few studies in this respect.[72] Almost all previous studies agreed on the higher concentration of ROS,[65, 73] NO, NOS, NO metabolites (especially in suicidal depression),[65, 66] XO,[66] 8-oxo-2′-deoxyguanosine (8-oxodG),[66] 8-OHdG and MDA[65] in depressed patients. The latter two were assumed to be associated with frequent episodes and recurrent disease, respectively. In contrast, there was a reduction in the function of the endogenous antioxidant system due to the lower levels of vitamin E, zinc,[66, 74] coenzyme Q10 (CoQ10), GSH, uric acid,[66] omega-3, PON1[74] and TAC among patients compared to control groups.[65] Data regarding alternation in the concentrations of CAT, SOD, and GP_x were controversial[65, 66, 74]; however, GP_x was a good indicator of the disease severity and progression. Given genetic variations, few single nucleotide polymorphisms (SNPs) were identified in SOD, CAT, PON1 and Bcl-2 gene coding antiapoptotic protein, with the latter correlated to treatment resistance in depressive disorder.[72] Additionally, few studies found DISC1,[68] TOMM40,[72] and CACNA1C[68] to be candidate genes in MDD. Some studies reported lower mtDNA copy number content, while some were not associated any significant changes.[64, 72] Hence, it needs more careful investigations. Mitochondrial dynamics were also reported to be altered during depression since the expression of Fis1/Drp1 and MFN1/MFN2/Opa1 was increased and decreased, respectively, in animal studies.[68] The detachment of hexokinase 1 (HK1),[72] a rate-limiting enzyme in glycolysis, from the outer mitochondrial membrane, especially in MDD patients, was assumed to be positively associated with a decline in glucose metabolism,[17] ATP production,[68] basal/maximal metabolic rate,[72] and OXPHOS dysfunction[17] in several brain regions in both animal and clinical evaluations. In line with this, the level of some mitochondrial-related metabolites, namely NAA, alanine, citrate, α-ketoglutarate, succinate, malonate, lactate, pyruvate, and fumaric acid, was changed in MDD models and patients.[75] As well, an alternation in mitochondrial morphology was demonstrated in former laboratory models.[68] Despite these, very few kinds of literature are available, specifically discussing alternations in each respiratory chain enzyme during MDD. Reports indicated that MDD was accompanied by suppressed complex I activity[68] and reduced ratio of complex I + II/complex IV and complex I + III/complex IV.[72]

What is clear is substantial deficits regarding the investigation of MDD relations with mitochondria-related genetic variations and OXPHOS activity. This issue should be addressed through studying living human brain regions rather than animal models before taking further steps.

3.4 Bipolar and related disorders

3.4.1 Bipolar disorder (BD)

Previously known as manic-depressive disorder, BD is considered a bridge between schizophrenia and MDD, where individuals face alternate stages of mania and depression. According to the DSM-5, BD is classified into three groups: type 1 (BD1), type 2 (BDII), and cyclothymic disorder, or cyclothymia. Although the intensity of symptoms may differ among subtypes, manic episodes are associated with very up, energetic feelings, decreased need for sleep, and appetite loss. Simultaneously, during depressed phases, patients experience sad, worried, restless, and hopeless feelings.[76] The global prevalence of BD is up to 4.8%,[11] which when accompanied with other mental (i.e., eating disorders, behavioral disorders, substance abuse, anxiety disorders, and personality disorders)[77] and medical comorbidities, puts BD in the sixth place of the leading cause of disability worldwide.[26] As with other psychological disorders, the role of several environmental and genetic risk factors has been identified in BD; however, the disease's exact nature remains unknown. This is the main reason justifying the lack of gold-standard diagnostic methods, leading to severe delays between the illness occurance, diagnosis and useful management. There has been a growing tendency toward identifying new specific targets better to understand BD's etiology and design potential molecules. Different mitochondrial aspects, including metabolism, oxidative stress, morphological changes and distribution, genetics, and internal apoptosis pathways, are promising candidates in this regard. The mentioned abnormalities ultimately alternate the level of lactate, PCr, NAA, adenosine diphosphate (ADP), ETC and intracellular pH, and expression of apoptotic, antioxidants, mitochondrial-related genes, and mtDNA. According to previous findings, some of the mitochondrial complex I genes, namely NDUFS1, NDUFS3, NDUFS7, NDUFS8, NDUFV1, and NDUFV2, were downregulated in BD.[26]

Moreover, identified polymorphisms in MT-ND1, MT-ND2, MT-ND3, MT-ND5, and MT-CYB were associated with ETC activity deficiency.[26,40] In line with this, the reduced levels of messenger RNA (mRNA), UQCRC2 (complex III), COX5A/COX6C/COX15 (complex IV), and ATPSC1/ATP5J/ATP5G3 (complex V) lessen the activity of other respiratory chain complexes.[11,17] Moreover, the mRNA levels of isocitrate dehydrogenase 3A and 3B, an essential enzyme-producing nicotinamide adenine dinucleotide (NADH) for OXPHOS, decreased in the dorsolateral prefrontal cortex and cerebral tissues of these patients, led to the accumulation of isocitrate in cerebrospinal fluid (CSF).[11] These factors together can result in two necessary alternations in mitochondrial biogenesis. First, the elevated release of cytochrome *c*, activation of caspases 3 and 9, programmed cell death, and further cell death in both manic and depressed episodes. As evidence, the expression of apoptotic genes, FAS, APAF, Bak, Bax, and BAD was elevated in patients' hippocampus and cerebral cortex.[11,13]

In contrast, the Bcl-2 family's antiapoptotic genes were downregulated, which may be due to the existence of polymorphism within this gene.[13] Second, deficiency in respiratory chain complexes was associated with the activation of anaerobic glycolysis pathway instead of OXPHOS, wherein the conversion of NADH to NAD^+ is accompanied by the reduction of pyruvate to lactate, accumulation of lactate, and a further decrease in pH, like what is seen in schizophrenia.[11,26] This alternation was observed in the frontal lobe, basal ganglia, and anterior cingulate regions in manic episodes. Since the role of haplotypes U, K, and UK has been proved in this regard, it seems that this feature may be inherited through mtDNA.[13] Of the other contributory factors for energy production, PCr, NAA, and ADP levels were as well altered during BD.[11] The level of PCr, a high-energy molecule, was decreased in the brain of type II BD patients both in manic and depressed episodes, which was demonstrated as a reduction in the PCr peak area percentage in the left frontal lobe (depressive state) and right frontal lobe (manic and euthymic states).[13,26] Additionally, the reduced level of inorganic phosphate in BD was associated with OXPHOS impairment within this population.[11] Both ADP and NAA followed the same pattern in BD patients. The lower level of ADP was identified in the anterior cingulate cortex (manic episode) and left ventrolateral prefrontal cortex (euthymic episode). The reduced level of NAA in basal ganglia, hippocampus, planum temporale, transverse temporal gyrus, and the dorsal prefrontal cortex, and the orbital frontal gray matter was the sign of neuronal degradation/loss. It undermined neuronal integrity and metabolism.[26]

As with mitochondrial function, there were also reports of mitochondrial biogenesis alternation in BD patients because of morphology and distribution, especially in the prefrontal cortex.[11,26] As a candidate susceptibility gene for BD and SCZ, G72 (DAOA) gene plays a significant role in mitochondrial fragmentation, which was thought to be responsible for the increased number of smaller mitochondria in the prefrontal cortex of BD patients.[13] DISC1 participating in mitochondrial transport and fusion, regulation of neuronal outgrowth, differentiation, and migration, was downregulated, which was linked to the high number of early manic episodes in BD.[13] It should be noted that polymorphisms in both DAOA and DISC1 were positively associated with the predisposition to BD.[78] Despite these, no changes were reported regarding other mitochondrial dynamics-related genes, including Drp1, MFN1, and OPA1 levels in BD.[26] Besides what is mentioned regarding genetic alternations in BD, polymorphisms in two other genes important in calcium hemostasis, CACNA1C, and TRPM2, can elevate the intracellular calcium level in BD, leading to apoptosis and death.[11,77] The expression of mitochondrial leucyl-tRNA synthetase and the level of mitochondrial UCP2 mRNA were increased and decreased, respectively.[11] Considering mtDNA copy number and deletions, former results are controversial[13,79]; highlighting the considerable need for further investigations.

The last aspect is oxidative stress playing a dramatic role in OXPHOS activity and dopaminergic metabolism. Previous assessments are evidence of increased oxidative markers, i.e., 4-hydroxynonenal (anterior cingulate cortex), 3-nitrotyrosine (prefrontal cortex), 8-oxo-dG, 8-oxo-7,8-dihydro guanosine (8-oxoGuo), MDA,[80] and NO independent of the phase of BD in most of the studies.[76,81] However, there is no consensus data regarding alternations in antioxidants, SOD, CAT, GP_x, and GST.[11,77,81] There were also some alternations in the expression of oxidative stress-related genes, i.e., GST, SOD, HO1, and GP_x.[82] These variations may be owing to the different methodology and state of the disease. Therefore, more well-designed studies are required in this regard.

3.5 Eating disorders

3.5.1 Anorexia nervosa (AN)

Being estimated to cause a 6%–8% mortality rate,[83] AN is a serious condition, especially among females and adolescence,[84] where patients express unreasonable fear of gaining weight. AN is proved to be closely linked to environmental (e.g., lifestyle, smoking, and childhood abuse) and genetic factors along with psychological and physical stressors.[84] It is more commonly seen among individuals suffering from drug-abuse, perfectionism, and obsessive-compulsive personality traits.[83, 84] Patients suffering from eating/purging AN subtype make themselves vomit or misuse laxatives and diuretics to lose weight. While restricting type, they bear massive starvation and take intense exercise; both can increase oxidative stress, which exerts further psychiatric complications and irreversible multiorgan damages.[83, 84] As with ADHD, the role of mitochondrial abnormalities in the pathogenesis of AN has not been well investigated. From former studies, it is only concludable that chronic starvation, low consumption of foods rich in antioxidants, and heavy physical activity are directly associated with significantly elevated levels of some pro-oxidants, NO and oxidized GSH, as well as CAT due to compensatory mechanisms, and apolipoprotein B, an oxidative stress indicator.

In contrast, AN patients had lower total/free GSH and free cysteine parameter levels; however, results regarding the alternation in the levels of SOD,[83, 84] uric acid,[85] and vitamins (i.e., vitamin E metabolites and vitamin A) were discrepant. No significant alternations in zinc, selenium, vitamin C, and serum albumin levels were reported as well.[84] Rather than oxidative stress, glycolysis and mitochondrial respiration were substantially diminished in AN patients' peripheral blood mononuclear cells following chronic caloric restriction. Besides, an upregulation in the expression of antioxidative genes, GP_X1, GP_X3, PRDX2, PRDX6, and DUOX2, was reported. In fibroblasts cultured with AN patient's sera, GP_X3 was also upregulated; however, there was a reduction in the expression of NOX5 and PTGS2.[86]

Weight gain is the primarily compensatory way in patients with anorexia nervosa. As it might include high-fat or high-meat diet regimens acting as strong pro-oxidants, doubts have arisen over this approach's potency toward enhancing oxidative stress. A recent meta-analysis revealed that weight gain decreased SOD activity while improved TAC, CAT activity, and the level of endogenous antioxidant, albumin.[87] Nonetheless, due to the cross-sectional design and small-scale studies, findings cannot be fully trusted. Another issue is the low body mass index of these patients, making data extrapolation quite impracticable to other psychiatric diseases.[84] Therefore, it is urgent to design further preclinical or clinical studies to discover the alternation of oxidative stress and OXPHOS and genetic variants in different stages of the disease re-alimentation; hence, paving the way for investigating AN pathology and developing specific pharmacological treatments.

3.6 Obsessive-compulsive and related conditions

3.6.1 Obsessive-compulsive disorder

OCD is one of the chronic, life-long mental disorders affecting about 2.3% and 3% of adults and children, respectively.[88] It is characterized by having compulsive actions and repeated obsessive thoughts or behaviors on several aspects of hygiene, religion, punctuality, etc. These all impose considerable impacts on both mental and functional states of one's life. Its significant association with other psychiatric disorders, including anxiety disorders, substance use disorders and mood disorders, as well as the unknown nature of the disease and low responsiveness to current therapies, highlight the unmet need to find new pharmacological targets. Regarding mitochondrial abnormalities, the existing literature is just confined to oxidative stress as with ADHD and AN. Genetic variations and environmental stressors have been found to change the neutral cell condition to increase some oxidant biomarkers, namely 8-OHdG and MDA. The latter may be related to the genetic variations in manganese SOD and mitochondrial UCP2 of OCD patients. As well, having a higher frequency of neuronal NOS allele justifies the increased level of NO. This is while, there are inconsistencies regarding the changing pattern of antioxidants. Precisely, the level of GSH, vitamin E, vitamin C, and vitamin A, PON, and TAC were reduced or remained unchanged while there were elevations in the states of GPx, SOD, and CoQ10. CAT level in OCD patients was also reported to be the same as healthy controls in two studies. Overall, there is a necessity in conducting more related in-depth evaluations of other mitochondrial aspects to control the disease better is apparent.[17]

3.7 Stress-related disorders

3.7.1 Posttraumatic stress disorder (PTSD)

In some cases, being exposed to severe trauma, including fatal incidents, appalling injury, and sexual assault, is linked to a psychiatric disorder called PTSD. Based on DSM-5, the affected patients experience four symptom clusters of intrusive memories, avoidance, negative cognitions and mood, and arousal (changes in physical and emotional reactions).[8, 89] The disease may also accompany other comorbidities like sleep disturbance, chronic pain, anxiety, major depression and neurocognitive, cardiovascular, respiratory, digestive, and musculoskeletal disorders.[20, 90] Several neurological systems have been proved to be at play in the pathogenesis of PTSD, with the noradrenergic system, serotonergic receptors (mainly 5-HT_{1A} and 5-HT_{1B}), endocannabinoids, and the cannabinoid CB_1 receptor, hypothalamic-pituitary-adrenal

(HPA) axis, and the opioid system being the most significant.[91] Additionally, alternations in the level of pro-oxidants/antioxidants and expression of oxidative-stress-related genes rooted from chronic stress can be the cause of both propagation and progression of PTSD, creating a vicious cycle. Despite normal conditions wherein glucocorticoids suppress inflammation, under chronic repeated stress, glucocorticoids increase inflammation and oxidative stress by inducing the expression of pro-inflammatory cytokines.[20] In this regard, former studies proved the elevated level of lipid peroxidation and reduced level of PON1 as an antioxidant in earthquake survivors. Depleted levels of thioredoxin reductase, SOD, and GST[20] along with increased GSH concentrations in the dorsolateral prefrontal cortex and anterior cingulate cortex of PTSD patients[20] are other manifestations of oxidative imbalance in this disease.

Moreover, oxidative stress brings down BDNF, making neuronal cells more susceptible to damage and death.[89] According to genetic-wide association studies (GWAS), the insufficiency of the antioxidant system for detoxifying pro-oxidants in PTSD is related to the existence of SNPs in RORA, ALOX12, and ALOX15 genes.[89] RORA gene SNP (rs8042149) in PTSD patients disturbs the normal function of the RORA gene in the prefrontal cortex, hypothalamus, and hippocampus, which is the protection of neuronal cells under oxidative stress through inducing the expression of peroxiredoxin 6 and GP_X1. Contracting to the RORA gene, ALOX12 and ALOX15 provoke oxidative stress by encoding lipoxygenase 12 and 15, respectively, which ultimately leads to neuronal cell death. SNPs in ALOX12 (rs1042357 and rs10852889) were shown to reduce the cortical thickness of PTSD patients' brain.[20] Except inducing oxidative stress, the overstimulation of glucocorticoid receptor caused by the repeated activation of the HPA axis can, directly and indirectly, impose detrimental effects on mitochondrial membrane resulting in the release of cytochrome *c* from mitochondria to cytosol, elevated caspases function, and further apoptosis.[92] For instance, an increase in Bax/Bcl-2 ratio was reported from previous PTSD models.[8, 92] Of the other responsible factors, DNA methylation,[20] alternation in the autophagy-related molecules (i.e., p62, Beclin-1 and microtubule-associated protein 1 light chain 3),[92] and telomere shortening[90] due to oxidative stress, play a crucial role in the pathogenesis of PTSD.

Since the diagnosis of PTSD is mainly based on the clinical symptoms,[89] and there are only few USFDA approved medications for PTSD (i.e., sertraline and paroxetine),[8] considering genetic variations and mitochondrial targeting might give rise to the better management of the disease.

4 Conclusion

Considering the extensive literature on the association of mitochondrial abnormalities with neuronal disorders such as schizophrenia, depressive disorders, and bipolar disorder, investigations should be focused on developing preventive approaches, diagnostic biomarkers, and precision medicines targeting different aspects of mitochondria, particularly respiratory chain and mitochondrial-related genes. However, more studies are required for other mental disorders, especially ones that have been thoroughly overlooked so far. Obtaining accurate data from living human brain regions usually requires invasive strategies. Most former studies are from postmortem assessment, where postmortem factors may also influence the brain. Therefore, obtaining practical methods to rule out these factors and finding alternative noninvasive techniques to study the living human brain is essential.

Acknowledgment

Authors acknowledge INSF general support.

References

1. Whiteford HA, Degenhardt L, Rehm J, et al. Global burden of disease attributable to mental and substance use disorders: findings from the Global Burden of Disease Study 2010. *Lancet (London, England)*. 2013;382(9904):1575–1586.
2. Trautmann S, Rehm J, Wittchen H-U. The economic costs of mental disorders: do our societies react appropriately to the burden of mental disorders? *EMBO Rep.* 2016;17(9):1245–1249.
3. Regier DA, Kuhl EA, Kupfer DJ. The DSM-5: classification and criteria changes. *World Psychiatry.* 2013;12(2):92–98.
4. World Health Organization. *Mental health atlas: 2011.* Geneva: World Health Organization; 2011.
5. Measuring universal health coverage based on an index of effective coverage of health services in 204 countries and territories, 1990–2019: a systematic analysis for the Global Burden of Disease Study 2019. *Lancet (London, England).* 2020;396(10258):1250–1284.
6. Atkinson C, Lozano R, Naghavi M, Vos T, Whiteford H, Murray CJL. The burden of mental disorders in the USA: new tools for comparative analysis of health outcomes between countries. *Lancet.* 2013;381:S10.
7. Executive B. *Global burden of mental disorders and the need for a comprehensive, coordinated response from health and social workers at the country level: report by the secretariat.* Geneva: World Health Organization; 2012.
8. Preston G, Kirdar F, Kozicz T. The role of suboptimal mitochondrial function in vulnerability to post-traumatic stress disorder. *J Inherit Metab Dis.* 2018;41(4):585–596.
9. Citrigno L, Muglia M, Qualtieri A, et al. The mitochondrial dysfunction hypothesis in autism spectrum disorders: current status and future perspectives. *Int J Mol Sci.* 2020;21(16).

10. Bansal Y, Kuhad A. Mitochondrial dysfunction in depression. *Curr Neuropharmacol.* 2016;14(6):610–618.
11. Scaini G, Rezin GT, Carvalho AF, Streck EL, Berk M, Quevedo J. Mitochondrial dysfunction in bipolar disorder: evidence, pathophysiology and translational implications. *Neurosci Biobehav Rev.* 2016;68:694–713.
12. Flippo KH, Strack S. An emerging role for mitochondrial dynamics in schizophrenia. *Schizophr Res.* 2017;187:26–32.
13. de Sousa RT, Machado-Vieira R, Zarate Jr CA, Manji HK. Targeting mitochondrially mediated plasticity to develop improved therapeutics for bipolar disorder. *Expert Opin Ther Targets.* 2014;18(10):1131–1147.
14. Norkett R, Modi S, Kittler JT. Mitochondrial roles of the psychiatric disease risk factor DISC1. *Schizophr Res.* 2017;187:47–54.
15. Waligóra A, Waligóra S, Kozarska M, Damasiewicz-Bodzek A, Gorczyca P, Tyrpień-Golder K. Autism spectrum disorder (ASD)—biomarkers of oxidative stress and methylation and transsulfuration cycle. *Psychiatr Pol.* 2019;53(4):771–788.
16. Picard M, McEwen BS. Psychological stress and mitochondria: a systematic review. *Psychosom Med.* 2018;80(2):141–153.
17. Iwata K. Mitochondrial involvement in mental disorders; energy metabolism, genetic, and environmental factors. *Methods Mol Biol.* 2019;1916:41–48.
18. Maia A, Oliveira J, Lajnef M, et al. Oxidative and nitrosative stress markers in obsessive-compulsive disorder: a systematic review and meta-analysis. *Acta Psychiatr Scand.* 2019;139(5):420–433.
19. Alvarez-Arellano L, González-García N, Salazar-García M, Corona J. Antioxidants as a potential target against inflammation and oxidative stress in attention-deficit/hyperactivity disorder. *Antioxidants (Basel).* 2020;9:176.
20. Miller MW, Lin AP, Wolf EJ, Miller DR. Oxidative stress, inflammation, and neuroprogression in chronic PTSD. *Harv Rev Psychiatry.* 2018;26(2):57–69.
21. Morris G, Stubbs B, Köhler CA, et al. The putative role of oxidative stress and inflammation in the pathophysiology of sleep dysfunction across neuropsychiatric disorders: focus on chronic fatigue syndrome, bipolar disorder and multiple sclerosis. *Sleep Med Rev.* 2018;41:255–265.
22. Adzic M, Brkic Z, Bulajic S, Mitic M, Radojcic MB. Antidepressant action on mitochondrial dysfunction in psychiatric disorders. *Drug Dev Res.* 2016;77(7):400–406.
23. Cikánková T, Fišar Z, Bakhouche Y, Ľupták M, Hroudová J. In vitro effects of antipsychotics on mitochondrial respiration. *Naunyn Schmiedebergs Arch Pharmacol.* 2019;392(10):1209–1223.
24. Strobbe D, Campanella M. Anxiolytic therapy: a paradigm of successful mitochondrial pharmacology. *Trends Pharmacol Sci.* 2018;39(5):437–439.
25. Nussbaumer M, Asara JM, Teplytska L, et al. Selective mitochondrial targeting exerts anxiolytic effects in vivo. *Neuropsychopharmacology.* 2016;41(7):1751–1758.
26. Cikankova T, Sigitova E, Zverova M, Fisar Z, Raboch J, Hroudova J. Mitochondrial dysfunctions in bipolar disorder: effect of the disease and pharmacotherapy. *CNS Neurol Disord Drug Targets.* 2017;16(2):176–186.
27. Mousavi T, Hadizadeh N, Nikfar S, Abdollahi M. Drug discovery strategies for modulating oxidative stress in gastrointestinal disorders. *Expert Opin Drug Discov.* 2020;15(11):1309–1341.
28. Frye RE. Mitochondrial dysfunction in autism spectrum disorder: unique abnormalities and targeted treatments. *Semin Pediatr Neurol.* 2020;35:100829.
29. Pangrazzi L, Balasco L, Bozzi Y. Oxidative stress and immune system dysfunction in autism spectrum disorders. *Int J Mol Sci.* 2020;21(9).
30. Castora FJ. Mitochondrial function and abnormalities implicated in the pathogenesis of ASD. *Prog Neuropsychopharmacol Biol Psychiatry.* 2019;92:83–108.

31. Lord C, Elsabbagh M, Baird G, Veenstra-Vanderweele J. Autism spectrum disorder. *Lancet (London, England).* 2018;392(10146):508–520.
32. Wang C, Geng H, Liu W, Zhang G. Prenatal, perinatal, and postnatal factors associated with autism: a meta-analysis. *Medicine.* 2017;96(18), e6696.
33. Hutton J. Does rubella cause autism: a 2015 reappraisal? *Front Hum Neurosci.* 2016;10:25.
34. Modabbernia A, Velthorst E, Reichenberg A. Environmental risk factors for autism: an evidence-based review of systematic reviews and meta-analyses. *Mol Autism.* 2017;8:13.
35. Sa-Carneiro F, Calhau C, Coelho R, Figueiredo-Braga M. Putative shared mechanisms in autism spectrum disorders and attention deficit hyperactivity disorder, a systematic review of the role of oxidative stress. *Acta Neurobiol Exp (Wars).* 2020;80(2):129–138.
36. Scassellati C, Bonvicini C, Benussi L, Ghidoni R, Squitti R. Neurodevelopmental disorders: metallomics studies for the identification of potential biomarkers associated to diagnosis and treatment. *J Trace Elem Med Biol.* 2020;60:126499.
37. Frustaci A, Neri M, Cesario A, et al. Oxidative stress-related biomarkers in autism: systematic review and meta-analyses. *Free Radic Biol Med.* 2012;52(10):2128–2141.
38. Hollis F, Kanellopoulos AK, Bagni C. Mitochondrial dysfunction in autism spectrum disorder: clinical features and perspectives. *Curr Opin Neurobiol.* 2017;45:178–187.
39. Legido A, Jethva R, Goldenthal MJ. Mitochondrial dysfunction in autism. *Semin Pediatr Neurol.* 2013;20(3):163–175.
40. Cruz ACP, Ferrasa A, Muotri AR, Herai RH. Frequency and association of mitochondrial genetic variants with neurological disorders. *Mitochondrion.* 2019;46:345–360.
41. Nguyen RL, Medvedeva YV, Ayyagari TE, Schmunk G, Gargus JJ. Intracellular calcium dysregulation in autism spectrum disorder: an analysis of converging organelle signaling pathways. *Biochim Biophys Acta Mol Cell Res.* 2018;1865(11 pt B):1718–1732.
42. LeClerc S, Easley D. Pharmacological therapies for autism spectrum disorder: a review. *P T.* 2015;40(6):389–397.
43. Lopresti AL. Oxidative and nitrosative stress in ADHD: possible causes and the potential of antioxidant-targeted therapies. *Atten Defic Hyperact Disord.* 2015;7(4):237–247.
44. Corona JC. Natural compounds for the management of Parkinson's disease and attention-deficit/hyperactivity disorder. *Biomed Res Int.* 2018;2018:4067597.
45. Verlaet AAJ, Maasakkers CM, Hermans N, Savelkoul HFJ. Rationale for dietary antioxidant treatment of ADHD. *Nutrients.* 2018;10(4):405.
46. Joseph N, Zhang-James Y, Perl A, Faraone SV. Oxidative stress and ADHD: a meta-analysis. *J Atten Disord.* 2015;19(11):915–924.
47. Pruett BS, Meador-Woodruff JH. Evidence for altered energy metabolism, increased lactate, and decreased pH in schizophrenia brain: a focused review and meta-analysis of human postmortem and magnetic resonance spectroscopy studies. *Schizophr Res.* 2020.
48. Ben-Shachar D. Mitochondrial multifaceted dysfunction in schizophrenia; complex I as a possible pathological target. *Schizophr Res.* 2017;187:3–10.
49. Ben-Shachar D. The bimodal mechanism of interaction between dopamine and mitochondria as reflected in Parkinson's disease and in schizophrenia. *J Neural Transm.* 2020;127(2):159–168.
50. Bergman O, Ben-Shachar D. Mitochondrial oxidative phosphorylation system (OXPHOS) deficits in schizophrenia: possible interactions with cellular processes. *Can J Psychiatry.* 2016;61(8):457–469.
51. Roberts RC. Postmortem studies on mitochondria in schizophrenia. *Schizophr Res.* 2017;187:17–25.
52. Rajasekaran A, Venkatasubramanian G, Berk M, Debnath M. Mitochondrial dysfunction in schizophrenia: pathways, mechanisms and implications. *Neurosci Biobehav Rev.* 2015;48:10–21.
53. Koga M, Serritella AV, Sawa A, Sedlak TW. Implications for reactive oxygen species in schizophrenia pathogenesis. *Schizophr Res.* 2016;176(1):52–71.

54. Fraguas D, Díaz-Caneja C, Ayora M, et al. Oxidative stress and inflammation in first-episode psychosis: a systematic review and meta-analysis. *Schizophr Bull.* 2018;45:742–751.
55. Fraguas D, Díaz-Caneja CM, Rodríguez-Quiroga A, Arango C. Oxidative stress and inflammation in early onset first episode psychosis: a systematic review and meta-analysis. *Int J Neuropsychopharmacol.* 2017;20(6):435–444.
56. Sussulini A, Hauser-Davis R. Metallomics applied to the study of neurodegenerative and mental diseases. *Adv Exp Med Biol.* 2018;1055:21–37.
57. Martins-de-Souza D, Harris LW, Guest PC, Bahn S. The role of energy metabolism dysfunction and oxidative stress in schizophrenia revealed by proteomics. *Antioxid Redox Signal.* 2011;15(7):2067–2079.
58. Gonçalves V, Cappi C, Hagen CM, et al. M93—a comprehensive analysis of nuclear-encoded mitochondrial genes in schizophrenia. *Eur Neuropsychopharmacol.* 2019;29:S1005–S1006.
59. Machado AK, Pan AY, da Silva TM, Duong A, Andreazza AC. Upstream pathways controlling mitochondrial function in major psychosis: a focus on bipolar disorder. *Can J Psychiatry.* 2016;61(8):446–456.
60. Hjelm BE, Rollins B, Mamdani F, et al. Evidence of mitochondrial dysfunction within the complex genetic etiology of schizophrenia. *Mol Neuropsychiatry.* 2015;1(4):201–219.
61. Gonçalves VF, Andreazza AC, Kennedy JL. Mitochondrial dysfunction in schizophrenia: an evolutionary perspective. *Hum Genet.* 2015;134(1):13–21.
62. Schulmann A, Ryu E, Goncalves V, et al. Novel complex interactions between mitochondrial and nuclear DNA in schizophrenia and bipolar disorder. *Mol Neuropsychiatry.* 2019;5(1):13–27.
63. Gonçalves VF, Cappi C, Hagen CM, et al. A comprehensive analysis of nuclear-encoded mitochondrial genes in schizophrenia. *Biol Psychiatry.* 2018;83(9):780–789.
64. Sharma S, Akundi RS. Mitochondria: a connecting link in the major depressive disorder jigsaw. *Curr Neuropharmacol.* 2019;17(6):550–562.
65. Bhatt S, Nagappa AN, Patil CR. Role of oxidative stress in depression. *Drug Discov Today.* 2020;25(7):1270–1276.
66. Wigner P, Czarny P, Galecki P, Su KP, Sliwinski T. The molecular aspects of oxidative & nitrosative stress and the tryptophan catabolites pathway (TRYCATs) as potential causes of depression. *Psychiatry Res.* 2018;262:566–574.
67. Pitsillou E, Bresnehan SM, Kagarakis EA, et al. The cellular and molecular basis of major depressive disorder: towards a unified model for understanding clinical depression. *Mol Biol Rep.* 2020;47(1):753–770.
68. Culmsee C, Michels S, Scheu S, Arolt V, Dannlowski U, Alferink J. Mitochondria, microglia, and the immune system—how are they linked in affective disorders? *Front Psychiatry.* 2018;9:739.
69. Lopresti AL, Maker GL, Hood SD, Drummond PD. A review of peripheral biomarkers in major depression: the potential of inflammatory and oxidative stress biomarkers. *Prog Neuropsychopharmacol Biol Psychiatry.* 2014;48:102–111.
70. Palta P, Samuel LJ, Miller 3rd ER, Szanton SL. Depression and oxidative stress: results from a meta-analysis of observational studies. *Psychosom Med.* 2014;76(1):12–19.
71. Rossetti AC, Paladini MS, Riva MA, Molteni R. Oxidation-reduction mechanisms in psychiatric disorders: a novel target for pharmacological intervention. *Pharmacol Ther.* 2020;210:107520.
72. Klinedinst NJ, Regenold WT. A mitochondrial bioenergetic basis of depression. *J Bioenerg Biomembr.* 2015;47(1–2):155–171.
73. Black CN, Bot M, Scheffer PG, Cuijpers P, Penninx BW. Is depression associated with increased oxidative stress? A systematic review and meta-analysis. *Psychoneuroendocrinology.* 2015;51:164–175.

74. Maurya PK, Noto C, Rizzo LB, et al. The role of oxidative and nitrosative stress in accelerated aging and major depressive disorder. *Prog Neuropsychopharmacol Biol Psychiatry.* 2016;65:134–144.
75. Nedic Erjavec G, Konjevod M, Nikolac Perkovic M, et al. Short overview on metabolomic approach and redox changes in psychiatric disorders. *Redox Biol.* 2018;14:178–186.
76. Fries GR, Zamzow MJ, Andrews T, Pink O, Scaini G, Quevedo J. Accelerated aging in bipolar disorder: a comprehensive review of molecular findings and their clinical implications. *Neurosci Biobehav Rev.* 2020;112:107–116.
77. Kim Y, Santos R, Gage FH, Marchetto MC. Molecular mechanisms of bipolar disorder: progress made and future challenges. *Front Cell Neurosci.* 2017;11:30.
78. Szczepankiewicz A. Evidence for single nucleotide polymorphisms and their association with bipolar disorder. *Neuropsychiatr Dis Treat.* 2013;9:1573–1582.
79. Yamaki N, Otsuka I, Numata S, et al. Mitochondrial DNA copy number of peripheral blood in bipolar disorder: the present study and a meta-analysis. *Psychiatry Res.* 2018;269:115–117.
80. Caldiroli A, Auxilia AM, Capuzzi E, Clerici M, Buoli M. Malondialdehyde and bipolar disorder: a short comprehensive review of available literature. *J Affect Disord.* 2020;274:31–37.
81. Brown NC, Andreazza AC, Young LT. An updated meta-analysis of oxidative stress markers in bipolar disorder. *Psychiatry Res.* 2014;218(1–2):61–68.
82. Benes FM, Matzilevich D, Burke RE, Walsh J. The expression of proapoptosis genes is increased in bipolar disorder, but not in schizophrenia. *Mol Psychiatry.* 2006;11(3):241–251.
83. Mascitti M, Coccia E, Vignini A, et al. Anorexia, oral health and antioxidant salivary system: a clinical study on adult female subjects. *Dent J.* 2019;7(2):60.
84. Solmi M, Veronese N, Manzato E, et al. Oxidative stress and antioxidant levels in patients with anorexia nervosa: a systematic review and exploratory meta-analysis. *Int J Eat Disord.* 2015;48(7):826–841.
85. Simeunovic Ostojic M, Maas J. Anorexia nervosa and uric acid beyond gout: an idea worth researching. *Int J Eat Disord.* 2018;51(2):97–101.
86. Omodei D, Pucino V, Labruna G, et al. Immune-metabolic profiling of anorexic patients reveals an anti-oxidant and anti-inflammatory phenotype. *Metab Clin Exp.* 2015;64(3):396–405.
87. Solmi M, Veronese N, Luchini C, et al. Oxidative stress and antioxidant levels in patients with anorexia nervosa after oral re-alimentation: a systematic review and exploratory meta-analysis. *Eur Eat Disord Rev.* 2016;24(2):101–105.
88. Smith L, Tracy DK, Giaroli G. What future role might N-acetyl-cysteine have in the treatment of obsessive compulsive and grooming disorders?: a systematic review. *J Clin Psychopharmacol.* 2016;36(1):57–62.
89. Kim TD, Lee S, Yoon S. Inflammation in post-traumatic stress disorder (PTSD): a review of potential correlates of PTSD with a neurological perspective. *Antioxidants (Basel).* 2020;9(2):107.
90. Miller MW, Sadeh N. Traumatic stress, oxidative stress and post-traumatic stress disorder: neurodegeneration and the accelerated-aging hypothesis. *Mol Psychiatry.* 2014;19(11):1156–1162.
91. Bailey CR, Cordell E, Sobin SM, Neumeister A. Recent progress in understanding the pathophysiology of post-traumatic stress disorder: implications for targeted pharmacological treatment. *CNS Drugs.* 2013;27(3):221–232.
92. Jia Y, Han Y, Wang X, Han F. Role of apoptosis in the post-traumatic stress disorder model-single prolonged stressed rats. *Psychoneuroendocrinology.* 2018;95:97–105.

CHAPTER 11

Mitotherapy, an attractive avenue for the future medicine

Jalal Pourahmad[a], Somayeh Handali[b], and Mohsen Rezaei[c]
[a]Department of Pharmacology & Toxicology, School of Pharmacy, Shahid Beheshti University of Medical Sciences, Tehran, Iran
[b]Medical Biomaterial Research Center (MBRC), Tehran University of Medical Sciences, Tehran, Iran
[c]Department of Toxicology, Faculty of Medical Sciences, Tarbiat Modares University, Tehran, Iran

Contents

1 Introduction

Mitochondria are considered the cell's energy generators for metabolic needs, proliferation, and differentiation, and also are involved in heat regulation, metabolites providing for biomacromolecules formation, calcium homeostasis, and control of apoptosis.[1–3] Mitochondrial dysfunction has been implicated in a range of pathological conditions, including diabetes, inflammation, cardiovascular diseases, neurodegenerative diseases, and cancer.[2, 4] Mitochondrial DNA is more vulnerable and sensitive to mutagens than nuclear DNA due to the absence of histones.[5] Some tissues and organs including cardiac and skeletal muscles, central nervous system (CNS), and also oocytes are rich in mitochondrial content while hepatocytes have nearly 800 mitochondria. Along with the mitochondrial number, cells may be different in their mitochondrial size, shape, and distribution. Mitochondria may be seen in various shapes that can be changed during the pathological conditions.[2] Normal and cancer cells are capable of releasing their mitochondria. Extracellular mitochondria are found in the form of free (free Mitos) inside the vesicles or platelets and as circulating cell-free mitochondrial DNA (cf-mtDNA). Circulating cell-free mitochondria with hereditary and nonhereditary

Mitochondrial Metabolism
https://doi.org/10.1016/B978-0-12-822416-8.00009-9

components are considered essential systemic messengers for cell-cell communication. Instead, these organelles may contain impaired or altered biomolecules with the potential for initiation of immune reactions and inflammation.[5] Their beneficial or deleterious roles are mainly dependent on the functional states of the extracellularly released mitochondria.[6] Extracellular vehicles (EVs) are also involved in the intercellular transferring of mitochondria. EVs protect mitochondrial components and other biological molecules of donor cells against the degradative enzymes during the intercellular movement. This is necessary for these trafficking components to be able to reprogram and adjust the recipient cell functions. Intercellular transferring of mitochondria is a promising strategy for restoration of bioenergetics and impaired metabolic pathways in the damaged recipient cells. Mitochondria from mesenchymal stem cells (MSCs) inhibited the release of proinflammatory cytokines, and increased the capacity of alveolar macrophages for bacterial clearance when applied in acute respiratory distress syndrome (ARDS). Exosomal release of mitochondria has also contributed to the throwing out of the partially damaged organelles prior to mitophagy. This process has been observed in Hela cells in response to mitochondrial reactive oxygen species (ROS).[7] Endocrine or paracrine responses can be altered by all the mentioned extracellular forms of mitochondria.[8] "Mitochondria know no boundaries"[9] and in recent years, their transplantation is considered an attractive therapeutic strategy with potential benefits in the treatment of metabolic, neurological, and cardiac diseases.[8] Related results showed that mitochondrial transplantation repaired damaged mtDNA and cell function. These advantages were reported for < 50 cell doublings in in vitro studies.[10] The concept of mitochondrial transplantation has come from the ability of mitochondria to transfer from one cell to another one. Their movement between cells is facilitated by various means including gap junctions, exosomes, cell fusion, and tunneling nanotubes (TNTs) formation. Another mechanism by which mitochondria internalize into the cells is actin-dependent macropinocytosis; however, the detailed molecular signaling for the pathway remained unclear. It has been suggested that mitochondrial internalization can be enhanced by some targeting strategies including peptide-mediated mitochondrial delivery (PMD) to improve the therapeutic outcome in patients with mitochondrial disease.[2]

Healthy mitochondria can inhibit tumor proliferation and abnormal mitochondria can change the respiration of cancer cells from the oxidative phosphorylation to aerobic glycolysis (Warburg phenomenon), which is implicated in the initiation of the carcinogenesis. These changes have been evident in melanoma cells and their rapid growth and increased hypoxia further fuel more proliferation and metastasis. It was found that in mitochondria-treated mice, the progression of tumor has been retarded and the metastatic melanomas per lung have been significantly reduced. Normal mitochondria considerably inhibited tumor growth and increased survival days of animals. Cancer cell mitochondria adjust their function and structure to the acidic and hypoxic microenvironment of tumor, but isolated mitochondria cannot tolerate these conditions inside the tumor and subsequently produce cytotoxic substances that may result in apoptosis. Glycolysis in cancer cells was increased with enhanced lactate production via the inhibition of pyruvate dehydrogenase (PDH), α-ketoglutarate dehydrogenase (αKDH) and activation of glucose-6-phosphate dehydrogenase (G6PD) and pyruvate kinase (PK), while the normal mitochondria treatment has considerable opposite effects. Normal mitochondria transplantation may also confer other synergistic effects for inhibiting tumor growth such as downregulating of glutaminolysis by decreasing the glutamate dehydrogenase (GDH) activity and lowering the contents of glutamine, glutamate, and ammonia. Organelle functionality is also important for obtaining proper consequences of mitotherapy. It was observed that when mitochondria have been isolated from the liver of young mice, the proliferation of tumor cells inhibited more efficiently than mitochondria from aged mouse. In this regard, mitophagy and mitochondrial dynamics including fission and fusion were also different in target tumor cells. Moreover, the mitochondrial content of tumor cells is less than the somatic cells and this reduced number of mitochondria can be considered an escape way from the apoptotic cell death commenced by mitochondria.[11]

2 Mitotherapy in diseases

Mitotherapy efficacy has been established in the liver, kidney, heart, spinal cord, and brain.[12–15] Mitotherapy has been evaluated in disease models in many in vitro and in vivo studies. In this regard, neurologic and cardiovascular disorders, cancer, and drug-induced toxicities have received more attention. In the following section, we will discuss some studies that have reported the effect of mitotherapy on neurological diseases, heart diseases, and cancer. Astrocytes, by providing neurotrophic factors, have long been known to have a neuroprotective role and maintain the function of neurons and protect them from endogenous toxic insults. However, they may contribute to irregular release of inflammatory mediators that leads to neuronal damage.[3] Mitochondrial transfer from astrocytes to damaged neurons plays a significant role in the neuro repairing function of astrocytes and apparently, this is not a one-way exchange. Transferring of mitochondria from neuron to astrocyte can also happen that allows neurons to stay away from damaged mitochondria. Furthermore, transferring of healthy mitochondria from specific progenitor cells to neurons and brain endothelial cells restored their mitochondrial function and improved their viability and blood-brain barrier activity.[3] Based on this intercellular mitochondrial trafficking, novel therapeutic approaches have been proposed to restore the function of damaged recipient cells.[6] In many studies, the use of normal functional mitochondria in a number of disease models has been evaluated and promising results have been obtained through these in vitro and in vivo studies. Beside this, modification of mitochondria for better therapeutic consequences has also been studied. Intact mitochondria and peptide-mediated allogeneic mitochondrial delivery have been studied in rat model of Parkinson's disease (PD) and PC12 cells exposed to 6-hydroxydopamine (6-OHDA). Following the nigrostriatal lesion induction by 6-OHDA and treatment with mitochondrial transplantation, lowered oxidative stress and apoptotic death in the rat PC12 cells were observed particularly in cells treated with modified mitochondria. Furthermore, restoration of mitochondrial dynamics and function was evident in PD rats.[13] Spinal cord ischemia (SCI) and mitochondrial dysfunction that may clinically be evident as postoperative paraplegia are considered the probable side effects of major vascular surgery. Transplantation of normal mitochondria to the ischemic area at the early stage in rats and in vivo imaging using a fluorescent MitoTracker revealed an improvement in the lower limb locomotion of rats with cord ischemia. Endoplasmic reticulum stress and proinflammatory cytokine levels including interleukin-6 and tumor necrosis factor-α, as well as caspase-3 activation, decreased in the mitochondrial transplanted animals.[16]

In ischemic heart tissue, mitochondrial transplantation decreased the cardiomyocytes death and improved energy availability and contractility. Administered mitochondria were able to enter the cardiomyocytes within minutes and were found in endosomes and lysosomes. Most of these administered mitochondria can escape from these compartments and once in the cytosol can fuse with native mitochondria to reinforce this endogenous network.[17] Modified exogenous mitochondria were administered to a myocardial ischemia-reperfusion (I/R) mouse model. Aldehyde dehydrogenase

2 (ALDH2) has a regulatory function in mitochondrial homeostasis and its activation has been demonstrated to improve the outcomes of mitochondrial transplantation in the model. In this way, oxygen consumption, ATP production, and viability of cardiomyocytes were noticeably increased.[18]

Mitotherapy has also been studied in rats with pulmonary hypertension, and ex vivo heart-lung block imaging showed a higher MitoTracker fluorescence in the pulmonary arteries. ATP levels and right ventricular performance were also improved. Moreover, the results indicated that the expression of actin and myosin in vascular smooth muscle cells of pulmonary arteries was upregulated and hyper responsiveness of pulmonary arteries to α-adrenergic stimulation was reduced. The improvement in the function of lung and heart in pulmonary hypertension indicated an appropriate efficacy of mitochondrial transplantation, suggesting a novel approach for treatment of pulmonary hypertension.[19]

Pediatric myocardial ischemia-reperfusion injury is accompanied by mitochondrial damage. Transplantation of viable autologous mitochondria isolated from nonischemic skeletal muscle into injured myocardium noticeably improved ventricular function within several days following the therapy.[20] Mitochondrial transplantation in other study indicated the protection against myocardial warm ischemia and improved myocardial function and oxygen consumption.[21] More investigation is needed to shed light on the safety, efficacy, and the appropriate dose of the treatment.

The transfer of mitochondria from cancer cells to other normal cells has been identified in the previous studies that can be used by tumor cells for cancer progression. This kind of mitochondrial trafficking supports the normal mitochondrial transplantation for cancer therapy. In vitro study of mitochondrial transplantation in breast cancer cells (MCF-7) resulted in considerable apoptosis, reduced cell growth, and oxidative stress. Furthermore, chemoresistance of MCF-7 cells to doxorubicin and paclitaxel has been reduced. Modification of mitochondria such as conjugation with cell-penetrating peptide Pep-1 improved the effectiveness of treatment and enhanced mitochondrial dynamic.[22]

3 Mitochondria internalization mechanisms and their functionality

Exogenous mitochondria can be injected directly or delivered via other routes including intravenous or intramuscular into the target organ. A number of mechanisms have been suggested for intercellular mitochondrial transfer, including TNTs, extracellular vesicles, gap junctions and cellular fusion.[3, 9] Various factors have been reported to trigger mitochondrial transfer, including glucose starvation, oxygen deprivation, UV irradiation, inflammatory signaling, and treatment with anticancer medications, including doxorubicin and topoisomerase II inhibitor etoposide.[23] On the other hand, glutathione (GSH) and *N*-acetyl cysteine (NAC) (as antioxidants) and decreased expression of NAPDH oxidase 2 inhibit mitochondrial transfer which indicates the possible role for ROS in both donor and recipient cells. A previous report indicated an increased ROS amount in leukemia that may be related to the enhanced mitochondrial uptake in acute myeloid leukemia (AML) compared with normal cells.[23]

TNTs (as membrane nanochannels) are the major delivery pathway for the transfer of mitochondria.[24] TNTs are F-actin-based structures that facilitate intercellular transfer of membrane components, lipid droplets, organelles (such as mitochondria and lysosomes), viruses, Ca^{2+} and signals molecules.[25, 26] The diameter of TNTs is in the ranges of 50–200 nm with the length of up to 150 μm.[24] These dynamic structures appear de novo within minutes and may remain functional for several hours.[3] Signaling pathways for TNT formation is overlapped with endosomal trafficking, in which exocyst complex components are shared. These components mediate the vesicular transport from Golgi apparatus to the plasma membrane. For TNT formation, M-Sec (as a part of the exocyst complex) interacts with Ras-related protein-A (RalA, small GTPase). For this reason, M-Sec is known as a TNT formation marker.[26]

In cancer cells, TNTs play a critical role in the intercellular communication which results in the progression of solid tumors including prostate cancer, glioblastoma, lung cancer, and colon cancer. It has been reported that transport of mitochondria via TNTs between endothelial and cancer cells such as breast and ovarian cancer cells leads to chemotherapy resistance in the recipient cancer cells.[25] Studies exhibited that decreased ATP production in the mitochondria, employing F-actin depolymerization drugs, inhibition of gluconeogenesis using metformin and blocking the mTOR pathway suppresses TNTs formation, which results in inhibition of organelles transfer in cancer cells.[25,27] However, in ex vivo studies, it was observed that hydrogen peroxide (H_2O_2) facilitates TNT formation and induces intercellular mitochondrial transfer.[23] Miro1, a mitochondrial Rho-GTPase that is a calcium-sensitive adaptor protein, assists mitochondrial movement through the microtubules across cytoplasmic bridges including TNTs. This protein, along with other accessory proteins including Miro2, TRAK1 (trafficking kinesin protein 1), TRAK2 (trafficking kinesin protein 2) and Myo19 (myosin XIX), contributes to the binding of mitochondria to the KIF5 (kinesin-related protein 5) motor proteins. Miro1 as an integral protein was found to be involved in mitochondrial transfer from MSCs to epithelial cells. However, it was observed that the absence of Miro1 did not interfere with the formation of TNT, but the rate of mitochondrial transfer through these tubes was significantly reduced.[4]

Hypoxic condition has been implicated in the enhancement of TNT formation in chemoresistant ovarian cancer cells (SKOV3, C200). Suppressing the mTOR pathway using either metformin (an antidiabetic drug) or everolimus (an mTOR inhibitor) inhibited the TNT formation in vitro, confirming that TNTs can potentially be targeted for cancer therapy.[28] Suppression of TNT formation in ovarian tumor-derived epithelial and ovarian adenocarcinoma cells by metformin and everolimus has also been demonstrated in other studies.[28]

Hypoxia may assist the formation of TNTs and also the propagation of hypoxia-inducible factor-1α (HIF-1α) and vascular endothelial growth factor (VEGF) between adjacent cells for stimulating of angiogenesis. Time-lapse imaging revealed the transferring of VEGF and even HIF-1α via TNTs. These findings can provide an explanation for reprogramming of other normal cells by cancerous cells that may lead to invasion and metastasis.[29]

Expression of P-glycoprotein (P-gp) is evident in many human cancers and it is related to the multidrug resistance (MDR) process. TNTs contributed to the transport of P-gp in MCF-7 so that they could be targeted in therapeutic strategies to cope MDR.[30]

Cytochalasin B as an inhibitor of actin polymerization prevented the TNT formation; however, it has had modest effect on the previously formed TNTs.[31] Moreover, tolytoxin, a cyanobacterial macrolide, at nM concentrations could inhibit polymerization of actin, which led to notable reduction in the number of connected cells via TNT and transfer of mitochondria in both neuronal cells (SH-SY5Y) and nonneuronal cells (SW13).[32]

Gap junctions (GJs) are intercellular channels that connect eukaryotic cells and provide a pass for direct transferring of metabolic intermediates and electrical signals between cells. Commonly, molecules up to 1000 Da can diffuse via gap junction channels (GJCs), including, metabolites, nutrients, second messengers (such as calcium, glucose, cAMP, cGMP, and IP3), amino acids, nucleotides, cations, anions, and mitochondria.[3, 33] Contribution of almost 20 connexin genes and 3 pannexin genes is necessary for gap junction formation in mice and humans.[3] Components of

GJs that are transmembrane hemichannels (or connexons) aggregate in the cell membrane and link to similar connexons on the adjacent cell with the intercellular distance that is around 2–3 nm.[33] Double-membrane vesicles are internalized via GJCs and connexins in clathrin-dependent endocytosis.[23] Miro1 is involved in the transfer of organelles and its overexpression leads to increased organelle transfer. Miro proteins contain two GTPase domains, two calcium-sensing domains, and a transmembrane component that stabilize them into the outer mitochondrial membrane. In the presence of calcium, binding of Miro1 to the KIF5 kinesin motor is inhibited. It was suggested that mitochondrial transfer is dependent on calcium exchange mediated by specific connexin containing GJCs.[23] Specific connexin containing GJCs has been implicated in an in vivo mouse model of acute lung injury that translocate mitochondria in microvesicles from bone marrow-derived stromal cells to reach the epithelia. Inhibiting of this specific connexin gap junction formation does not affect the cytoplasmic transferring; however, the mitochondrial relocation is reduced.[34]

Extracellular vesicles (EVs) are stable structures released by prokaryotic and eukaryotic cells that contain biomolecules (RNAs, lipids, and proteins) and based on their size, contents, and mechanism of biogenesis, they are classified as apoptotic bodies (ABs), microvesicles (MVs), and exosomes. EVs are observed in different body fluids and are involved in many physiological functions through the carrying of essential components between cells. EVs may also contain mitochondria or mtDNA for the maintenance of cellular viability.[8] MVs (50–1000 nm) are larger particles than exosomes (40–120 nm). However, they are different in other characteristics rather than size. Recently, annexin A1 was recognized as a specific marker that exists in both small (< 150 nm) MVs and large (< 1000 nm) EVs.[7] The presence of mitochondria has been evident in exosomes derived from T-cell regulators that are proinflammatory HLA-DR$^+$ subsets of airway myeloid-derived regulatory cells (MDRCs) in asthma. Florescence probe-tagged exosomes were observed in cocultured recipient peripheral T cells which confirmed a direct exosome-mediated cell-cell transfer. These translocated mitochondria are eventually colocalized with recipient mitochondrial network and are capable of producing ROS within those T cells. This pattern of intercellular communication displays a novel mechanism for mitochondria transfer and also for the redox or bioenergetic regulation of target cells.[35]

EVs secretion is necessary for therapeutic effect of MSCs. This kind of intercellular communication contributes to reprogramming and changes in other cells in terms of their fate and metabolism. Therefore, EVs may be involved in the development of chronic diseases including tumor progression and neurological diseases even though the EVs secretion can principally be originated by cells that are not directly associated with nervous system such as erythrocytes.[8] As a result, inhibition of secretion and uptake of exosomes are considered a promising strategy for metastasis prevention. Decrease in the tumor-derived exosomal uptake by heparin significantly reduced the metastasis in oral squamous carcinoma.[36] Moreover, an antibody was designed to reduce the generation of tumor-derived exosomes, which resulted in decreasing of metastasis in breast cancer in a mouse model.[37] These results show a potential therapeutic effect for designed antibodies against metastasis-related exosomes. Cancer cell exosomes production has been shown to be influenced by extracellular acidity. In this regard, culturing of melanoma cell lines at low pH (~6.7) led to an increased exosomes generation compared with physiological conditions (pH 7.4) of culturing for the same cells.[38, 39] These results were confirmed in other tumor cell lines including colon cancer, breast cancer, prostate cancer, and osteosarcoma. Acidic environment's contribution to higher exosome secretion has not been mechanistically understood, but it seems that a higher secretion of exosomes at this condition reduces the accumulation of intracellular metabolites that could be toxic at high concentrations.[40] Subsequently, specific proton-pump inhibitors have been shown to decrease the exosomes in the plasma of xenograft models. In parallel, alkalinization of tumor environment may help the cancer patients with tumor metastasis.[41]

A suppressive effect of exosomes on immune system has also been suggested. This effect is in favor of tumor metastasis since secreted exosomes were found to interfere with NK cells' cytotoxicity and dendritic cells' (DC) differentiation. They also contribute to the apoptotic cell death induction in cytotoxic T cells and macrophage M2 polarization. Early evaluation of exosome therapy in metastatic melanoma patients revealed an acceptable adverse effect profile and partial effectiveness that promote their use for therapeutic purposes. Exosomes from autologous DC and MHC class I peptides showed high efficacy and improved survival in advanced nonsmall-cell lung cancer.[42]

Exosomes have been used as nanocarriers for drug and gene delivery. For this purpose, they exhibited biocompatibility, acceptable safety profile, low immunogenicity, ease of production and storage, efficient loading capacity, and longevity.[43] Most of the cells produce exosomes, which are stable in biological fluids and are capable of passing the blood-brain barrier due to their small sizes. Exosomal uptake of cancer cells is nearly 10 times higher than similar-sized liposomes that confirm their higher specificity for tumor cell targeting.[43] Some anticancer drugs including siRNAs and chemotherapeutic compounds successfully delivered by bioengineered exosomes to cancer cells and a new generation of exosome-based cancer vaccines showed promises in early phases of clinical trials.[42] Enhanced permeability and retention (EPR) effect due to abnormal blood vessels formation facilitates the accumulation of exosomes in tumors as compared to normal tissues that make them efficient carrier for drug delivery. Furthermore, modification of exosomes with targeting peptides or antibodies can even make them one of the best carriers for therapeutic drug delivery in cancer therapy.[43] Doxorubicin loaded in exosomes (ExoDOX) significantly increased drug accumulation in cancer cells and prevented its off-target cardiotoxicity compared with free DOX. P-glycoproteins (P-gp) as cell membrane transporters are active pumps for expelling out of chemotherapeutic agents. ExoDOX treatment on P-gp-overexpressing cells indicated noticeable accumulation of DOX compared to free drug or a combination of P-gp-inhibitor (verapamil) and free drug. The IC_{50} values that were obtained for ExoDOX were 50-folds lower than that of free DOX in the above-mentioned multidrug-resistant cell lines. Moreover, paclitaxel-containing exosomes (ExoPTX) derived from macrophage in comparison with Taxol significantly reduced tumor growth in a murine model of metastatic lung carcinoma.[42]

Numerous evidences indicated that isolated mitochondria entered the cells through the endocytosis; however, the mechanism has yet to be known. Coincubation of human astrocytes with starved human glioma cells (U87) resulted in transferring of mitochondria into the later. Rate of glycolysis decreased in U87 cells under starvation that eventually reduced the conversion of NAD^+ to NADH, so that remaining NAD^+ accumulated in the extracellular space. Extracellular NAD^+ then catalyzes and undergoes cyclization by CD38 (a member of the NAD^+ glycohydrolase family) to intracellular cyclic ADP-ribose (cADPR) that triggers the release of Ca^{2+}. Cytoskeleton remodeling and invagination of plasma membrane (internalization) can be initiated as the consequence of intracellular Ca^{2+} release. Therefore, internalization of isolated mitochondria via endocytosis can be mediated by NAD^+-CD38-cADPR-Ca^{2+} signaling. A number of cell functions including increased gene and protein expression associated with the tricarboxylic acid (TCA) cycle, enhanced aerobic respiration, decreased glycolysis, triggering of the mitochondrial apoptotic pathway, and inhibition of the U87 cells growth can be achieved by mitochondrial transfer. Injection of isolated mitochondria into U87 xenograft tumors resulted in the internalization of mitochondria and inhibition of the growth of glioma in nude mice. In vitro and in vivo evaluations of mitochondrial transplantation revealed a higher radiosensitivity of glioma cells and starvation-induced NAD^+-CD38-cADPR-Ca^{2+} signaling can play a major role in mitochondria endocytosis to restore aerobic respiration and weakening of the Warburg effect.[44]

Internalization of EGFP-labeled mitochondria was evident in human osteosarcoma (HOS) cells when they were exposed to inhibitors of different endocytic pathways, which indicated that macropinocytosis or an actin-dependent endocytosis is involved in this processes. Time-lapse microscopy revealed a large number of motile cellular vesicles contained mitochondria. Furthermore, confocal imaging showed EGFP-labeled mitochondria colocalization with other macropinocytic molecules during internalization into HOS cells.[45]

Cell fusion is another mechanism by which mitochondria are being transferred between two independent cells. In this process, membranes of the two cells are fused together and organelles and cytosolic compounds are shared, but their nuclei remain intact.[9] The underlying mechanism of this phenomenon has not been fully understood and its physiological significance in terms of permanent vs partial fusion and also hybrid cells characterization remained unclear. It was found that for successful reprogramming of cardiomyocytes, besides the transferring of stem cell mitochondria into them, intercellular biomolecules containing F-actin and microtubules are also required.[46]

Routes of mitochondria internalization into the recipient cells are shown in Fig. 1.

Following the internalization of exogenous mitochondria, their fusion with endogenous mitochondria occurs by using several proteins. In mitochondria transplantation of human-induced pluripotent stem cell-derived cardiomyocytes (iPS-CMs) to human cardiac fibroblasts (HCFs), mitofusin contributed to tethering and fusing of outer mitochondrial membranes, while optic atrophy 1 (OPA1) proteins display a significant role in inner mitochondrial membrane fusion. At least five OPA1 isoforms are found in cardiomyocyte mitochondria; among them, longer OPA1s are membrane-associated proteins with major contribution to inner membrane fusion and shorter ones are soluble proteins involved in mitochondrial fusion and some other functions.[17]

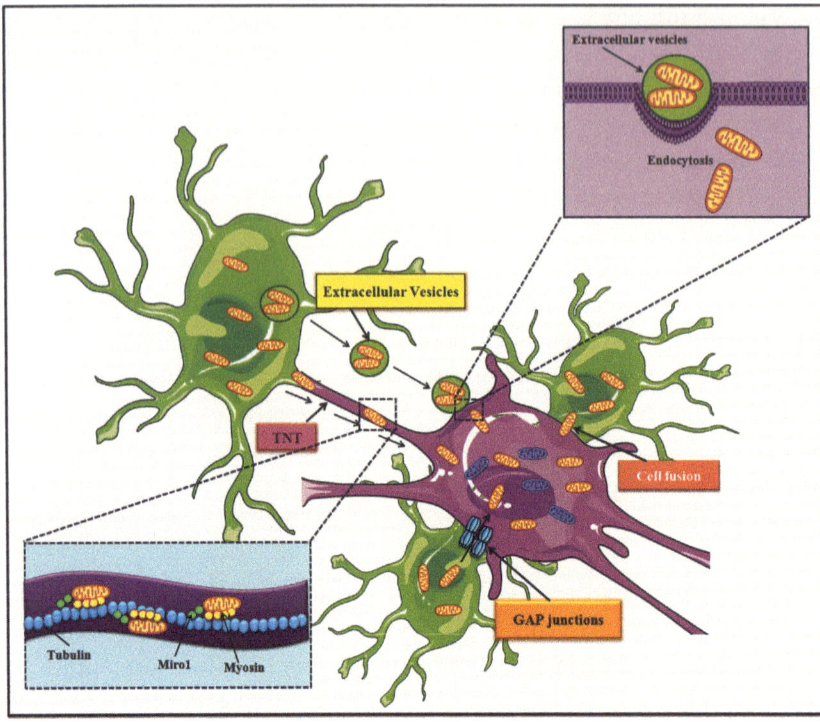

Fig. 1 Routes of mitochondria internalization into the recipient cell.

4 Immune reactions in mitotherapy

cf-mtDNA presence in the blood contributes to the autoimmune disease, sepsis, malignancy, tissue injury, and organ dysfunction. CpG motifs of mitochondrial DNA are capable of stimulating proinflammatory pathway via the endosomal, nucleotide-sensing Toll-like receptor, TLR9. It was found that red blood cells (RBCs) bind to the CpG motifs of cf-mtDNA through TLR9 to scavenge these particles and prevent lung damage. This protective mechanism is highlighted when TLR9-dependent RBC-mediated CpGDNA scavenging is inhibited that leads to lung damage in vivo. This finding indicates the critical role for RBCs in regulating systemic inflammation by scavenging of cf-mtDNA.[47]

Mitochondrial transplantation appears to be safe in autologous conditions, while for therapeutic purposes, allogeneic transplantation would be more practical.[48] Activation of immune system by mitochondria or their contents is expected to happen following the allogeneic mitochondrial transplantation. However, this reaction was not observed in a recent study in which isolated mitochondria from normal cells were used in cardiovascular diseases. Platelets are being activated at the beginning stage of injury and are involved in immune response. At this condition, their mitochondria are also activated and release ROS and apoptotic mediators. These activated mitochondria are highly immunogenic once outside the platelets. However, nonactivated mitochondria from normal platelets exhibited regulatory effect when in contact with immune cells. Apparently, mitochondrial condition and type are contributed to their immunogenicity and regulatory properties.[8] Mitochondrial transplantation in rabbits has not been associated with the inflammatory biomarker appearance and proinflammatory cytokines elevation in blood. Furthermore, intraperitoneal administration of mitochondria to mice did not precipitate an immune reaction, although a local response at the site of administration has been reported which may contribute to the local effects of immune cells and their released mediators.[49]

Transplantation of autologous mitochondria in a porcine model of I/R did not trigger inflammation and serum cytokine levels' elevation. Intraperitoneal injection of single or multiple doses of allogenic mitochondria in mice was used for immune response evaluation. In this study, inflammatory mediators including immunoglobulin M (IgM), interferon-γ (IFN-γ), and interleukin-2 (IL-2) were not significantly increased following the transplantation.[48]

Ramirez-Barbieri et al. showed that single or serial intraperitoneal administration of syngeneic or allogeneic mitochondria in mice did not produce a detectable immune or autoimmune response. They did not report any direct or indirect acute or chronic alloreactivity, allo-recognition or damage-associated molecular patterns (DAMPs) reactions to this treatment. Also, histopathological evaluation demonstrated no evidence of tissue damage, necrosis, and inflammation in heart and lung.[10]

Some opposite results were obtained by other studies, which indicated an immune reaction following mitochondrial transplantation. Exogenous mitochondria-induced DAMPs and increased levels of proinflammatory cytokines were confirmed in the recipient organs.[50] Previous studies reported that circulating free mtDNA induced DAMPs and triggered an immune response and cell damage, which led to significant injuries to heart and lung.[51–53] Transplanted mitochondria were linked to the activation of vascular endothelial cells and releasing of proinflammatory cytokines.[54]

5 Mitochondrial modification for transplantation

As mentioned previously, viable mitochondria can be isolated from patient's own muscle (e.g., rectus abdominis) and transplanted into the damaged organ during mitochondrial transplantation process. Mitochondria following the isolation are usually stored on ice for a while and they gradually lose their activity over time. Therefore, a rapid transplantation procedure is required for obtaining better outcomes. Rapid isolation and usage of mitochondria during transplantation are known as one of the major limitations for this procedure and some cold storage and cryopreservation protocols have been introduced to maintain the function of isolated mitochondria for longer periods.

Besides the searching for optimum condition for preservation of isolated mitochondria, a number of studies have been carried out to efficiently deliver exogenous mitochondria into the target organ. Transferred mitochondria can be coated with specific peptides to target the diseased site. For instance, mitochondria were mixed with AsOR-PL (asialoorosomucoid-polylysine, an asialoglycoprotein-based carrier) for targeting hepatocyte and LLO (listeriolysin O, an endosomolytic peptide) to form AsORLLO for facilitating the release of internalized mitochondria from endosomes. These positively charged peptides are mixed with negatively charged mitochondria to form the desired complexes. Results showed that these targeted mitochondria were efficiently delivered to rat liver and the modification led to better outcomes in terms of endosomal escape of internalized mitochondria.[55]

MDR is considered one of the main obstacles for the efficacy of chemotherapy. To cope with MDR, a siRNA-containing nanomaterial-coated mitochondria complex successfully delivered siRNA and mitochondria into the MDR cells. In this approach, the functional siRNA decreased the expression of pump resistance-related proteins, while exogenous mitochondria triggered apoptotic signaling and improved intracellular metabolism network.[56] Some biocompatible polymer including dextran can enhance the delivery rates of transplanted mitochondria to cells. Binding of a conjugate composed of dextran and triphenylphosphonium to exogenous mitochondria preserved the organelle functionality and enhanced cellular internalization compared with intact mitochondria.[57]

References

1. Angajala A, Lim S, Phillips JB, et al. Diverse roles of mitochondria in immune responses: novel insights into immuno-metabolism. *Front Immunol.* 2018;9:1605.

2. Roushandeh AM, Kuwahara Y, Roudkenar MH. Mitochondrial transplantation as a potential and novel master key for treatment of various incurable diseases. *Cytotechnology.* 2019;71:647–663.
3. Shanmughapriya S, Langford D, Natarajaseenivasan K. Inter and intracellular mitochondrial trafficking in health and disease. *Ageing Res Rev.* 2020;62:101128.
4. Ahmad T, Mukherjee S, Pattnaik B, et al. Miro1 regulates intercellular mitochondrial transport & enhances mesenchymal stem cell rescue efficacy. *EMBO J.* 2014;33:994–1010.
5. Al Amir Dache Z, Otandault A, Tanos R, et al. Blood contains circulating cell-free respiratory competent mitochondria. *FASEB J.* 2020;34:3616–3630.
6. Nakamura Y, Park J-H, Hayakawa K. Therapeutic use of extracellular mitochondria in CNS injury and disease. *Exp Neurol.* 2020;324:113114.
7. Liu D, Dong Z, Wang J, Tao Y, Sun X, Yao X. The existence and function of mitochondrial component in extracellular vesicles. *Mitochondrion.* 2020;54:122–127.
8. Miliotis S, Nicolalde B, Ortega M, Yepez J, Caicedo A. Forms of extracellular mitochondria and their impact in health. *Mitochondrion.* 2019;48:16–30.
9. Torralba D, Baixauli F, Sánchez-Madrid F. Mitochondria know no boundaries: mechanisms and functions of intercellular mitochondrial transfer. *Front Cell Dev Biol.* 2016;4:107.
10. Ramirez-Barbieri G, Moskowitzova K, Shin B, et al. Alloreactivity and allorecognition of syngeneic and allogeneic mitochondria. *Mitochondrion.* 2019;46:103–115.
11. Fu A, Hou Y, Yu Z, Zhao Z, Liu Z. Healthy mitochondria inhibit the metastatic melanoma in lungs. *Int J Biol Sci.* 2019;15:2707.
12. Masuzawa A, Black KM, Pacak CA, et al. Transplantation of autologously derived mitochondria protects the heart from ischemia-reperfusion injury. *Am J Physiol Heart Circ Physiol.* 2013;304:H966–H982.
13. Chang J-C, Wu S-L, Liu K-H, et al. Allogeneic/xenogeneic transplantation of peptide-labeled mitochondria in Parkinson's disease: restoration of mitochondria functions and attenuation of 6-hydroxydopamine–induced neurotoxicity. *Transl Res.* 2016;170. 40–56.e3.
14. McCully JD, Cowan DB, Pacak CA, Toumpoulis IK, Dayalan H, Levitsky S. Injection of isolated mitochondria during early reperfusion for cardioprotection. *Am J Physiol Heart Circ Physiol.* 2009;296:H94–H105.
15. Gollihue JL, Patel SP, Mashburn C, Eldahan KC, Sullivan PG, Rabchevsky AG. Optimization of mitochondrial isolation techniques for intraspinal transplantation procedures. *J Neurosci Methods.* 2017;287:1–12.
16. Fang S-Y, Roan J-N, Lee J-S, et al. Transplantation of viable mitochondria attenuates neurologic injury after spinal cord ischemia. *J Thorac Cardiovasc Surg.* 2019;161:e337–e347.
17. Cowan DB, Yao R, Thedsanamoorthy JK, Zurakowski D, Pedro J, McCully JD. Transit and integration of extracellular mitochondria in human heart cells. *Sci Rep.* 2017;7:1–11.
18. Sun X, Gao R, Li W, et al. Alda-1 treatment promotes the therapeutic effect of mitochondrial transplantation for myocardial ischemia-reperfusion injury. *Bioact Mater.* 2021;6:2058–2069.
19. Hsu C-H, Roan J-N, Fang S-Y, et al. Transplantation of viable mitochondria improves right ventricular performance and pulmonary artery remodeling in rats with pulmonary arterial hypertension. *J Thorac Cardiovasc Surg.* 2020. In press.
20. Emani SM, Piekarski BL, Harrild D, Pedro J, McCully JD. Autologous mitochondrial transplantation for dysfunction after ischemia-reperfusion injury. *J Thorac Cardiovasc Surg.* 2017;154:286–289.
21. Guariento A, Doulamis IP, Duignan T, et al. Mitochondrial transplantation for myocardial protection in ex-situ–perfused hearts donated after circulatory death. *J Heart Lung Transplant.* 2020;39:1279–1288.

22. Chang J-C, Chang H-S, Wu Y-C, et al. Mitochondrial transplantation regulates antitumour activity, chemoresistance and mitochondrial dynamics in breast cancer. *J Exp Clin Cancer Res.* 2019;38:1–16.
23. Griessinger E, Moschoi R, Biondani G, Peyron J-F. Mitochondrial transfer in the leukemia microenvironment. *Trends Cancer.* 2017;3:828–839.
24. Sahinbegovic H, Jelinek T, Hrdinka M, et al. Intercellular mitochondrial transfer in the tumor microenvironment. *Cancer.* 2020;12:1787.
25. Jash E, Prasad P, Kumar N, Sharma T, Goldman A, Sehrawat S. Perspective on nanochannels as cellular mediators in different disease conditions. *Cell Commun Signal.* 2018;16:76.
26. Nawaz M, Fatima F. Extracellular vesicles, tunneling nanotubes, and cellular interplay: synergies and missing links. *Front Mol Biosci.* 2017;4:50.
27. Lou E, O'Hare P, Subramanian S, Steer CJ. Lost in translation: applying 2D intercellular communication via tunneling nanotubes in cell culture to physiologically relevant 3D microenvironments. *FEBS J.* 2017;284:699–707.
28. Desir S, Dickson EL, Vogel RI, et al. Tunneling nanotube formation is stimulated by hypoxia in ovarian cancer cells. *Oncotarget.* 2016;7:43150.
29. Lou E, Zhai E, Sarkari A, et al. Cellular and molecular networking within the ecosystem of cancer cell communication via tunneling nanotubes. *Front Cell Dev Biol.* 2018;6:95.
30. Pasquier J, Galas L, Boulangé-Lecomte C, et al. Different modalities of intercellular membrane exchanges mediate cell-to-cell p-glycoprotein transfers in MCF-7 breast cancer cells. *J Biol Chem.* 2012;287:7374–7387.
31. Bukoreshtliev NV, Wang X, Hodneland E, Gurke S, Barroso JF, Gerdes H-H. Selective block of tunneling nanotube (TNT) formation inhibits intercellular organelle transfer between PC12 cells. *FEBS Lett.* 2009;583:1481–1488.
32. Senol AD, Pepe A, Grudina C, et al. Effect of tolytoxin on tunneling nanotube formation and function. *Sci Rep.* 2019;9:1–15.
33. Zefferino R, Piccoli C, Di Gioia S, Capitanio N, Conese M. Gap junction intercellular communication in the carcinogenesis hallmarks: is this a phenomenon or epiphenomenon? *Cells.* 2019;8:896.
34. Li C, Cheung MK, Han S, et al. Mesenchymal stem cells and their mitochondrial transfer: a double-edged sword. *Biosci Rep.* 2019;39.
35. Hough KP, Trevor JL, Strenkowski JG, et al. Exosomal transfer of mitochondria from airway myeloid-derived regulatory cells to T cells. *Redox Biol.* 2018;18:54–64.
36. Sento S, Sasabe E, Yamamoto T. Application of a persistent heparin treatment inhibits the malignant potential of oral squamous carcinoma cells induced by tumor cell-derived exosomes. *PLoS One.* 2016;11, e0148454.
37. Nishida-Aoki N, Tominaga N, Takeshita F, Sonoda H, Yoshioka Y, Ochiya T. Disruption of circulating extracellular vesicles as a novel therapeutic strategy against cancer metastasis. *Mol Ther.* 2017;25:181–191.
38. Logozzi M, Angelini DF, Iessi E, et al. Increased PSA expression on prostate cancer exosomes in in vitro condition and in cancer patients. *Cancer Lett.* 2017;403:318–329.
39. Parolini I, Federici C, Raggi C, et al. Microenvironmental pH is a key factor for exosome traffic in tumor cells. *J Biol Chem.* 2009;284:34211–34222.
40. Logozzi M, Mizzoni D, Angelini DF, et al. Microenvironmental pH and exosome levels interplay in human cancer cell lines of different histotypes. *Cancer.* 2018;10:370.
41. Zhao H, Achreja A, Iessi E, et al. The key role of extracellular vesicles in the metastatic process. *Biochim Biophys Acta Rev Cancer.* 2018;1869:64–77.
42. Syn NL, Wang L, Chow EK-H, Lim CT, Goh B-C. Exosomes in cancer nanomedicine and immunotherapy: prospects and challenges. *Trends Biotechnol.* 2017;35:665–676.
43. Wang J, Zheng Y, Zhao M. Exosome-based cancer therapy: implication for targeting cancer stem cells. *Front Pharmacol.* 2017;7:533.

44. Sun C, Liu X, Wang B, et al. Endocytosis-mediated mitochondrial transplantation: transferring normal human astrocytic mitochondria into glioma cells rescues aerobic respiration and enhances radiosensitivity. *Theranostics.* 2019;9:3595.
45. Patel D, Rorbach J, Downes K, Szukszto MJ, Pekalski ML, Minczuk M. Macropinocytic entry of isolated mitochondria in epidermal growth factor-activated human osteosarcoma cells. *Sci Rep.* 2017;7:1–13.
46. Acquistapace A, Bru T, Lesault PF, et al. Human mesenchymal stem cells reprogram adult cardiomyocytes toward a progenitor-like state through partial cell fusion and mitochondria transfer. *Stem Cells.* 2011;29:812–824.
47. Hotz MJ, Qing D, Shashaty MG, et al. Red blood cells homeostatically bind mitochondrial DNA through TLR9 to maintain quiescence and to prevent lung injury. *Am J Respir Crit Care Med.* 2018;197:470–480.
48. Yamada Y, Ito M, Arai M, Hibino M, Tsujioka T, Harashima H. Challenges in promoting mitochondrial transplantation therapy. *Int J Mol Sci.* 2020;21:6365.
49. Chernyak B. Mitochondrial transplantation: a critical analysis. *Biochemistry (Mosc).* 2020;85:636–641.
50. Pollara J, Edwards RW, Lin L, Bendersky VA, Brennan TV. Circulating mitochondria in deceased organ donors are associated with immune activation and early allograft dysfunction. *JCI Insight.* 2018;3.
51. Kuck JL, Obiako BO, Gorodnya OM, et al. Mitochondrial DNA damage-associated molecular patterns mediate a feed-forward cycle of bacteria-induced vascular injury in perfused rat lungs. *Am J Physiol Lung Cell Mol Physiol.* 2015;308:L1078–L1085.
52. Wang L, Xie L, Zhang Q, et al. Plasma nuclear and mitochondrial DNA levels in acute myocardial infarction patients. *Coron Artery Dis.* 2015;26:296.
53. Cloonan SM, Choi AM. Mitochondria in lung disease. *J Clin Invest.* 2016;126:809–820.
54. Lin L, Xu H, Bishawi M, et al. Circulating mitochondria in organ donors promote allograft rejection. *Am J Transplant.* 2019;19:1917–1929.
55. Liu X, Khouri-Farah N, Wu CH, Wu GY. Targeted delivery of mitochondria to the liver in rats. *J Gastroenterol Hepatol.* 2020;35:2241–2247.
56. Chen W, Shi K, Chu B, Wei X, Qian Z. Mitochondrial surface engineering for multidrug resistance reversal. *Nano Lett.* 2019;19:2905–2913.
57. Wu S, Zhang A, Li S, et al. Polymer functionalization of isolated mitochondria for cellular transplantation and metabolic phenotype alteration. *Adv Sci (Weinh).* 2018;5:1700530.

Index

Note: Page numbers followed by *f* indicate figures and *t* indicate tables.